Domestic Animal Behavior

for Veterinarians and Animal Scientists

Second Edition

for **Veterinarians and**

Domestic Animal Behavior

Animal Scientists / SECOND EDITION

KATHERINE ALBRO HOUPT, V.M.D., Ph.D.

 IOWA STATE UNIVERSITY PRESS / AMES

Katherine Albro Houpt is a professor of physiology at the New York State College of Veterinary Medicine, Cornell University, Ithaca.

© 1991 Iowa State University Press, Ames, Iowa 50010

♾ Printed on acid-free paper in the United States of America

First edition, 1982
Second printing, 1986
Third printing, 1989
Second edition, 1991

Library of Congress Cataloging-in-Publication Data

Houpt, Katherine A.
 Domestic animal behavior for veterinarians and animal scientists / Katherine Albro Houpt. — 2nd ed.
   p.   cm.
 Includes bibliographical references and index.
 ISBN 0-8138-1062-0 (alk. paper)
 1. Domestic animals — Behavior.   I. Title.
 [DNLM: 1. Animals, Domestic. 2. Behavior, Animal. SF 756.7 H839d]
SF756.7.H68   1991
636′.001′9 — dc20
DNLM/DLC   91-7030
for Library of Congress

Contents

Preface

I WROTE THIS SECOND EDITION TO TAKE ADVANtage of the increase in knowledge of domestic animal behavior. In it, I have paid considerable attention to specific topics, such as location of the teat by the neonatal ungulate. In areas such as learning, I have shifted emphasis from the whole animal to the cellular mechanism; a similar shift is affecting most areas of biology.

The behavior of free-ranging cats was virtually unknown a decade ago, and now dozens of studies have revealed that cats have a more flexible social environment than was originally believed. The field of equine behavior also has grown tremendously owing in part to the efforts of my students, Sharon Crowell-Davis and Lee Boyd.

I particularly desired this book to serve not only as a source of information concerning normal domestic animal behavior but also as a guide to the diagnosis and treatment of problem behavior. Another decade of experience in treating behavior problems has led us to conclude that truly abnormal behavior is rare. Most behavior problems are normal behaviors that are inconvenient or dangerous to humans. The problem animal's behavior usually must be changed for it to remain in the home or on the farm. Changes can be effected by simple management, by behavior modification, by medication, or by surgery.

I dropped two subjects that were covered in the first edition: the human–companion animal bond and cruelty. This omission is not because these subjects lack importance. On the contrary, they are so important that conferences and a considerable literature are devoted to them. Several books have been written on the human-animal bond and many on the subject of welfare, some of which deal specifically with farm animals. Because so much recent information is available elsewhere and because of page constraints in this book, the subjects were omitted.

This edition is authored solely by me because Thomas Wolski has returned to private veterinary practice and is helping to raise his family as he participates in community activities.

Domestic
Animal
Behavior

for Veterinarians and Animal Scientists

Second Edition

1

Communication

RELATIVE SENSORY ACUITY. Communicating with animals, in particular, learning to understand the messages the animal is sending, is a most important part of veterinary medical diagnosis. Communication is a vital part of animal husbandry and the art of veterinary medicine and a very useful adjunct to the science of veterinary medicine. For example, before ordering a complete blood count and liver function tests, the astute clinician will already know that a dog is suffering from abdominal pain because it assumes an abnormal posture with its rump high and head low, or that a horse that paces its stall and kicks at its belly may be suffering from the abdominal pain of colic. A rectal examination will locate the area involved more specifically.

Another important aspect of communication between veterinarian and patient or between handler and stock is assessment of an animal's emotional state or temperament. Adequate restraint or, preferably, a quiet tractable patient is necessary for thorough examination and diagnosis. Most practitioners eventually learn to recognize animals that will be aggressive or fearful and, therefore, will require tranquilization, muzzling, or more stringent restraints. It would be most helpful for agriculture and veterinary students to learn in advance how to recognize animals' moods. Learning by experience to recognize behavior problems may be at the expense of a badly bitten hand or a kicked leg. For their own safety, as well as for diagnostic acuity, clinicians should learn to listen to and watch for the messages their patients are transmitting to them and to each other. Farmers also will prevent injury to themselves and to their stock if they can interpret the animals' messages.

Animals communicate not only by auditory signals as humans do but by visual and olfactory signals as well. Many olfactory messages cannot be detected by humans, although male pheromones, for example those in the urine of tomcats and the very flesh of boars and buck goats, can be discerned. Everyone is aware of vocal communication by animals, but many of these calls remain to be decoded. It is the visual signals made by ears, tail, mouth, and general posture that most benefit the veterinarian who is gauging the temperament and health of the patient. Lately, emphasis has increased on the same type of communication between humans. Now researchers should concentrate not on body English but on body canine, and so on.

Perception

Visual Acuity. Communication between animals depends on their ability to perceive messages. The sensory abilities of domestic animals, with the exception of dogs and cats, have not been systematically studied. Perception in animals is almost always compared with that in humans. Cats can discriminate illumination at one-fifth the threshold of humans, but their resolving power is only one-tenth that of humans (Ewer 1973). Siamese cats do not have stereoscopic vision; other cats do (Packwood and Gordon 1975). Environmental conditions affect visual acuity; it has been shown that free-ranging cats are hypermetropic whereas caged cats are myopic (Belkin et al. 1977). When the relative visual acuity of some other species was compared, the pig ranked highest, followed by sheep, cattle, dogs, and horses (Hebel 1976). A reflective layer of the choroid, the tapetum, functions to exploit any incoming light and is possessed by cats, dogs, horses, and cattle.

COLOR VISION. A question often put to a behaviorist is whether animals have color vision. All domestic animals have been shown to possess color vision in that they will discriminate based on color, but color is probably not as relevant to these animals as it is to birds, fish, and primates. For example, it is very difficult to teach cats to discriminate between colors, although they learn other visual discriminations with ease and have two types of cones that absorb green and blue. Nevertheless, cats (Sechzer and Brown 1964), dogs (Neitz et al. 1989), horses (Grzimek 1952), cattle (Gilbert and Arave 1986; Dabrowska et al. 1981), pigs (Klopfer 1966), goats (Buchenauer and Fritsch 1980), and sheep (Munkenbeck 1983) can also discriminate based on color alone.

Therefore, all the species of domestic animals have been shown to possess color vision. It is interesting that ruminants can discriminate medium and long wavelengths (yellow, orange, and red) better than short wave lengths (violet, blue, and green). Bulls can indeed perceive the matador's red cape (Riol et al. 1989)

EYE PLACEMENT. Eye placement in the skull affects vision. Horses' eyes are set quite laterally and can, therefore, see to the side and well to the rear, but they cannot see well right in front of their heads. Lateral vision is necessarily monocular, and horses see binocularly only in the 70° arc directly in front of their heads. Contrary to many popular and scientific sources, horses do not have a ramped retina (Sivak and Allen 1975). Figures 1.1 and 1.2 illustrate the fields of vision of the cat and horse.

Fig. 1.1. Fields of vision of the cat. A large binocular area results from the forward position of the eyes (J. H. Prince, reprinted from *Duke's Physiology of Domestic Animals*, M. J. Swenson, ed., copyright © 1970, 1977 by Cornell University, with permission of Cornell University Press).

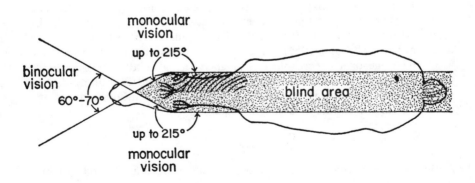

Fig. 1.2. Fields of vision of the horse. The area of binocular vision is small but the field of monocular vision is very wide (Waring et al. [1975], copyright © 1975, with permission of Bailliere Tindall and W. B. Saunders Co.

Auditory Acuity. In hearing, as in vision, cats and dogs appear to perceive over a greater range than humans. Humans can hear 8.5 octaves; cats can hear 10, but cats, despite their mobile pinnae, can only discriminate between sounds 5° apart whereas humans can distinguish sounds 0.5° apart. The pinnae in cats significantly lower their auditory thresholds. The absolute upper limit of hearing in cats and dogs is 60–65 kHz (kilohertz = kilocycles per second). Dogs and cats can discriminate ⅛–¹⁄₁₀ tones (Ewer 1973; Neff and Diamond 1958). Detection of higher frequencies requires a disproportionate increase in cochlear nerve fibers; the number of fibers needed per octave is not constant but rises with the rise in frequency (Rose 1968). Cats have 40,000 cochlear fibers and humans have 30,000, although the range of hearing in cats is only 1.5 octaves greater than that of humans. Cats can discriminate frequency to 8 kHz. Sheep also appear to respond to higher frequencies than humans (Ames and Arehart 1972). The auditory acuity of dogs enables them to respond to silent dog whistles and distracting devices ultrasonic to humans, but not dogs (Campbell 1989). Dogs are not known to produce ultrasonic calls, but rodents do make ultrasonic noises (Allin and Banks 1972), which carnivores use to locate them.

Olfactory Acuity. Olfaction is to animals what writing is to humans—transmission of a message in the absence of the sender. For auditory or visual signals to be sent, the sender must be present, but an odor persists for minutes (or days) after the sender has gone. Olfactory acuity is probably the most important sense of domestic animal species because individual odor recognition and pheromonal release are an important part of their communication. Dogs probably have the greatest olfactory acuity of the domestic species in that they can detect substances at a very low concentration, and this macrosmatic species is the one most investigated. Dogs can detect aliphatic acids at one-hundreth the concentration detectable by humans (Moulton et al. 1960); they can distinguish between the odors of identical twins (Kalmus 1955) and detect the odors of fingerprints 6 weeks after the fingerprints were placed on glass (King et al. 1964). Dogs are trained to sniff out drugs and natural gas leaks, and bloodhounds have been used for centuries to track people; apparently no modern invention is as reliable as the canine olfactory mucosa.

DOGS

Auditory. The common vocal communications of dogs are the bark, whine, howl, and growl.

Bark. Barking is a territorial call of dogs used to defend a territory and to demarcate its boundaries. Stray dogs, whose resting places may be quite temporary, rarely bark (Beck 1973). However, as a stray dog passes the yards of owned dogs, it precipitates territorial barking. The observant owner can recognize various types of barks. The bark to be let in the house differs from that directed at human intruders, which may differ from that directed at canine intruders. Barking occurs in wild canids; a wolf in a seminaturalistic pen will bark at an intruder, but barking has been a trait selected for in domesticated dogs. Dogs function well as burglar alarms, but the barking trait can become a problem in a highly urbanized environment. A barking dog can achieve a noise level that is considered a nuisance; 2200 complaints about barking are made in Los Angeles per year (Senn and Lewin 1975). For this reason, owners may request that their dogs be debarked (vocal cordectomized). Various procedures are described for vocal cordectomy (Archibald 1974) (see Behavior Problems, this section, for alternative treatments). Dogs are not rendered silent, but the strength and pitch of their voices are lowered. A more acute problem is the barking of kenneled or caged dogs in a veterinary clinic. The noise level generated by barking can exceed the 90-decibel level set by the Occupational Safety and Health Act (Occupational Safety and Health Administration 1972). Animal hospitals must be constructed with very good sound insulation for that reason.

Whine and Howl. Whining is an et-epimeletic or care-soliciting call of the dog. It is first used by puppies to communicate with the mother who provides warmth and nourishment. Mature dogs will whine when they want relief from pain or even in a mildly frustrating situation, such as when they want to escape outdoors or to reach a rabbit for which they are digging.

Howling is a canine call that has not been deciphered well. It occurs more frequently in wild canids, coyotes, and wolves and in some breeds of dogs, such as huskies, malamutes, and to a lesser extent, hounds. Harrington and Mech (1978) found that the incidence of howling in wolves increased tenfold during the denning season. As the year's pups mature the pack becomes more dispersed, and the howling apparently takes the place of scent marking in coordinating pack-member spacing and activity.

Harrington (1986) has also found that wolves can discriminate strange adult howls from strange pup howls and answer only the former. A different component, lower in frequency, occurs in the answering howls of wolves approaching the source of a strange howl. Whether this is true in dogs as well remains to be determined.

Growl. Growling is an aggressive call in dogs, as is commonly known. From an evolutionary standpoint, it is interesting that even the most placid dog can be induced to growl if one threatens to take a bone away from it. A scarcity of food in general can increase aggression (see Chap. 2), but bones seem to have a particular value even for the satiated dog.

Visual. A dog's emotional state can be determined by observing its ears, mouth, facial expression, tail, the hair on its shoulders and rump, and its overall body position and posture (Fig. 1.3). The calm dog stands with ears and tail hanging down. The alert dog's tail and ears point upward and it may point with one front foot. As the dog becomes more aggressive, the hair on the shoulders (hackles) and rump rises, and the lips draw back. The ears remain forward, and the tail may wag slowly. With increasing aggression, the lips retract and expose the teeth in a snarl. The dog stands straight. As the dog becomes frightened, the ears go back until they flatten against the head, and the tail descends until it is between the legs.

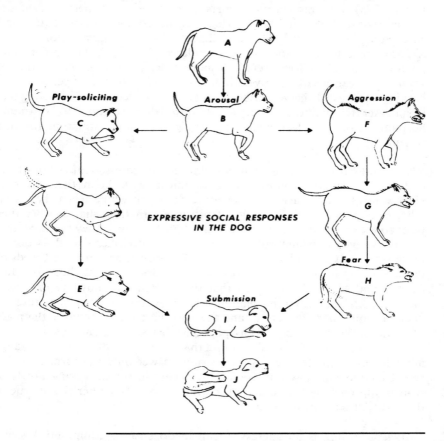

Fig. 1.3. Body postures of the dog. (*A, B*) Neutral to alert attentive positions; (*C*) play bow; (*D, E*) active and passive submissive greeting—note tail wag, shift in ear position and in distribution of weight on fore and hindlimbs; (*F, H*) gradual shift from aggressive display to ambivalent fear-defensive aggressive posture; (*I*) passive submission; (*J*) rolling over and presentation of inguinal-genital region (Fox and Bekoff [1975], copyright © 1975, with permission of W. B. Saunders Co.).

Posture. The posture of the fear-biting dog, the one most likely to injure a veterinarian, is that of the frightened dog, with tail and ears down and the body leaning away from the source of fear. It will have raised hackles and lips retracted in a snarl. When approaching a dog, care must be taken to notice any lifting of the lip because this may be the only prediction of defensive aggression or fear biting. The fear biter will escape, if possible, but if it is approached within its critical distance, which may be only 1 or 2 feet from it, it will attack.

More common, fortunately, is a dog in which fear is not mixed with aggression. The fearful dog crouches with its tail between its legs and its ears flattened. If the dog is abjectly submissive, it will lie on its side and lift its hindleg, displaying the inguinal area. It may also make licking movements. Finally, it may urinate. This behavior probably represents a reversion to puppy habits in which the puppy lies on its side and presents the inguinal area to the dominant mother, and allows her to lick and clean it.

Submissive urination is a frequent behavioral problem that occurs more often in young dogs and small dogs. It may be difficult to live with a dog that is dominant over its owner, but it is fairly messy to live with a dog that urinates submissively. Punishing the dog for urinating in fear or excitement aggravates the problem. The dog is already afraid of its owner, and punishment will only confirm and reinforce that fear. The wisest course is to avoid overexciting the dog.

Overenthusiastic greetings and too-harsh punishments should be avoided. If the person in the household who most often elicits the submissive behavior generally ignores the animal, the problem may be minimized. Submissive urination often declines as the dog matures.

Dogs greet their owners as they did their mothers, by licking their faces. As puppies, dogs lick their mothers' faces to beg for regurgitated food. Although wild canids frequently regurgitate food for their pups, domestic dogs do not do so very often; nevertheless, the begging behavior is shown by domestic puppies. The behavior persists in the adult dog who either licks the owner, or, if prevented by discipline or its small stature, makes licking intention movements. Mouthing the owner's hand is another greeting that is a submissive gesture.

Dogs have a play signal. It is necessary to signal that the action to follow will be play; otherwise, the recipient of the playful act will consider it genuine aggression or sexual activity and respond in kind. Bowing with the forequarters lowered and the hindquarters elevated and topped by a rapidly wagging tail is the signal for canine play (Bekoff 1977). Often one paw is waved or rubbed at the dog's own muzzle.

Genetic and Surgical Alteration. Ear, tail, and hair position are all important in visual communication between dogs, but communication tends to break down in breeds that have been modified either genetically or surgically. Dogs with dependent ears like hounds can only hint at attentiveness or fear. It is hard to detect piloerection on a long-haired dog. Can an Afghan raise its hackles? Hair also prevents many breeds from seeing the signals of other dogs. The Old English sheepdog is a good example: hair covers its face so that it cannot see, and, because of coccygeal amputation, its tail position must be left to the imagination. Simply cutting the hair that obstructs a dog's vision can improve its temperament. Dogs with docked tails learn to wag their whole hindquarters to express pleasure, if not fear.

Olfactory. The legendary olfactory acuity of dogs has already been mentioned. Because dogs can smell some olfactants in a concentration 0.01 to 0.00001 that of the threshold for humans (Becker et al. 1962), it is not surprising that dogs use odors to communicate.

The importance of olfactory communication to dogs is exemplified by the diligence with which male dogs scent mark vertical objects by urinating on them. It is believed that the species, sex, and even individuals can be identified by dogs from the odor of the urine. It is worth noting that dogs scent mark much more frequently in areas where other dogs have marked. The record may be that observed by Sprague and Anisko (1973) of 80 markings by one dog in 4 hours. Even though male dogs rarely empty their bladders completely, such efforts exhausted this dog's supply, and the last urinations were dry.

Elimination Postures. Elimination postures in male and female dogs are a subject of interest to owners. Owners are often concerned because their young males do not lift their hindleg but still squat. Although standing and lifting the hindleg are typical innate male behaviors mediated by testosterone (Berg 1944), males urinate in other positions 3% of the time. Urinating bitches assume not only the squatting position (68% of the urination) but may lift their hindlegs (2%), or use various combinations of the two postures.

Urine marking is the most common form of scent marking in dogs, but vertical objects may also be marked with feces, as any kennel cleaner has observed. Again, males are more likely to mark with feces than females (Sprague and Anisko 1973) (Fig. 1.4). Hart (1974a) found that castration reduces scent marking in male dogs. Dogs that cannot smell (anosmic) and that cannot, therefore, identify other dogs' urine, mark less frequently and do not urinate on the urine of other dogs as intact dogs do. When dogs scratch after defecation, they are not making rudimentary burying movements but are spreading the scent and possibly adding the odor of secretions from interdigital sebaceous glands.

Urine. The most powerful means of olfactory communication in the canine species is the urine of an estrous bitch. Doty and Dunbar (1974) have shown that male dogs are more strongly attracted to the urine of an estrous bitch than to vaginal or anal sac secretions, although Goodwin et al. (1979)

Fig. 1.4. Elimination postures of the dog (Sprague and Anisko 1973 copyright © 1973, with permission of E. J. Brill Publishers).

present strong evidence that it is the vaginal secretion methyl p-hydroxy-benzoate that induces the actual mating behavior sequence in the male (see Chap. 4). Dunbar (1978) also demonstrates the marked preference by an estrous bitch for male urine compared with either estrous or nonestrous urine. The urine contains pheromones, substances secreted by one animal that affect the behavior of another animal. In estrous urine, these compounds are probably estrogen metabolites. The urine of a bitch in heat can attract males from great distances. The attractant effect of the bitch's pheromone is usually considered a nuisance, but there are practical applications. For instance, the pheromone could be used to attract stray dogs that could then be easily captured. One might expect that male dogs inevitably would be attracted to the urine of a receptive female, but Beach and Gilmore (1949) noted that a dog without mating experience did not investigate estrous urine in preference to anestrous urine like sexually experienced dogs did.

Anal Secretions. Urine is not the only olfactory means by which dogs communicate. The anal sacs and anal gland did not evolve simply to be clinical problems for veterinarians. (Expressing impacted anal sacs is probably the least glamorous procedure in veterinary medicine.) The anal gland secretions normally are eliminated with the feces and, no doubt, give them a unique odor (Doty and Dunbar 1974). Dogs, on meeting, usually sniff under each other's tails. The purpose of this behavior probably is one to identify the individual by its smell. A very excited dog can express its anal sacs forcefully; the resulting odor is pungent enough to be smelled by humans and may function as a fear pheromone (Donovan 1967). Otic secretions are also believed to function in individual identification (Fox and Bekoff 1975), and investigation of one another's ears is a common greeting behavior of dogs.

Behavior Problems

Excessive Barking. Some of the problems associated with canine communication, such as submissive urination, have been discussed; barking can be a serious enough problem to warrant debarking, but before surgery is undertaken it is important to determine where and when the dog is barking. If it is barking only when out in the yard alone, the solution is simple — walk the dog on a leash and keep it inside otherwise. Owners often need to be persuaded that this is not unkind to the dog. Barking is probably not something dogs enjoy; they are defending their territories, not exchanging friendly greetings. In other cases, the dog is barking when left alone, which can be treated as a separation problem (see Chap. 8). A more common problem is the dog that barks at guests. The dog is exhibiting territorial behavior just as it would if a burglar were attempting to enter the house. The owner's task is to teach the dog to stop inappropriate barking. Behavior modification methods can be used. It is usually better to positively reinforce a desirable habit than to punish an undesirable habit. This is particularly true of undesirable behavior that might have originated in fear. When the dog barks, the owner should call it and reward it for coming or command it to sit and reward it for sitting. The owner should never yell at the dog for barking, for loud noises will only encourage it to bark more. Dogs should not be encouraged to bark as puppies by praising or by adding to their excitement by saying, "What's there?" Teaching a dog to "speak" for food is, of course, the best way to create a problem barker at meal times. Mild punishment in the form of a water gun or lemon juice sprayed on the dog's mouth can be used. If mild behavioral methods do not work, conditioning with either an ultrasonic or an electric shock collar could be used to teach the dog that barking should be kept to a minimum. As a final resort vocal cordectomy (debarking) may save the life of a dog that has been a barking problem.

Urine Marking. The communication by pets that is least appreciated is urine marking. Male dogs rarely urine mark in the house, but the marking of every vertical object in a city block is another source of pollution that may not only kill the trees sprayed but also spread such urine-borne diseases as leptospirosis. Urine marking in the house can be treated by castration or progestins. Smelling the urine of other dogs does, no doubt, excite a dog, whether or not it might lead to aggression directed toward humans. For a number of reasons, therefore, dogs should be encouraged to urinate within their own territories.

CATS

Auditory. Moelk (1944), in an early study of domestic cat vocalization, listed an extensive vocabulary that may not be recognized by every cat owner. McKinley (1982) has used sonographic analysis to study feline vocalizations and divided feline vocalizations into pure calls in which the vocalizations are homogeneous and complex calls made up of two or more pure types.

Pure calls

MURMUR. A murmur is a soft rhythmically pulsed vocalization given on exhalation. Other murmurs are the request or greeting call, which can vary from a coax to a command, and the acknowledgment or confirmation call, a short, single murmur with a rapidly falling intonation. A soft buzz-like vocalization, the purr, is easy to recognize. It occurs only in social situations and may indicate submission or a kittenlike state. Remmers and Gautier (1972) have shown that purring is associated with rapid (25/sec) disynchronous contraction of the muscles of the larynx and diaphragm.

GROWL. A harsh, low-pitched vocalization (Busnell 1963), a growl is usually of long duration and given in agonistic encounters.

SQUEAK. A high-pitched raspy cry, the squeak is given in play, in anticipation of feeding and by the female after copulation.

SHRIEK. A shriek is a loud, harsh, high-pitched vocalization given in intensely aggressive situations or during painful procedures.

HISS. Hissing is an agonistic vocalization produced while the mouth is open and teeth are exposed. This vocalization is probably defensive and can be used to gauge whether a cat is being defensive or offensive.

SPIT. Spitting is a short, explosive sound given before or after a hiss in agonistic situations.

CHATTER. Some cats chatter their teeth while hunting; especially when unable to reach their prey.

Complex Calls

MEW. A mew is a high-pitched, medium-amplitude vocalization. Phonetically it sounds like a long "e." It occurs in mother-kitten interactions and in the same situations as the squeak.

MOAN. A moan is a call of low frequency and long duration. The sound is "o" or "u." It is given before regurgitating a hair ball or in epimeletic situations such as begging to be released to hunt.

MEOW. The characteristic feline call "ee-ah-oo," the meow, is given in a variety of greeting or epimeletic situations just as the mew and squeak are.

Visual

Posture. The postures and facial expressions of the cat are shown in Figure 1.5A,B. A cat carries its tail high when greeting, investigating, or frustrated. It depresses its tail and wags the tip during stalking. When walking or trotting, the cat holds its tail out at a 40° angle to the back, but as the pace increases, it holds its tail lower (Kiley-Worthington 1976). A relaxed cat, like a relaxed dog, usually stands with tail hanging, but its ears are usually forward. When the cat's attention is attracted, it raises its tail, points both ears forward and holds them erect. Raising the tail might be considered a greeting signal. The aggressive cat walks erectly on tiptoe with its head down. Because the cat's hindlegs are longer than its front legs, it appears to be slanting downwards from rump to head. It holds its partially piloerected tail down but arched away from the hocks and its ears erect and swiveled so that the openings point back. It rotates its whiskers forward and protrudes its claws.

Expression. The frightened cat crouches with ears flattened to its head, salivates, and spits. The pupils of the aggressive cat are constricted; as the animal becomes more defensive, the pupils dilate. The light-colored iris of the eye offers an especially prominent signal of the cat's mood; it is probably an important intraspecific signal and should be used to advantage by the veterinarian. The eyes of an excited cat appear red because the retinal vessels can be seen through the dilated pupils. Contrary to popular belief, the "Halloween cat" is not the most aggressive one; this cat with the arched back, erect tail, and flattened ears, which is piloerected and hissing, corresponds to the fear-biting dog. The cat is fearful but will become aggressive if its critical distance is invaded. One clue to the cat's emotions is that the hind feet appear to be advancing while the front feet retreat; the paws are gathered close together under the cat.

The gape, a response to a strange smell, is the expression most commonly seen when the cat smells a strange cat's urine; this may be the equivalent of the *Flehmen* response of the ungulates (see Horses). The mouth is

Fig. 1.5A. Body postures of the cat. Aggressiveness is increasing from A_0 to A_3, fearfulness from B_0 to B_3. $A_3 B_0$ is the most aggressive cat, $A_0 B_3$ the most fearful, and $A_3 B_3$ the frightened-aggressive cat (Leyhausen 1975, English ed., Katzen—eine Verhaltenskunde, 6th ed., copyright © 1982, illus 65, p. 146 with permission of Paul Parey, Belin and Hamburg).

opened, the tongue is flicked behind the upper incisors where there is an opening in the hard palate that communicates with the vomeronasal organ (Kolb and Nonneman 1975) (Fig. 1.6). At the same time, an autonomic response to the odor is occurring in which the urine brought to the hard palate is aspirated into the vomeronasal organ during parasympathetic stimulation. Fluid is flushed from the vomeronasal organ during sympathetic stimulation (Eccles 1982).

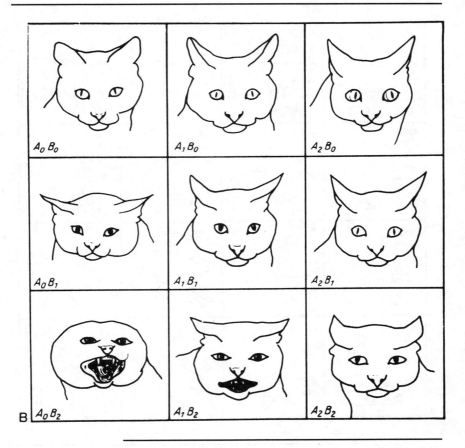

Fig. 1.5B. Facial expressions of the cat (Leyhausen, Katzen—eine Verhaltenskunde, 6th ed., copyright © 1982, illus. 65, p. 146, [1975], with permission of Paul Parey, Berlin and Hamburg). A_2B_0 is offensively aggressive; A_0B_2 is defensively aggressive.

Olfactory

Scent Marking. Male cats also scent mark. Cats defend only a small home range, and spraying by tomcats is a common behavioral complaint. Owners often present the animal for castration for this reason. Spraying nearly always ceases if the tomcat is castrated before he is one year old. The result in mature males is more variable, but 87% of older males also abandon the habit after castration (Hart and Barrett 1973). Occasionally, a castrated male or a female will begin to spray. Spraying is most likely to occur when a strange cat is brought into the home, but cats probably also use scent

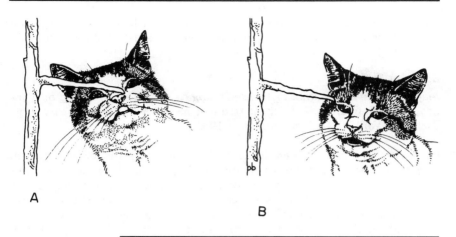

A

B

Fig. 1.6. Gape expression of a cat. (*A*) The cat touches the investigated object with its nose and may lick its nose; (*B*) then opens its mouth while gazing in a preoccupied fashion (drawing by Priscilla Barrett, Cambridge, UK).

marking to arrange their activities temporally with other cats, as has been suggested by Eaton (1970) for the cheetah. De Boer (1977) showed that much of the signal value of urine is lost within 24 hours by comparing male cats' interest in fresh and older urine marks; this may explain why Eaton's cheetah groups were spaced approximately 1 day's travel apart and why tomcats seem to make the rounds of their ranges daily. Cats can apparently distinguish the urine of familiar cats from that of strange cats (Natoli 1985). Spraying can be triggered by the addition of a new family member, a move to a new home, or a change in the owner's and, thus, the cat's daily routine. The smell of male cat urine is quite detectable by humans and usually objectionable. For this reason, an enterprising and behaviorally oriented practitioner, with some success, placed another noxious scent (mothballs) around the boundaries of a castrated male cat's territory in an attempt to drive away marauding tomcats.

Anal Secretions. Cats are well known for fastidiously covering their feces, but in some situations, such as at the boundaries of their territories, cats may leave their feces uncovered. Cats probably use fecal and anal sac odor for communication; two strange cats spend considerable time circling one another attempting to smell the perianal area. If the cats are not too antagonistic, they will eventually permit each other to sniff.

Bunting. Bunting (cheek-rubbing) behavior may also be a form of olfactory communication in that glandular secretion from the cat's face is deposited on the object bunted. Cats bunt the objects to which they respond with a gape. Urine up to 3 days old can elicit these responses (Verberne and De Boer 1976).

Behavior Problems

Inappropriate Elimination or House-Soiling by Cats. House-soiling or inappropriate elimination is the most frequent behavior problem of cats (Borchelt and Voith 1981; Hart and Hart 1985).

DIAGNOSIS. The first problem in treating elimination problems in cats is to determine exactly what type of elimination is occurring because the various forms of house-soiling, urinating or defecating, are treated differently. If they are urinating, they may be either spraying or squatting. The owner can usually differentiate between spraying and squatting, even if the cat is not observed, by the location of the urine and the amount of urine produced. Cats usually spray small amounts on vertical objects, usually in areas with some social significance, such as windows through which the cat sees other cats, beds, or chairs, or, possibly in response to the sounds produced by microwave ovens. Cats urinating in the usual elimination posture, the squatting position, rather than marking produce larger volumes of urine, usually at the edges of a room, on rugs, in sinks or bathtubs, or on beds, especially water beds or mattresses lying directly on the floor. Cats may defecate just outside the litter box, on rugs, in fireplaces, or in the soil of flowerpots; rarely a cat will neither urinate nor defecate in the litter box. Because the causes of the different types of elimination problems are different, the treatments differ; so it is important to determine in which way the cat is house-soiling.

CAUSES. Spraying is a normal and frequent behavior of intact male cats and is the reason most male cats, in contrast to most male dogs, are castrated. The primary cause of spraying in castrated males or females is the presence, particularly the odor, of other cats in the environment. For this reason, a cat in a household with more than one other cat is more apt to spray than one in a single cat household. Male cats with female housemates are more likely to spray than those with male housemates (Hart and Cooper 1984). The proportion of cats spraying increases with the number of cats in the household, so that homes with a dozen or more cats almost inevitably have problems with spraying (Olm and Houpt 1988). If the owner is unsure which cat is spraying, it is possible to administer a capsule containing sodium fluorescein (0.3 ml of 10% solution, 100 mg/ml) to the most likely offender and then examine the house with an ultraviolet light to detect fluorescence. If no fluorescence is detected, the next most likely cat can be given fluorescein 2 days after the first (Hart and Hart 1985).

Spraying may occur when a new cat is introduced or when a cat is introduced to a new house and gradually subsides as the novelty of the new cat or new home declines; in other cases the problem may persist. Incidence appears to be seasonal, with more cats spraying in early spring and late summer than at other times of the year.

Nonspraying or squatting urination is apt to be a sequela of feline urological syndrome. The cat apparently associates the painful urination with the litter box and avoids it, or it may postpone urination until the urge is too great for it to reach the litter pan. Cats that use sinks or bathtubs are often suffering from feline urological syndrome. Other cats are reacting to some aspect of their litter — the smell or texture, the location of the box, the litter box container. Cats that like to perch on the edge of the litter pan while urinating rather than squat in the litter may prefer the floor because it is more stable. A marking aspect may be present in nonspraying urination, too, especially when the cat urinates on a bed or couch or when there are many cats in the household. In multicat households, one dominant cat can keep another cat from using the litter pan.

Defecation problems are less frequent and are more apt to be a response to the litter than to the social environment. Constipation may cause the cat to avoid the litter because of the pain associated with straining in that location.

TREATMENT. Any animal with a behavior problem should be examined for and treated for a concurrent medical problem. This is particularly true of feline house-soiling problems because of the association of feline urological syndrome with failure to urinate in the litter box.

Spraying is best treated by castration of intact males and by reduction of the number of cats in the household. Most owners are unwilling to reduce the number of cats because it is difficult to find homes for adult cats. In 30% of the cases, synthetic progestins, medroxyprogesterone acetate or megestrol, reduce spraying (Hart 1980; Hart and Hart 1985), but the side effects of the drugs, particularly diabetes mellitus, preclude their chronic use. Diazepam (2.5 mg BID [twice a day] or as needed) is also successful in reducing spraying.

Cats seem to depend most on their sense of smell to recognize other cats. For this reason, olfactory tractotomy is successful in eliminating spraying in most female cats and half the male cats that spray. Appetite and food intake is not adversely affected (Hart 1981). Lesions created surgically in the medial preoptic are also successful, but unlike a tractotomy, which is a reasonably simple procedure requiring no specialized equipment, the procedure to produce brain lesions requires specialized equipment and training (Hart and Voith 1978). In male cats ischiocavernosus myectomy has proven successful (Bali and Hoörmeyer 1986).

Nonspraying elimination should be treated by changing the characteristics of the litter. Different types of litter should be presented: clay, sand, corncob, scented, alfalfa, paper, and so forth. The litter most preferred by cats is made of fine particles similar to sand. Two boxes should be provided for each cat, one for urine and one for feces, and the litter should be changed daily. Several locations and container types should be tried. A tray can be substituted for a box, a covered litter pan substituted for an open one, or vice versa. If a social cause is suspected, the same drugs used for spraying can be used.

Defecation problems should also be treated by changing the litter substrate or location. A carpeted frame or "catwalk" around the litter box on which the cat can sit while defecating is often successful.

In general, punishment is not successful; the cat learns only to avoid house-soiling in the owner's presence. Furthermore, if stress is part of the underlying cause of the misbehavior, punishment will only increase the stress. Retraining the cat by restraining it in a cage or bathroom may lead the cat to use the litter box while restrained but seldom affects its behavior when it is free to roam the house. One method of gradually reintroducing the cat to freedom is to keep it on a leash whenever the owner is home. The owner can prevent the cat from soiling and take it to the litter if it starts to squat or to spray. If a discrete area such as a bathtub is used, keeping a few inches of water in the tub will prevent the cat from using it.

Repellents may effectively discourage a cat from using a particular area to eliminate, but it will usually just choose another area. Effective repellents are citrus-scented air fresheners and strong-scented soaps, both of which are effective longer than commercial animal repellents. When dealing with this problem, as when dealing with any behavior problem, the best solution is to change the animal's motivation. It is important to remove the odor from a soiled area. Various commercial preparations are available, however, soap, water, and white vinegar are usually sufficient. Owners should be discouraged from using ammonia because the odor is part of the typical urine odor and might actually encourage the cat.

A few zealous owners may be willing to train the cat to eliminate on command, that is, to housebreak the cat the way most dogs are. The owner should be aware of the times the cat usually eliminates and call the cat to the litter box at those times, encourage it to eliminate by stirring the (clean) litter, and praise it for eliminating. Toilet training the cat has also been suggested.

Clawing and Scratching. Clawing or scratching may be considered grooming behavior because the cat is loosening old layers of the claw but seems to be primarily a form of marking behavior. Whatever the motivation for scratching may be, it is often undesirable, especially if the new sofa or draperies are the chosen site. Many cats are declawed to eliminate the problem, but bad scratching habits can be prevented. If kittens are encouraged to use a scratching post, they usually will not abuse furniture. A good scratching post should have loosely woven material, not a firm rug, on it because the purpose of the scratching post is to allow the cat to hook its claws in the fabric. Because cats scratch more often when they awaken, as they stretch, the post should be placed near the cat's usual sleeping place. The best teacher of a kitten is its mother, so kittens should be obtained from queens that use a scratching post (Hart 1978).

If all else fails, the cat should be declawed rather than euthanized or sent to a shelter for eventual euthanization. There is no question that the cat will experience some pain at the time of declawing, so every effort should be made to improve analgesia, but there do not appear to be any long-lasting behavior sequelae to onychectomy (Bennett et al. 1988; Morgan and Houpt 1989)

HORSES

Auditory. Spectrograms of equine vocalizations are shown in Figure 1.7. Vocal communication appears to be important in maintaining herd cohesion.

Neigh. The neigh (or whinny) is a greeting or separation call. It is most often heard when horses or a mare and foal are separated. A separated mare and foal will neigh repeatedly. These appear to be nonspecific distress calls, which the mare but not the foal may recognize individually (Wolski et al. 1980). Some horses will call to their owners, but usually only when they are in sight.

Nicker. The soft nicker is a care-giving, epimeletic, or care-soliciting, et-epimeletic, call given by a mare to her foal upon reunion and is probably recognized specifically by each (Tyler 1972). A horse may also nicker to its handler, and a stallion may nicker to a mare in estrus.

Snort, Squeal, and Roar. Nickers or neighs usually elicit a reply; other equine vocalizations, such as snorts, squeals, and roars, do not. The roar is a high-amplitude vocalization of a stallion, usually directed toward a mare. A sharp snort is an alarm call in horses. More prolonged snorting or sneezing snorts appear to be a frustration call given when horses are restrained from galloping or forced to work. Snorts and nickers emanate from the nostrils when the mouth is closed; other calls are given with the mouth open.

When two strange horses meet, or when horses have been separated for some time, they greet one another by putting their muzzles together nostril to nostril (Fig. 1.8). The nostrils are flared, but if any vocal signals are given they are inaudible to humans. Usually one, the other, or occasionally both horses will squeal and strike or jump back, although neither has been bitten or threatened. The squeal is, therefore, a defensive greeting. It is heard frequently when horses are forming a dominance hierarchy and also when many bites are being exchanged. Mares that are not in estrus squeal and strike when a stallion approaches too closely. A squeal may also be heard in response to pain.

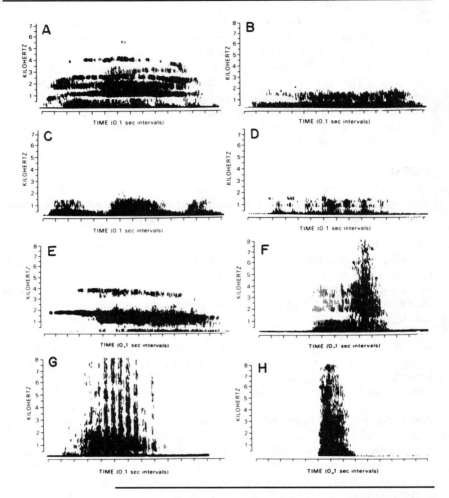

Fig. 1.7. Sound spectrograms of the common vocalizations of the horse. (*A*) Squeal, (*B*) nicker (horse awaiting food), (*C*) nicker (stallion courting), (*D*) quiet nicker (mare to foal), (*E*) whinny, (*F*) groan with final exhale, (*G*) snort, (*H*) blow (Waring et al. 1975, copyright © 1975, with permission of W. B. Saunders Co.).

Visual

Expression. The horse's ears are probably the best indication of its emotions. The alert horse looks directly at the object of interest and holds its ears forward. Ears pointed back indicate aggression, and the flatter the ears are to the head, the more aggressive the horse is (Trumler 1959) (Fig. 1.9). Veterinarians are frequently called upon to examine a horse for soundness

Fig. 1.8. Equine greeting. Nostril to nostril investigation by a horse and pony (Houpt 1977a).

for a prospective buyer. If the horse reacts to examination, or even saddling, by swiveling its ears back, it may not be a desirable purchase, even if perfectly sound physically.

Other facial expressions of the horse are more subtle; nevertheless, they can be profitably used to understand a horse's mood. A submissive horse turns its ears outward. Young horses (less than 3 years old) have a more dramatic display, snapping, also called champing or tooth clapping,

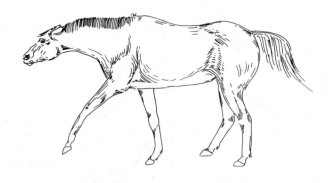

Fig. 1.9. Aggressive posture of a horse. The ears are back, the horse is striking with its front leg and lashing its tail (Houpt 1977b).

in which the lips are retracted, exposing the teeth that are sometimes clicked together (Fig. 1.10). This expression will be shown by a yearling colt to an approaching stallion or toward a mare with a newborn foal.

The sexually receptive mare shows a unique expression, the mating face, in which her ears are swiveled back and her lips hang loose (Fig. 1.11). The *flehmen* response, or curled upper lip of the courting stallion, will be discussed in the next section. A horse that sees but cannot reach food, or is anticipating food, makes chewing movements and sticks out its tongue (Fig. 1.12). More difficult to identify is the horse in pain. A horse that is exhausted but in pain will show loose lips but clenched masseter muscles. Before a horse is in such pain with colic that it kicks at its belly, it will repeatedly swivel its ears back as if attending to its abdomen. The various facial expressions of horses are illustrated by Schafer (1975).

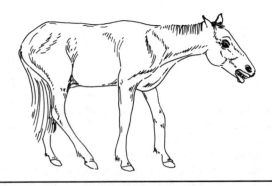

Fig. 1.10. The submissive posture of a horse. The tail is tucked in and the ears are turned outward. The horse is also snapping (opening and closing its mouth while retracting the lips) (Houpt 1977a).

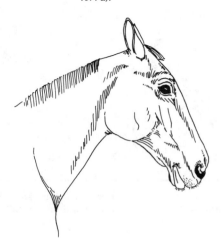

Fig. 1.11. Mating expression of the mare (Houpt 1977a).

Fig. 1.12. Food-anticipating expression of the horse (Houpt et al., 1978a, copyright © 1978, with permission of Elsevier Scientific Publishing).

Visual Field. Horses have a very wide visual field because of the lateral placement of their eyes (see Fig. 1.2). They tend to position their ears in the same direction that they are looking. Thus, when the horse's ears are pointed straight ahead, it is looking ahead. This can be a clue that the horse is about to shy at an object. Usually the rider can identify the frightening object by looking where the horse's ears are pointing. The horse can then be coaxed to investigate and conquer its fear of the object. When the horse turns its ears to the side and back, it is looking to the side. It can see in a wider range to the side and well behind its head. The horse can see best when it is grazing and only its legs impede its vision. Then it can see in all directions and detect an approaching predator easily. Horses have blind spots, including the area just ahead of them when their heads are high. As mentioned previously, horses appear to have color vision.

Posture. The posture and bodily actions of the horse are also useful indicators of its moods. The relaxed horse stands quietly, whereas its nervous counterpart prances and chafes at the least restraint. The aggressive horse, which is threatening to kick, lashes its tail and may even lift one of its hindlegs. The frightened horse tucks its tail tightly against its rump and stands with its feet close together. Muscle guarding is seen, especially if the animal anticipates pain. The stallion moving his mares assumes a unique posture, called herding, driving, or snaking, with head down, nearly touching the ground, and ears flattened (Fig. 1.13). Horses paw the ground, not in aggression but in frustration, if they are eager to gallop or, more commonly, if they want to graze and are restrained by rope or reins. Pawing to eat may be a behavior derived from pawing through snow for grass and may be considered a form of displacement behavior.

Fig. 1.13. Driving posture of the horse. The stallion, on the left, drives a mare. This behavior is also called snaking, herding or rounding (Houpt 1977a).

Olfactory

Scent Marking. Olfactory communication plays an important part in the sexual behavior of horses. Stallions curl their upper lip in the flehmen position, or "horse laugh," when they smell the urine of an estrous mare or are near an estrous mare (Fig. 1.14). Estrous urine alone does not stimulate more episodes of flehmen by stallions than nonestrous urine (Marinier et al., 1988; Stahlbaum and Houpt, 1989), but the frequency of flehmen toward a particular mare increases as she approaches estrus, perhaps because the mare urinates more frequently. After the stallion investigates urine by putting his lip in it, the flehmen position carries the urine into the nasal cavity. When his lips are raised in the flehmen position ,the nostril opening is partially blocked, and the horse, by breathing deeply, carries the urine into the vomeronasal organ. Although stallions flehmen most frequently, geldings and mares also exhibit the behavior in response to olfactory or gustatory stimuli. Cough medicine or a new bit often cause the "horse laugh," obviously not a sign of amusement.

Horses also use olfactory cues, especially from their own or other horses' manure, to find their way home. Wild stallions use manure piles along well-used pathways, possibly to scent mark (Feist and McCullough 1976). These piles may separate bands of horses both spatially and temporally. Even in a pasture, stallions select one place to defecate and back into the pile to eliminate so that the pile does not grow much wider. On the other hand, mares and geldings face outward, gradually increasing the diameter of the pile. Because horses will not eat grass contaminated with feces, a pasture containing mares and geldings rapidly becomes "horsed out" or inedible (Odberg and Francis-Smith 1977). Despite the discrimination of older horses, foals show coprophagia, which will be discussed in more detail in Chapter 6.

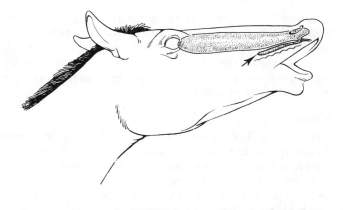

Fig. 1.14. Flehmen response or lip curl. The location of the vom-eronasal organ is indicated by the arrow.

PIGS

Auditory. Vocal signals are probably the most important means of communication in pigs; 20 calls have been identified (Hafez and Signoret 1969), and 6 are easily recognizable to humans. Kiley (1972) has analyzed the vocalizations of ungulates in depth.

Grunt, Bark, and Squeal. The common grunt is 0.25 to 0.4 seconds long and is given in response to familiar sounds or while a pig is rooting. The staccato grunt or short grunt is, as the name implies, shorter (0.1–0.2 sec) and is given by an excited pig; it may precede a squeal. A crescendo of staccato grunts is given by a threatening sow and may precede an attack on anyone who disturbs her litter. The long grunt (0.4–1.2 sec) appears to be a response to pleasurable stimuli, especially tactile ones. The bark is given by a startled pig. The squeal is a more intense vocalization, and the pig that is actually hurt screams.

The various grunts and combinations do not appear to have specific meanings, but the intensity of the vocalization varies with the intensity of the situation. A common sequence is to proceed from common grunts to staccato grunts to repeated grunts without interruption to grunt squeals to screams as the animal is approached, chased, picked up, and injected. Staccato greeting grunts are given by pigs that are reunited after a separation, and a series of 20 grunts with no pause may be given by the hungry pig. Nursing calls will be described in Chapter 5.

Isolation in a strange place causes pigs to vocalize. Short grunts are followed by screams. At the same time the rate of defecation increases (Fraser 1974b). Mature pigs often react to restraint by tantrum behavior accompanied by very loud calls but, interestingly enough, with no increase in heart rate (Marcuse and Moore 1944). When disciplining a subordinate pig, a dominant pig will give a sharp bark as it feints with its snout. The pig in chronic pain grinds its teeth.

Visual

Posture. Possibly because the vocabulary of swine is so large, visual signals do not appear to be as important. However, one can learn something about pig thermoregulatory problems, if not their moods, by observing their posture. Newborn pigs are relatively deficient in fur or fatty insulation and their surface/volume ratio is large; therefore, maintaining body temperature is difficult. Pigs have compensated for their poor physiological abilities with several behavioral strategies to reduce heat loss. A warm piglet lies sprawled out, but a cold one crouches with its legs folded against the body, which minimizes the surface area in contact with a cold floor.

Group Behavior. Group behavior is even more important. Pigs, especially newborn piglets, huddle when they are cold. They thereby convert several small bodies into one large one, both decreasing their surface area and using one another for insulation. Pigs can select an optimal temperature when a gradient is present, both in the laboratory and on the farm. Therefore, heat lamps are provided, and newborn pigs, except those that were brain damaged by anoxia at birth, stay under the lamp at a comfortable 85°F. Adult pigs still huddle when they are cold, but their thermoregulatory problem is more apt to be hyperthermia. Pigs do not sweat, and although they pant, it is not sufficient for cooling. Again, behavioral thermoregulation takes over and pigs wallow in mud, which promotes evaporative heat loss more effectively than plain water (Mount 1979).

Tail Position. Of the visual signals in pigs, the tail, particularly in piglets, is a good index of general well being. A tightly curled tail indicates a healthy pig, and a straight one indicates some distress. The pig's tail is elevated and curled when greeting, when competing for food or chasing other pigs, and during courting, mounting, and intromission. The tail straightens when the pig is asleep or dozing but curls again when the pig arouses unless the animal is isolated, ill, or frightened. The tail will twitch when the skin is being irritated. Amputation of pigs' tails removes a valuable, if crude, diagnostic aid. See Chapter 2 for other visual signals of swine.

Olfactory. Boars may use behavioral signs more than pheromones to determine the sexual receptivity of the sow. However, because females can identify intact males, probably by the strong odor produced by the androgen metabolites present in both the saliva and preputial secretions (Signoret et al. 1975), it seems likely that boars are able to perceive estrogen compounds with either the olfactory or vomeronasal system.

Olfactory stimuli serve to identify pigs individually, for pigs can distinguish conspecifics by odor (Meese et al. 1975). Pigs investigate any newcomer or any pig that has been temporarily removed by nosing it. The ventral body surface is a preferred site for sniffing. The fact that pigs can form a dominance hierarchy while blindfolded indicates that olfactory and auditory rather than visual signals are important to pigs (Ewbank et al. 1974).

CATTLE AND SHEEP

Auditory and Visual. Despite the intimate association of humans and ruminants for thousands of years, very little is known about communication in these species. Kiley (1972) has analyzed cattle vocalization phonetically and according to the motivation of the animal. The "mm" call is of low amplitude and is usually detectable only within 7 meters. It is a common call given by a cow to her calf, or while waiting to be fed or milked. A "mm(h)" call is given in a slightly more frustrating situation, for example, when a cow is isolated. A threatening bull roars at high amplitude with a "(M)enh" sound. A very hungry calf will give a high-intensity "menh" call. During copulation, grunting sounds occur. Some humans can recognize cows by voice, so it would not be surprising if cattle were able to recognize one another. Cattle appear to respond to a vocalization with a vocalization of similar intensity; thus an excited call is answered by excited calls. Calves have a special moo, almost a baa, or play call (Brownlee 1954).

Vocal communication in a prey species like cattle may be most important in transmitting information about general safety or danger. It may have been more important for cattle (and horses) to be alert and ready to flee than to communicate more precise information in their calls. Typical aggressive and submissive postures of cattle will be described in Chapter 2.

If domestic animal communication is studied in as great depth and with the same ingenuity as bird communication has been studied, it may be found that vocal communication is more precise in domestic animals. Careful analysis of the situation in which a call is given, recording the call, and playing back the call to conspecifics in a naturalistic setting may help to break the code of domestic animal languages.

The only vocalizations of sheep that have been studied in detail are those of the ewe and lamb that are involved in mutual recognition (see Chap. 5). Adult sheep continue to use vocalizations as contact calls. A flock of a dozen or less sheep is usually quiet, but large flocks are noisy as the smaller flocks, of which the large one is made, attempt to stay together. Sheep stamp as a visual signal for aggression. The other visual signals used in courting behavior are discussed in Chapter 4. Sheep are also able to distinguish conspecifics by means of olfaction (Baldwin and Meese 1977a).

Olfactory. Olfactory communication is very important for sexual activity in ruminants. Goats and cattle can distinguish conspecifics by means of urine. Male urines are more easily distinguished than are female urines (Baldwin 1977). The Flehmen response is shown by all male ruminants in response to a female in estrus.

GROOMING BEHAVIOR

Pigs. Mutual grooming is well known as a social activity of primates, but its importance in domestic animals is rarely recognized. Subordinate pigs groom dominant ones. The dominant pig lies on its side while the subordinates nibble at its belly. The areas, such as the flanks and back, that the pig cannot reach with its own snout or hind feet (Fig. 1.15) are groomed by other pigs. Singly penned pigs scratch themselves on inanimate objects instead. If scratching seems particularly prolonged or intense, skin parasites may be present.

Horses. Horses mutually groom one another. Licking of the foal is only seen in the short period after the foal's birth, but horses will stand shoulder to shoulder and nibble at each other's withers and back. This behavior is more pronounced in the spring when heavy winter coats are being shed. Rolling, which serves to scratch the horse's back, is also more pronounced at this time. Horses rub their rumps and tails against fences. This appears to have erotic properties; males show penile erection. Rubbing can also be a

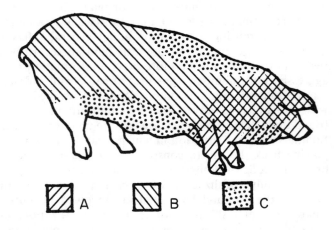

Fig. 1.15. Grooming behavior of the pig. (A) The area scratched by the hindlegs, (B) the area rubbed on vertical objects, (C) the area licked and nosed by other pigs (Hafez and Signoret 1969, copyright © 1969, with permission of W. B. Saunders Co.).

sign of perianal pruritus due to pinworm (*Oxyuris* spp.) infestation.

Summer weather brings irritating flies, many of which are biting species. To escape flies, horses spend many daylight hours in the shade (Tyler 1972), in grassless areas, or if available, in the snow or surf (Keiper and Berger 1982–83). Some horses have a particularly effective way to deal with flies by standing side by side, nose to tail, keeping the flies off one another's faces with their tails. Not all horses form pairs, despite the obvious advantages, but will stand closer to one another during times of high fly density (Duncan and Cowtan 1980). Horses tend to groom animals close to their own rank in the dominance hierarchy, which are also the horses in closest proximity to them (Clutton-Brock et al. 1976). Horse owners assume the role of grooming partner when they curry their horses.

Cattle. Cattle spend about 1 hour a day licking themselves. In addition, mutual grooming occurs, but this only occupies a few minutes of each day. Subordinate animals lick dominant ones around the head and neck (A. Fraser 1974); older and larger cattle receive and give more grooming than younger cattle. Milk production and milking order (order of entrance into the milking parlor) are also correlated with the amount of grooming received (Wood 1977).

Cats. Although licking is a very important part of maternal behavior in cats and dogs and self-grooming occupies a great deal of their time as adults, mutual grooming, or allogrooming, is not common. Cats sometimes lick one another; this is most likely to occur when a mother continues to groom her adult offspring, but long-term associates also allogroom.

Feline grooming is an important part of daily activities. One of the simplest types of grooming is licking the nose and lips. These are two distinct motions that rarely overlap. Licking the nose occurs after gaping, for example, and the tongue moves dorsally on the midline and then is pulled immediately vertically and into the mouth. A common licking problem is an exaggeration of this behavior in which the nose is chronically irritated by the abrasive tongue. Licking the lips moves the tongue along the edge of the upper lips to the corners of the mouth, usually after eating or drinking.

Feline face washing is a stereotyped behavior. The sitting position cat applies saliva to the medial aspect of the front leg, which is held horizontally. The paw is rubbed from back to front over the nose with a circular upward motion. This motion is repeated a few times, with the paw reaching out a little farther each time until it arrives behind the ear (only after three rubs) and then travels downward over the backside of the ear, forehead, and eye. Other areas of the body are cleaned, but not in the stereotyped order that the face is washed. The tongue is drawn over the coat in long strokes, mostly in the direction of the hair. During normal grooming the saliva applied is licked up again, but in a hot environment the saliva is allowed to remain to aid in thermoregulation. The neck, chest, shoulders, and front paws receive the most grooming. The stomach, rear legs, back, croup, tail, and anal areas receive less attention. All the former regions are licked as the cat sits. The cat lies to lick the sides, stomach, rear legs, tail, and front paws (Leyhausen 1979) (Fig. 1.16). A cat may sit like a bear on its haunches to lick the penis, which can be a sign of urethral obstruction.

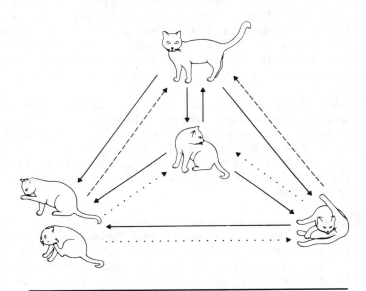

Fig. 1.16. Grooming postures of the cat. (*Top*) Non-grooming. (*Center*) Flank grooming. (*Clockwise*) Grooming of hindquarters; scratching ears; first step of face grooming, licking the front leg. Solid arrows indicate grooming sequences of normal cats. Dotted and dashed arrows indicate sequences in cats with tectal lesions (Swenson and Randall 1977, copyright © 1977, with permission of *J. Comp. Physiol. Psych.*).

DETECTION AND DIAGNOSIS OF DISEASE BY BEHAVIORAL OBSERVA- TION. This is not intended to be a textbook of clinical diagnosis, but a few examples will emphasize the importance of behavioral knowledge in veterinary medicine and animal husbandry.

Much is often made of the fact that veterinarians cannot ask their patients where the pain is localized, but the good clinician can get a clear answer from their behavior. Manipulation of a painful area is met with escape attempts or even attacks. If pain is localized in the abdomen, the animal will tense its abdominal muscles (muscle guarding). An animal with thoracic pain will abduct its elbows. A specific indication of pain is seen in the cow that has a foreign body penetrating the reticular wall (hardware disease). When its withers are pinched, it will flex its front legs and sink down, emitting a grunt of pain at the same time. This procedure is called *the grunt test,* because a normal cow will lower itself, but silently. The rabid cow is usually not vicious but bellows, a very unusual vocalization for the cow.

Lameness and colic are the two most common ailments of horses. The diagnosis and localization of the more subtle lamenesses are skills that take many years to acquire, but even the novice can recognize the signs of a recently foundered horse. It keeps shifting its weight and picking up its hooves and, if forced to move, appears to be walking on eggs. A horse in severe pain of any kind sweats profusely. Horses with abdominal pain look at their bellies, attempt to roll, and pace their stalls.

Pressing the head against a wall or fence is a sign of intracranial pathology and probably head pain in ruminants. Sudden changes in temperament, especially changes in aggression, are also signs of intracranial abnormalities.

A marked decrease in activity or rapid tiring are signs of heart disease, most commonly congestive heart failure in older dogs or congenital cardiac malformations in young dogs. Straining to urinate is a sign of urethral obstruction, an all too common malady of male cats. The head tilt of the dog with otitis or vestibular problems is pathognomonic. Anorexia is a general sign of illness in all species, and it is a good prognostic sign if an animal resumes feeding during the course of treatment.

2

Aggression and Social Structure

Aggression is not a unitary phenomenon but serves a variety of functions in an animal's life. Aggression is used to obtain food, to facilitate access to a sexual partner, or to establish an animal's place in a social hierarchy. Aggression takes place during formation of a dominance hierarchy, but once the dominance hierarchy is formed, it provides a means by which animals in the group can minimize additional serious combat. Within an established hierarchy, subtle threats replace physical violence in competitive situations. For animal practitioners, the problem is not to eliminate all aggression but to determine what type of aggression they are dealing with; only then can they cope with the problem of control effectively. For example, castration may stop a tomcat from fighting with neighborhood cats, but it may have little effect upon his hunting behavior.

One can discern at least eight functional categories of aggression. This classification, based upon the work of Moyer (1968), is not absolute but will show the variability in aggression.

CATEGORIES OF AGGRESSION

Social Aggression. Social aggression occurs when animals live in groups. It serves to establish the peck order, that is, who will be dominant over whom. When adult animals that have never been penned together are brought together for the first time, intense aggressive encounters may occur for several days until each animal has established its position in what generally turns out to be a hierarchy of dominant-submissive relationships. In this type of social grouping there is an alpha animal, which is seldom challenged by subordinates, a beta, or second-ranked animal, which is only challenged by the alpha animal, and so on. Within this organization, the type of aggressive encounter changes once the rank of each animal has been determined. No longer is an attack and subsequent fight needed for an alpha animal to establish its rights over a beta animal. Now, a direct stare, or the threat of a charge, will generally deter the beta animal from further confrontation. This assertion of dominance in the absence of physical combat is called ritualized aggression. Although perception of an extensive hierarchy is questionable in domestic species the relationship between any two animals certainly is recognized. The scarce resource over which social dominance is expressed can be food, a comfortable place to rest, or any action by one animal that is perceived as a threat or challenge by the other.

Territorial Aggression. Territorial aggression keeps other animals out of a particular geographical area; for example, when the domestic dog becomes a snarling menace to the delivery man. In essence, the dog is defending a territory that it considers its own, and strangers, canine or human, simply are not welcome.

Pain-induced Aggression. Pain-induced aggression develops directly from induced pain or fear of pain. The function, of course, is to reduce the pain by eliminating the source. When an animal breaks a leg, it does not discriminate between the pain that comes from the break and the unavoidable pain induced by the veterinarian who tries to set the broken leg. A defense reaction of many species, including dogs and cats, is to attack the cause of pain.

The veterinarian must expect to encounter pain-induced aggression frequently, for any animal will attempt to retaliate if it is suffering acute pain. Some species and some individual animals are more stoic than others, but most will bite or kick if the pain is severe. To judge when aggression may be induced by pain, the practioner must consider the anatomical area involved. A wound on the face will be more painful to the animal than a similar wound on the back because more receptors per unit surface are present on the face. Other areas that appear to be highly sensitive are the ears, when afflicted with otitis, and the rectum, especially when the anal sacs are infected. Any animal will be in pain if a bone is fractured or if the animal has been subjected to surgery.

Injections usually cause pain, and the veterinarian must be prepared for the animal's reaction. Subcutaneous injections of a nonirritating liquid are virtually painless if placed in a loose-skinned area, but intramuscular injections of an irritating fluid like tetracycline or ketamine are very painful, even if no nerve is struck. Horses can become very "needle shy" and quite unmanageable. Equine practitioners would be well advised to infuse local anesthetic with a small-gauge needle before placing a large-gauge needle (>18) in the jugular vein of a horse. A horse that can be repeatedly injected is worth the few extra minutes and few extra cents to render the procedure painless. Horses, dogs, and cats will usually direct their aggression at the veterinarian, if they cannot escape, but cows will direct their aggression at the painful area. A cow that can scratch its ear with a hind foot is quite capable of kicking someone standing at its shoulder, so one would do well to take advantage of the cow's pain-directed aggression. If possible, injections should be given on the side opposite that on which the injector is standing. The teats are probably the most sensitive area of the cow, with the possible exception of the muzzle, and teats should be handled from the opposite side of the udder if the procedure is to be painful or if the cow is nervous.

A cat with no previous history of behavioral problems offered an example of fear-induced aggression when it became markedly aggressive when admitted for treatment of a retro-orbital abscess. The cat was aggressive not only toward the hospital personnel but towards its owner as well. The aggressive behavior, almost fury, did not subside until the cat had been in its home environment for several days. The fact that pain increases aggression should help to explain why corporal punishment of an aggressive dog may exacerbate rather than attenuate its undesirable behavior.

Fear-induced Aggression. This aggression can be related to pain, but in other cases it is neophobia, fear of the unknown, or fear of a particular person or animal with no apparent cause. It is particularly important to differentiate fear from dominance.

Irritable Aggression. Irritable aggression develops when an animal is hungry, fatigued, or sick, and has less tolerance for disturbing situations. This is also the type of aggression seen in some old animals. It gives the organism rest by getting rid of disturbing stimuli.

Maternal Aggression. Maternal aggression is directly related to the protection of young. Although the male is generally considered the more aggressive of the two sexes, maternal aggression can equal the ferocity of any male attack.

Sexual Aggression. In some species, such as the cat or the mink, mating behavior is accompanied by some very severe biting by the male. Although this is an integral part of the mating pattern, it appears to be a biting attack that functions as a very convincing courtship pattern to restrain the female while the male achieves vaginal penetration. Sexual aggression also includes competition between animals for a sexual partner.

Predatory Aggression. Predatory aggression might be better called food-getting behavior rather than aggression per se. The cat stalking a bird or mouse is usually interested in eating the object of the attack; hence, the term aggression becomes difficult to apply except that the behavior is characterized by an attack and biting. However, because cats that are fully satiated will often hunt and not eat the catch, predatory aggression behavior is not entirely governed by hunger.

With this classification scheme in mind, aggression will be analyzed species by species. In subsequent sections, the major forms of aggression in domesticated species will be discussed. Some species, such as the canids, show a much wider spectrum of aggression than other species, or perhaps one observes a wider spectrum because of close association with humans. Pain-induced aggression is omitted because it occurs in all species.

BIOLOGICAL BASIS OF AGGRESSION

Genetic Factors: Breed Differences. Breed differences in aggression and the dominance of one breed over another have been discussed under the appropriate species. For example, dominance is clearly influenced by heredity, for twin cattle are often of equal dominance (Ewbank 1967a). Dairy cattle exhibit breed differences in dominance. Beef breeds, which do not differ markedly in weight, also show definite differences in temperament (Wagnon et al. 1966). Another clear indication of hereditary influence upon aggression is the difference in temperament among dog breeds. These differences, resulting from the selective breeding of dogs, have provided some very truculent breeds. Dogs are the best examples of genetic influences on behavior because there are so many breeds, and their short generation time makes it possible to select for behavioral characteristics easily. Mackenzie et al. (1986) have reviewed the numerous studies on heritability of temperament in dogs. See Prevention of Aggression (dogs) for more examples of heritability of aggression.

Environmental Control of Aggression. Various environmental factors can increase aggression. Hunger and crowding are the primary ones. Almost every study has found that decreasing enclosure size increases the rate of aggression in dairy cattle (Metz and Mekking 1984), beef cattle (Tennessen et al. 1985), and pigs (Jensen 1984). Most aggressive interactions occur at the time of feeding. Unpredictability of feeding also increases the rate of aggression (Carlstead 1986).

Neuroanatomical Control of Aggression. Most current understanding of how the brain regulates aggressive behavior comes from experimental work with the domestic cat. Here, experiments using electrical stimulation of the brain and lesions of specific brain areas as reviewed by Moyer (1968) have indicated the central importance of the hypothalamus for aggression. Although aggression can be categorized by function as it is here, neuroscientists have identified only two major types of aggression: predatory attack and an affective defensive reaction that can result in either attack or flight. These two general reactions can be distinguished by the brain systems that regulate them and the overt behavioral responses themselves.

Predatory attack is characterized by silent stalking in which the body is held close to the ground and by an attack patterned to achieve a quick kill of the prey. In contrast, the affective attack is characterized by extensive autonomic involvement in which there is pupillary dilation, piloerection, salivation, hissing, and often severe arching of the back. Whereas the predatory attack is preceded by a low profile and silent approach, the affective attack accentuates the size and ferocity of the animal.

Predatory attack can be elicited by electrical stimulation of the lateral hypothalamus. If cats are placed with rats, most cats will not attack. However, electrical stimulation of the lateral hypothalamus will induce a quiet stalking attack. If the electrically stimulated cat is confronted with a bowl of food instead of a rat, it will "attack" the bowl of food but generally will not eat. Instead, it will continue to prowl the cage with food dripping from its mouth.

Electrical stimulation of the medial hypothalamus will induce the affective defensive reaction that results either in retreat or, if this is not possible, in an affective attack. In this attack the claws and mouth are used as weapons, but the cat does not attempt to kill with a quick neck bite as it does with lateral hypothalamus stimulation.

Lesions in the Ventromedial Hypothalamus. Lesions in the ventromedial hypothalamus result in increased aggression that begins about two weeks after the lesion. The attack seen in these ventromedial hypothalamus–lesioned animals lacks much of the affective display seen with ventromedial hypothalamus stimulation. It has been suggested that the ventromedial hypothalamus may inhibit aggression that is normally expressed by activation of the lateral hypothalamus. Hence, lesions may eliminate this inhibition. Consistent with this suggestion is the finding that lesions of the lateral hypothalamus abolish all aggression.

A concept emphasizing interplay between the lateral and ventromedial hypothalamus has also been suggested for eating behavior in which the lateral hypothalamus is seen as a focal point for control of eating, and the ventromedial hypothalamus is regarded as a satiety center that inhibits eating. A parallel between aggression and eating exists in that many of the points in the lateral hypothalamus that will induce aggression when stimulated electrically will induce eating when stimulated at a lower electrical intensity. That aggression (predatory) is not controlled by neural systems that are identical to those for eating is indicated by the finding that some points in the lateral hypothalamus will induce eating but fail to induce attack at any intensity, and some points that induce attack fail to influence eating.

Attack behavior is also modified by electrical stimulation in the temporal region, particularly the amygdala. Whereas electrical stimulation in this region alone does not induce attack in the cat, amygdalar stimulation reduces the amount of hypothalamic stimulation needed to induce attack. Lesions of the amygdala reduce aggression, and animals with such lesions become quite placid (Zagrodzka and Fonberg 1977).

From these studies of neural control of aggression emerges a concept of two general forms of aggression: one concerned with food-getting behavior and the other a form of affective defense that can lead to a clawing, noisy attack. The primary control mechanisms seem to be located in the hypothalamus, but these are modulated by cortical limbic structures possibly through thalamic pathways (Bandler and Flynn 1974; Siegel et al. 1977).

Knowledge of the neuroanatomical basis of aggression can be used to diagnose and to locate a brain lesion and to determine surgical treatment (see Treatment of Aggression). A 7-year-old German shepherd guard dog was presented for treatment because it was no longer aggressive but rather docile and lethargic. Although no abnormalities of postural reflexes were present, the site of the lesion could best be determined by the change in the animal's temperament. Postmortem examination revealed a tumor in the frontal lobe. Psychomotor epilepsy characterized by bouts of rage is rare in veterinary medicine but does occur (McGrath 1960). An abnormal electroencephalogram will help confirm the diagnosis, and anticonvulsants (phenobarbital, for example) may control the rage.

Neurochemical Control of Aggression. In studies of cats, injection of minute amounts of acetylcholine into the hypothalamus results in the affective defensive reaction, but no particular locus appears to be more effective than others. In the rat, application of carbachol (a congener of acetylcholine) to the lateral hypothalamus induced a high level of mouse killing; norepinephrine and serotonin were ineffective in this regard. In the cat, injections of norepinephrine into the hypothalamus will reduce carbachol-induced aggression, and arecoline, a muscarinic drug, will increase prey attack whereas nicotine will not (Berntson et al. 1976). Reis (1974) has argued for an aggression facilitatory role of norepinephrine, but data on norepinephrine are often contradictory and difficult to interpret (Myers 1964).

Serotonin may act to inhibit aggression in cats (Katz and Thomas 1976) and provide some rationale for the use of low-protein diets that might allow more tryptophan, the amino acid precursor of serotonin, to traverse the blood brain barrier and thereby increase brain serotonin levels. Oral tryptophan may also be administered.

Hormonal Control of Aggression. In many species the male is more aggressive than the female, both at the interspecies and intraspecies levels. There are, however, some notable exceptions, such as the female with young. In such cases, the female can be a formidable foe to anyone approaching her offspring. At this time, the female is most likely to engage in affective interspecies aggression that is highly damaging but not likely to kill if the intruder leaves the area. At other times, female aggression is generally limited to discouraging male suitors when the female is not sexually receptive and to maintaining the female's place in a female hierarchy if she lives in a group, such as a dairy herd. Female dogs do show territorial defense but usually not with the gusto that males do. Through training or experience, however, the female can become an adequate watchdog.

Testicular Hormones. In species exhibiting a marked sex difference in aggression levels, testicular hormones appear to play two distinct roles in the control of aggression. During very early development, the presence of testicular hormones establishes a heightened potential for aggression. In the absence of testicular hormones, this aggression fails to develop. Hence, in the male, the presence of androgens during sexual differentiation enhances the potential for aggression, whereas the female escapes this androgen influence.

The effect of androgens on aggression has been thoroughly investigated in rodents but is not limited to them. Cats show similar sex differences in aggressiveness that depend on the neonatal hormonal environment (Inselman-Temkin and Flynn 1973). Female puppies treated with testosterone in utero and after birth were, as adult dogs, more successful in competing for a bone than normal females but still were less successful than males (Beach et al. 1982). Testosterone administration increased dominance rank in cows (Bouissou 1978; Bouissou and Guadioso 1982) and aggression in sheep (Parrott and Baldwin 1984). Castration has been practiced for centuries to improve tractability as well as to prevent breeding. Bulls, for example, are more aggressive than steers; the difference increases with age. Bulls also mount one another, which may be an expression of dominance rather than homosexuality (Klemm et al. 1983).

In addition to the developmental effects of androgens, there are the well-known concurrent effects of androgens upon aggression. Exposure to androgens during adulthood increases the probability that the male will show various forms of affective aggression (territorial, sexual, and irritable). Androgens have little to do with predatory aggression.

SOCIAL STRUCTURE AND AGGRESSION IN DOMESTIC ANIMALS
Cattle
Social Order in Free-ranging Cattle. In contrast to most other domestic species, cattle are not often found in the feral state. Their large size and nutritional requirements may be reasons. One group of cattle have remained relatively unmanaged on an estate in England for over 500 years. These animals, the Chillingham cattle, form cow and calf herds, but the bulls live separately in groups of two or three. More aggression occurs

among the cows than among the bulls or between cows and bulls when they are artificially fed (Hall 1989). The bulls maintain a hierarchy through displays with little overt aggression, but bulls from different home ranges rarely breed the same group of cows.

Social Aggression. The dominance hierarchy in cattle is known as the bunt order in polled cattle and the hook order in horned cattle. Dominance can be determined by observing the stances of the two cows involved. The dominant cow, when threatening the submissive one, will stand with its feet drawn well under and with its head down but perpendicular to the ground. The ears will be turned back with the inner surface pointing down and back. The submissive cow also stands with lowered head, but its head is parallel to the ground and its ears are turned so that the inner surface points to the side (Fig. 2.1). Aggressive bulls will turn perpendicular to the opponent and display their full height and length. Some may paw and horn the ground. Aggression is expressed by bunting or striking the opponent with the head.

Determinants of Dominance. The determinants of dominance in cattle appear to be height, weight, presence or absence of horns, age, sex, and territoriality, with age and weight being most important. The cow with horns will dominate a polled animal. In general the heavier animal will be dominant over the lighter one (Bouissou 1965, 1972), but in one study height was found to be negatively correlated with dominance (Collis 1976). In an established herd, the older cows tend to be dominant, probably because initially the older cows were larger than the younger ones; once the hierarchy is formed it remains stable (Schein and Fohrman 1955). Strange cattle added to the herd will tend to be subordinate even if they are older and heavier presumably because the cattle on their own territory have an advantage (Schein and Fohrman 1955). Bulls were dominant over cows in a study of Holstein cattle (Soffie et al. 1976).

Dominance hierarchies can be observed by simply noting all agonistic interactions between cattle. However, this can be a slow process. Clutton-Brock et al. (1976) noted only 0.1 agonistic encounters per hour in free-ranging Highland cattle, whereas they found that ponies engaged in agonistic behavior 1.9 times per hour. Providing food to hungry cattle almost always provokes aggressive behavior and can be a technique for determining food-related dominance quickly. The cow that delivers the most blows and spends the most time controlling the food is dominant. Physical contact is necessary for dominance to be determined, but vision is not. When two cows are in separate pens with the bucket anchored between the pens, both will attempt to eat from it; neither will retreat. If the cows are in the same pen, one will defer to the other, with or without a struggle. The same process occurs even if the cows are blindfolded (Bouissou 1971). A unique way to determine dominance is to put the animals facing one another in a

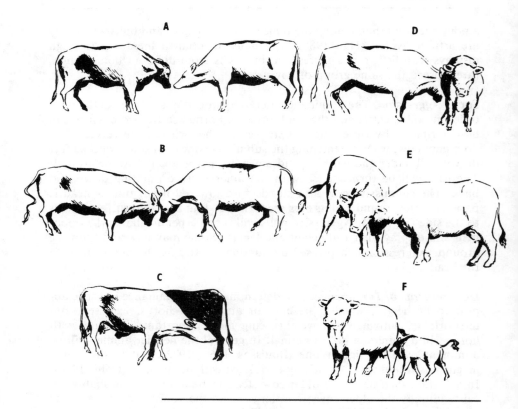

Fig. 2.1. Patterns of agonistic interactions in cattle. (*A*) Cows meeting after an active approach. The one on the left is threatening, whereas the one on the right has assumed a submissive posture. Note the head and leg position of each cow. (*B*) Physical combat—a fight. The cows bunt or push against each other head to head, each striving for a flank position. (*C*) The clinch. One contestant of an evenly matched pair slips alongside the other; the head of one is pushed between the legs and udder of the other. In unusually prolonged contests, the cows rest briefly in the clinch between bouts of bunting. (*D*) Flank attack. The animal that gains a flank position holds a decided advantage over the other. The flanked animal either submits and flees or strives to regain the head-to-head position. (*E*) The butt. The dominant animal directs an attack against the neck, shoulders, flank, or rump of the subordinate, which in turn submits and avoids the aggressor. (*F*) Play fighting. The calf butts the mother (Hafez et al. 1969, copyright © 1969, with permission of W. B. Saunders Co.).

passage that is too narrow for them to turn around. The animal that is forced to back out of the passage is the subordinant one (Stricklin et al. 1985).

Dominance hierarchies are similar whether determined by aggressive interactions in many situations or just in feeding situations. Dominance does not seem to be related to access to resting places in a free stall situation (Friend and Polan 1974), nor is entry into the milking parlor, although there is a tendency for middle-ranking cattle to come first, dominant cows in the middle, and low-ranking cattle last (Kilgour and Scott 1959; Gadbury 1975). In most cases there is no correlation between milk production and dominance (Dickson et al. 1967), but in at least one study, high-ranking cattle tended to be good producers (Brakel and Leis 1976). Purcell et al. (1988) have found that herds in which the average behavior in the milking parlor is better, that is, less restless and aggressive, have higher milk production than herds with poor behavior. Handling the cows when they are calving appears to improve behavior at milking (Hemsworth et al. 1987).

There are many triangular relationships in large herds because the number of cows is so much larger than that of a natural herd of cattle (Hall 1986). Furthermore, hierarchies fluctuate because dairy herds have so many changes in composition owing to culling poor producers, adding replacement heifers, and moving dry and calving cows to separate facilities. Cattle apparently remember one another; when a cow leaves her herd and then returns within a few weeks she will assume the same rank (Collis et al. 1979). Sick cows and heavily pregnant ones will withdraw from the herd and, therefore, show a change in status—a fact that the stock handler and veterinarian should note (Beilharz and Zeeb 1982).

When cattle are driven, the least dominant animals will be first and the dominant animals will be in the middle of the herd (Beilharz and Mylrea 1963b), although when grazing freely the dominant animals will be the furthest from an observer (Beilharz and Mylrea 1963a). During undisturbed grazing, the dominant cattle tend to be the leaders, but which individual cow leads is variable (Sato 1982). When feed is available from a stall where only one animal can eat at a time, animals that are dominant in other situations do not supplant subordinate animals (Stricklin and Gonyou 1981). Perhaps having protection on three sides allows the subordinates to maintain their positions.

Problems can arise when crush gates are used to speed entry into milking parlors because subordinate cows that would enter last are pushed into dominant cows. Aggressive interactions may result. Nonpregnant cows precede pregnant ones in the crush order (Donaldson and James 1963). The individual distance between grazing cattle is about 20 meters, and intrusion into this area may be met with threats or bunts (Clutton-Brock et al. 1976). Cattle high in the hierarchy have an interanimal distance greater than that of cattle low in the hierarchy. Bulls tend to have greater interanimal distance than steers (Hinch et al. 1982–83). Interanimal preferences may also be related to dominance because animals found close to one another in a field are also close in rank (Syme et al. 1975).

The dominant cow will not be the first into the milking parlor, but it will be the first to a feeding area. Once at a feeding area, the dominant cow spends more time eating (McPhee et al. 1964) and less time moving from place to place than a low-ranking cow (Albright 1969), both in the feedlot and on pasture. There are several interactions of reproductive status and dominance. Dominance increases with estrus and decreases with pregnancy (Schein and Fohrman 1955; Beilharz and Mylrea 1963a). Prolactin levels are negatively correlated with dominance (Arave and Albright 1976).

There are breed differences in dominance. Angus cattle are usually dominant over Herefords and shorthorns (Wagnon et al. 1966). Brahma-Hereford crosses are dominant over Herefords (McPhee et al. 1964). Among the dairy breeds, Ayrshires are dominant over Holsteins, and these two breeds are dominant over the smaller breeds, such as Jerseys (Brakel and Leis 1976). These breed differences indicate that dominance may be inherited and heritability of dominance has been calculated to be 0.4 to 0.5 (Beilharz et al. 1966; Dickson et al. 1970).

Stage of lactation and adaptation to a challenging environment can destabilize dominance hierarchies. For example, when pastured in the Alps, Holsteins were subordinate to the native Swiss breeds (Oberosler et al., 1982), although one would have expected the larger Holsteins to be dominant.

Once formed, dominance hierarchies reduce overt aggression; only the lowest ranking animals may suffer food deprivation when supplies are scarce or feeding space is limited. Most aggression is seen during the initial stages of hierarchy formation. Stock managers should mix strange animals with care and avoid putting hungry animals together. When a previously unacquainted group of cattle is formed, it is 24 to 28 hours before the hierarchy is formed. In addition to physical injury, cattle may also suffer from lack of rest because of the general turmoil. The normal pattern of standing and lying as a group does not emerge for at least 48 hours after the group is formed. The stress resulting from lack of rest and lack of rumination is added to the stress of transportation that usually precedes the formation of a new group (Wieckert 1971). These considerations may explain why cattle are more susceptible to diseases such as the shipping fever complex when new groups are formed.

Measuring Temperament. In laboratory animals, emotionality is determined by using an open field test in which the area of the field traversed, the defecations, and the vocalizations are measured. Kilgour (1975) applied the open field test to cattle, but there was no correlation between emotionality as measured in the open field and the score the cow received for temperament by the dairymen who milked her. Tulloh (1961a,b) has developed a temperament score for beef cattle and related it to the ease with which cattle can be driven into a chute or scale or stanchion. He found, using Herefords, shorthorns, and Angus, that Angus were most excitable, shorthorns intermediate, and Herefords most docile. Nervousness was inversely proportional to body weight; small cattle were more nervous than large ones. Although Herefords were docile, they were actually more difficult to drive because they would not respond quickly by moving away from the driver.

Postures of Aggression. The aggressive behavior of the bull as it confronts the threat of the matador is typical of aggression in cattle. The bull usually turns sideways, displaying his height and length to the challenger. The animal paws and often horns the ground, even dropping to its knees to do so. Intraspecies aggression in cattle is characterized by head-to-head combat. In the absence of horns, cattle use their heads as battering rams, pummeling each other's heads and shoulders until one can reach the more vulnerable flanks or simply inflict too much punishment on the other. Equally matched cows may fight for long periods, interrupting active aggression to rest in clinches; one cow will put its muzzle between the hindquarters and the udder of the other, effectively immobilizing it (see Fig. 2.1).

In cattle the greatest problem is the notorious and unpredictably aggressive behavior of bulls. Dairy bulls are generally more aggressive, as well as larger, than beef breed bulls. One reason that artificial insemination of cows has been so enthusiastically accepted, despite the lowered fertility that has resulted, is because it is no longer necessary to keep bulls on the farm. A teaser bull with a deviated penis or a vasectomy can detect cows in estrus, but not impregnate them, and will be as dangerous as a fertile bull. The same hormones that motivate him to mount the cow will also induce him to charge his owner. Removal of those hormones will reduce both mounting and charging. Bulls are sometimes kept in groups; their behavior in this situation is described by Dalton et al. (1967) and Kilgour and Campin (1973).

Clinical Cases of Aggression. Aggression toward people, including butting, kicking, and crushing, is most apt to be a problem in dairy cattle that are handled several times a day. Dangerous animals are usually culled, but a high-producing cow may be kept. She will pose the most danger to those unfamiliar with her temperament, that is, veterinarians. Some cows can be handled only from one side, so the astute clinician should try to work from the side of the cow where she is milked.

Veterinarians can also be the victims of bovine aggression when their treatment is most successful. A cow recumbent with hypocalcemia may, when treated with calcium intravenously, leap to her feet and attack because, whereas severe hypocalcemia results in muscular weakness, mild hypocalcemia can cause irritability.

Cattle will attempt to protect their young and caution is always warranted when a mother is with her offspring. The disappearance of bulls from dairy farms, as artificial insemination replaced natural service, reduced the risk of death from bovine aggression. Nevertheless, every precaution is taken to provide escape routes for the handlers at the artificial insemination centers where large numbers of bulls are kept.

An unusual case of bovine aggression demonstrated the importance of visual cues to cattle. A herd of Hereford cattle was bred to a shorthorn bull. Most of the calves were red with a little white, but one was mostly white with a few red spots. When the calf and its mother were released into the pasture a few days after the calf's birth, all of the other cows attacked the calf. Altering olfactory cues had no effect, but the problem was dealt with immediately by putting the calf and its dam in a corral where they could be seen but not injured, and eventually the problem was solved by adding more cows with white calves to the herd.

Horses

Social Aggression

WILD HERDS. A herd of horses is defined as those sharing the same general range. In a free-ranging horse herd, each stallion is associated with a band of mares numbering from 2 to 20 ($\bar{x} = 6$) (Tyler 1972; Keiper 1976; Waring 1983; Berger 1986). An old mare is apt to be the most dominant of the mares, and it is she who leads the herd in flight and in daily journeys to rest or to a new grazing area. The stallion drives the herd from behind, only going to the front to confront another stallion. Nevertheless, the stallion is usually, but not always, dominant over his harem (Berger 1977; Wells and von Goldschmidt-Rothschild 1979; Houpt and Keiper 1982). Mares apparently choose the stallion and the herd that they ultimately join. Fillies usually leave their natal band at puberty and join another band or a bachelor band. The dominant stallion within the bachelor band will be the one to form a new harem band with the young mare. Fillies avoid incestuous breeding with their sire by leaving their natal band (Duncan et al. 1984). The small percentage that do remain have very low foaling rates, indicating that inbreeding depression of reproduction does occur (Keiper and Houpt 1984).

When a new mare joins an established band, the stallion will protect her from the other mares, especially if she is in estrus. Later she will work her way into and up the hierarchy. The older mares who have been in the band longer will be dominant. Expressed in order of dominance, the typical hierarchy is adult male, adult female, juvenile male, juvenile female, male foal, female foal (Stebbins 1974). Within a herd the mares appear to have preferred associates, "friends" anthropomorphically, with whom they mutually groom, especially when their winter coats are shedding. Preferred associates share resources without competing. When the relationships of the mares are known, the preferred associates often are mother and daughter or siblings (Tyler 1972) and are usually animals close in social rank (Clutton-Brock et al. 1976). The dominance hierarchies remain stable in undisturbed feral herds (Keiper and Sambraus 1986), and age appears to be the most important determinant of dominance.

Colts may remain in the herd until they challenge the stallion by attempting to breed the mares. At this time, they will be driven off and may join a bachelor herd of stallions. Large bands may contain more than one stallion (Miller 1981; Berger 1986). The dominant stallion in the herd does most of the breeding; the subordinate stallion does most of the fighting with any other stallions that approach the band. These large multimale bands tend to supplant smaller bands at water sources (Miller and Denniston 1979).

Although facial expressions in horses have not been well studied, a submissive "snapping" has been noted (Trumler 1959) (see Fig. 1.10). In interspecies relationships, horses dominate cattle (Tyler 1972).

DOMESTIC HERDS. Although the strict order by sex and age described above may be observed in wild horses, quite a different picture emerges when herds of domestic horses are studied. Dominance hierarchies tend to be linear unless the group is large, in which case triangular and more complex relationships appear (Fig. 2.2). Montgomery (1957) studied one herd of 11 horses and found that dominance, as determined from observation of interactions of the whole herd, was determined by weight, not length of residency, but Clutton-Brock et al. (1976) did not find any correlation be-

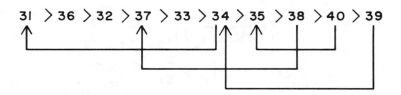

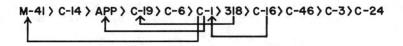

Fig. 2.2. Examples of dominance hierarchies in mares. (*A*) A herd of 10 Thoroughbred mares. (*B*) A herd of 11 mares of various breeds. The > symbol indicates that the horse on the right is submissive to the horse on the left. Arrows indicate direction of dominance in triangular relationships (Houpt et al. 1978a, copyright © 1978, with permission of Elsevier Scientific Publishing).

tween size and dominance in Highland ponies. In a larger study of 11 herds of horses and ponies, Houpt et al. (1978a) found that age and weight were not statistically correlated with rank in dominance hierarchies. All possible pairings of the herd members were made in food dominance tests so that both lower and higher rankings could be determined. Over an 18-month period the hierarchies in domestic horses were stable. The most aggressive horse was the dominant one. Horses under age three are never dominant over adult horses and, in fact, display little aggression toward one another even when vying for food, as Grzimek (1949) noted previously. The position of the stallion in the hierarchy is variable. In three herds of ponies of mixed size and age, geldings and mares were dominant over the stallions (Fig. 2.3), but Arnold and Grassia (1982) found that two horse stallions were dominant to the mares of their herds. Both studies used food competition as a measure of dominance. Mothers are not necessarily dominant over their daughters. The daughters of dominant mares tend to be dominant within their own groups (Houpt and Wolski 1980). The determinants of dominance in horses appear to be more closely related to the animal's temperament and the position of its mother in the herd than to physical characteristics.

Fighting can include a variety of responses in horses: running, either chasing or fleeing; circling; neck wrestling; biting; and kicking (Klingel 1974). Horses neck wrestle and nip at one another even in play, but biting or biting attempts with ears flattened to the head and lips retracted is a sign of serious aggression (Fig. 2.4). Kicking is considered the horse's most aggressive act (Tyler 1972). Although some have hypothesized that kicks

$$\text{Herd IV} \quad F_{58} > F_{55} \geq F_{56} \geq C_{57}$$

$$\text{Herd V} \quad F_{62} > F_{59} > C_{61} \geq C_{60}$$

$$\text{Herd VIII} \quad G_{75} > M_{78} > S_{81} > M_{76} > S_{77} > G_{80} > G_{79}$$

$$\text{Herd IX} \quad G_{83} > M_{84} > M_{85} > M_{87} > S_{88} > G_{86}$$

$$\text{Herd VII} \quad S_{71} > G_{72} > G_{73} > F_{74}$$

$$\text{Herd XI} \quad G_{90} > M_{94} > M_{98} > M_{95} > M_{91} > M_{93}$$

Fig. 2.3. Horse and pony herds in which mares (*M*) or fillies (*F*) are dominant over stallions (*S*), colts (*C*), or geldings (*G*).

Fig. 2.4. An aggressive expression in a horse. A threat to bite.

are defensive, that is, directed up the dominance hierarchy (Wells and von Goldschmidt-Rothschild 1979), it is more likely that kicking occurs when either the challenge or the danger is from the rear (Fig. 2.5). Rearing while striking at another with the front legs, and attempting to bite the other's neck or legs, appears to be a sex-specific type of aggression indulged in mostly by stallions. This is to be contrasted with the rump-to-rump kicking battles of mares. Although wild horses probably do not inflict much damage upon one another, despite this biting and kicking, the shod hoof of the domesticated horse is a formidable weapon, and a newly introduced horse may suffer severe injuries. Mares are hobbled prior to breeding to avoid serious injury to the stud; although the mare may be in full estrus, she may kick at a strange stallion that attempts to mount her without the days of courting that occur in the wild (Tyler 1972).

Horses can be severely injured in the process of forming a hierarchy, especially if they are so confined that the loser cannot escape. Horses that have been stalled separately for a few months may show much more aggression than when they have been together on a daily basis. Because neither age, weight, nor sex appear to be important determinants of dominance, one should hesitate to predict a hierarchy in horses. It is, therefore, a good management practice to introduce (or reintroduce) horses to one another

Fig. 2.5. Threat to kick. The horse on the right is threatening to kick the horse on the left. Note the lashing tail of the threatening horse and the tucked tail of the threatened horse.

across a fence (Houpt and Wolski 1980). The horses can investigate and threaten each other but will be able to escape easily without being kicked. Paradoxically, horses that are very aggressive toward other horses and tend to be high in the equine hierarchy are usually not difficult for a human to manage.

Sexual Aggression. Stallions defend their mares, not a fixed territory. Two herd stallions, upon meeting, usually prance toward one another. Once they are close enough to do so they investigate one another with their nostrils. They will then defecate and sniff at the manure. Feist and McCullough (1976) noted that within bachelor herds, the most dominant animals defecated last. These displays between stallions can lead to a fight, but aggression is much more likely to result in the absence of displays when a bachelor stallion tries to abduct a mare.

The aggression shown by nonestrus females of most species when mounted by a male might be termed antisexual aggression. It is best exemplified by the kicks of the mare when approached by a teaser stallion or by the bites and growls of the nonreceptive bitch when the male places a tentative paw on her back. Mares, however, do show sexual aggression. A dominant mare may prevent a subordinate mare from being bred by biting her until she runs rather than stands for the stallion.

Treatment of Aggression in Horses

AGGRESSION TOWARD PEOPLE. A simple way to acquire dominance over a horse is to flex its forelimb and strap it in that position for a few minutes. The horse is, in effect, three legged and should be urged to walk so that it is aware of its helpless situation. Of course, this exercise should be done only on a soft surface so the horse will not injure itself if it should fall. When the horse has been restrained for five minutes, the limb should be freed and the horse walked again. This process is repeated several times until the horse has learned that the person can give him the use of his leg. This technique is most effective for a person with whom the horse has had no prior experience.

For simple cases of aggression (i.e., of a horse toward its owner), rewarding nonaggression is usually effective. Alternatively, aggression can be punished, but many owners of pleasure horses are unable or unwilling to inflict appropriate punishment. In addition, the punishment must follow the misbehavior within seconds. If a horse threatens its owner, but the owner must run around the paddock to catch the horse before whipping it, the horse will not learn to stop threatening, it will learn to avoid the owner. The same principle applies to punishing a horse for misbehaving in the show ring when the class is over.

Rewards must also be carefully timed, but the timing of rewards is not as crucial as that of disciplinary action. The best reward for horses is food, and grain fed in many small portions can be used to "shape" certain responses. For example, if the horse tends to swing its rump toward anyone who enters the stall, the following course of treatment should be used. On the first day, no grain should be poured into the horse's bucket until it turns 45° or more toward the front of the stall. By feeding grain in measurements of 1 cupful or less, the owner will give the horse plenty of opportunities to learn that its movement toward the front of the stall will be rewarded. The next day, the criterion for reward should be raised to a 90° turn toward the front of the stall. This process can be continued until the horse learns that it must turn and face the front of the stall before it will receive its grain. The owner must be rigid about enforcing this, even if it means several days without grain for the horse.

A similar method can be used for treating a horse that lays back its ears in a threatening manner. The horse should be fed only when it puts its ears forward. It may be necessary to elicit the desired response by whistling or throwing pebbles, and then the grain should be given only as long as the horse's ears remain forward. Each day a longer duration of this behavior should be demanded before the horse is fed. Once the horse is responding well to that, the owner can assume dominance by standing over the feed, starting with hay and waving the horse off. This is the way that dominance is expressed between horses, but care must be taken to insure that the owner is the "winner."

These simple behavior modification exercises should be performed by the owner because the horse must learn that it cannot threaten the owner. The horse must be rewarded and punished by each person affected by its behavior.

For severely aggressive horses, a more drastic treatment program must be employed. The effectiveness of this treatment depends on three factors: a stall or barn that is virtually lightproof, only one source of food, and the absence of other horses.

The horse should be put in the dark alone and be hand-fed, receiving food and light only from people. Food should be withheld as long as the horse approaches aggressively. The animal should be given as many opportunities as possible to earn food and light, but initially the lights should be on for only a few minutes each day. If the horse refuses to eat for several days, hay may be provided, but grain should continue to be used as a reward for good behavior.

The presence of another horse will give the aggressive horse companionship, and the appearance of a person will, therefore, not be as rewarding. Solitude is a form of punishment for the herd-loving horse, and success is near when the horse nickers as people approach.

A final treatment of aggression is pharmacological and much more dangerous to the horse. The horse is immobilized with succinylcholine so that voluntary movement, but not respiration, is prevented. Treatment should be administered by a veterinarian who has resuscitation equipment on the premises. The principle depends on rendering the horse helpless, and once the horse has fallen to the ground, the owner and caretakers should handle the horse thoroughly, including its ears and feet. Its mouth should be opened and closed, and slightly frightening stimuli, such as flapping saddle blankets, should be introduced. The horse is to experience all this while fully conscious but unable to move.

In addition to the usual risk of using a muscle blocker, there is also the possibility of intensifying fear in a horse that is already frightened. The only type of horse considered for treatment with succinylcholine should be one that is not afraid of people—a dominant horse.

AGGRESSION TOWARD OTHER HORSES. Aggression between two or more horses can be treated by changes in management—separation of individuals. In other cases, widely spacing feed buckets is the easiest way to prevent aggression between horses by reducing competition over resources; the resource is usually food. Holmes et al. (1987) have shown how wire partitions along a feed trough allow a subordinate horse to eat in the presence of a dominant. If aggression occurs in other circumstances, it is more difficult to treat, particularly if it occurs at pasture. Such behavior may be treated hormonally with progesterone (Depo-Provera®: Upjohn) or medroxyprogesterone (Ovaban®: Schering) 65–85 mg/day orally for a 300 kg horse. The pharmacological basis for the effectiveness of this treatment is not known. It is hypothesized that progestin administration inhibits those areas of the hypothalamus that induce aggression, especially sex-specific types of aggression. At any rate, the benefits achieved by such treatment must be balanced against the side effects. For example, long-term use may affect fertility in stallions.

A shock collar operated by a remote transmitter may also be used to treat aggression at pasture. Shock, like any punishment, must be delivered at the proper time to be effective; specifically, the moment the horse threatens, kicks, or bites another horse. To ensure that the horse does not associate the punishment with the collar, a dummy collar should be worn by the horse for several days before the shock collar is worn. If possible, the transmitter operator should be hidden, and although the signal can be transmitted over great distances, it cannot be transmitted through metal (i.e., wire fences). Commercially available collars are made for dogs but are sufficient for use on horses, and a longer collar or two collars buckled together should fit most animals.

Care must be taken to avoid overusing shock, and only one unwanted behavior should be punished. Shock can also teach place avoidance: if a horse is shocked while crossing a stream, it subsequently may be afraid of water.

Aggression toward other horses that occurs under saddle or in harness is more easily punished, and most of these problems can be solved by a competent horse trainer.

Aggression in horses can be treated in a variety of ways. The method chosen should be determined by the type and severity of the aggression, as well as the circumstances under which it occurs. Owners should participate in treatment and should be urged not to breed vicious or unmanageable horses.

Pigs. Populations of feral swine are numerous. They form groups of approximately eight, most commonly three sows and their offspring. The males are solitary for much of the year, but may form all male groups in the late winter (Graves 1984). The males travel farther than the females. Young pigs do not leave the sow until they weigh 60 to 70 pounds. The pigs have overlapping home ranges of 300 to 2000 acres (Kurz and Marchinton 1972).

Social Aggression. Pigs have the most intriguing hierarchies because the ranks are formed soon after birth, not by the uncoordinated pushing for a nipple exhibited by puppies but by vicious blows with the appropriately named needle teeth piglets possess at birth. To reduce the injury and infection from snout lacerations during the neonatal period, most swine husbandmen clip the teeth to the gumline (Fraser 1975d). The resource over which the piglets are fighting is the preferred pair of teats, usually the most anterior pair that produces the most milk and has the lowest incidence of mastitis. In addition, pigs sucking at these teats are much less likely to be kicked by the sow's hindlegs. The hierarchy is formed within the first two days after birth; the heaviest and firstborn pigs are usually dominant (McBride 1963). Because the anterior teats produce the most milk, the pigs that suckle these teats grow fastest and remain dominant (McBride et al. 1964, 1965). Once formed, the teat order of hierarchy remains stable, especially the top and bottom ranks (Ewbank and Meese 1971). By the 6th day after birth the same teat is suckled 90% of the time (Hemsworth et al. 1976). Removal of pigs for as long as 25 days does not affect their rank upon return to the group (Ewbank and Meese 1971).

When strange weanling or older pigs are mixed, a hierarchy must be formed (Rasmussen et al. 1962; McBride et al. 1964), and this process takes several days (Meese and Ewbank 1973a). Boars are presumably dominant over sows, but when barrows (castrated males) and sows are penned together, the males may or may not be dominant (Meese and Ewbank 1973a). Once a hierarchy is formed, mounting behavior may be observed among pigs. The dominant pig will lie down and its belly will be nosed by the other pigs. This behavior, the function of which remains unknown, is seen most often when the dominant pig has returned to its group after a separation (Fraser 1974a). There are also breed differences in dominance by sex: more Hampshire males are dominant over females than are Durocs (Beilharz and Cox 1967). Small pigs and newcomers to an established group are usually subordinate (Fraser 1974a).

As in other species, once the hierarchy is formed, fighting is replaced by threats that consist of a sharp, loud grunt and feint with the snout by the dominant pig. Aggressive behaviors include thrusting the head upwards or sideways against the head or body of the opponent. These activities may be accompanied by biting. Levering, in which the snout is put under the body of the opponent, usually from behind, also occurs. The submissive gesture in pigs consists of twisting the head away from the opponent (Jensen 1980). The subordinate pig quickly gives ground (see Fig. 2.6). Leadership on a novel pasture is not correlated with dominance (Meese and Ewbank 1973b).

Vision is not necessary for dominance hierarchy formation in pigs because pigs temporarily blinded with opaque contact lenses form a hierarchy, although overt aggression is reduced (Ewbank et al. 1974). Olfactory bulbectomy reduces the expression of aggression by pigs, perhaps because anosmic pigs have difficulty in discriminating among pigs (Meese and Baldwin 1975a).

Most aggressive behavior is seen in relation to food. A subordinate sow may produce small litters, not because she lacks the genetic potential, but because she cannot obtain adequate nourishment. Individual feeding arrangements can alleviate the problem. However, when pigs of disparate size must be mixed, smaller pigs fare better if the larger pigs are added to the pen last so that the small pigs have the advantage of an established territory. There are breed differences in aggression. Yorkshires are more aggressive than Berkshires (McBride et al. 1964).

Crowding increases aggression in most species and pigs will show more aggression when the stocking rate is increased (Bryant and Ewbank 1972).

Sexual Aggression. A classic example of porcine aggression is the confrontation of two boars. Pigs tend to use loud vocal communications in general, but two boars threatening one another are eerily quiet. They strut shoulder to shoulder champing their jaws, from which fall clumps of thick white saliva containing an androstenol pheromone (Fig. 2.6). When they face one another they often paw the ground, a sign of aggression in many artiodactyls. The animals meet in frontal assault. They slash at each other's shoulders with their well-developed tusks, inflicting severe lacerations. The stronger pig will achieve a flank attack and, consequently, victory. The winner of a conflict usually chases the loser. Aggression between sows and barrows is similar; only champing and strutting are restricted to intact males.

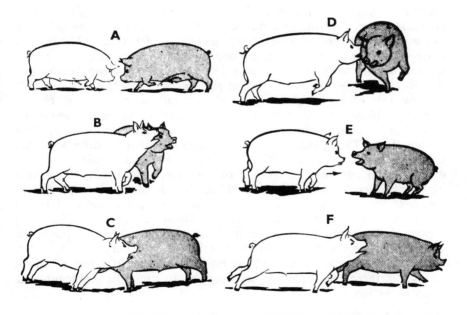

Fig. 2.6. Agonistic behavior in boars. (*A*) Pawing the ground during initial encounter, (*B*) strutting, (*C*) shoulder-to-shoulder contact and slashing, (*D*) perpendicular biting attack, (*E*) submission of pig on the right, (*F*) pursuit of the loser (Hafez and Signoret 1969, copyright © 1969, with permission of W. B. Saunders Co.).

Preventing Aggression

AMONG NEWLY MIXED PIGS. There are three methods for minimizing aggression among newly mixed pigs: tranquilization, provision of shelters and boar pheromone. The aggression can be reduced by tranquilizers such as azaperone 2.2 mg/kg (Symoens and van den Brande 1969) or amperozide 1 mg/kg (Björk et al. 1988). McGlone and Curtis (1985) used hides, small recesses in the pen into each of which a pig could put its head. It is not clear why hiding the head should reduce aggression. Perhaps the sight of the head stimulates aggression in the more aggressive pig. A simpler explanation is that the pig's weapons are the teeth and snout and if they are in the hides they are not being used on other pigs. Spraying or dabbing the boar pheromone 5-alpha-androstenol on the pigs reduces aggression among young pigs (McGlone 1985a, b; McGlone 1986) but not older ones. Perhaps young pigs are always submissive to adult boars, whereas an older pig may challenge an adult male or a pig that smells like one. Lithium has also been used to reduce aggression but probably does so by causing malaise (McGlone et al. 1980).

TAIL BITING. Aggressive tail biting is mostly seen in penned pigs housed on artificial floors. Crowding encourages the outbreak of tail biting as it does aggression in all species (Scott 1958a; Ewbank and Bryant 1972), but the main cause appears to be lack of opportunity for oral stimulation in a species that normally spends 7 hours a day rooting on pasture (Hafez and Signoret 1969). Quite possibly, bored pigs begin to nibble on each other's tails for lack of anything else to do. Once a tail has been bitten severely enough to bleed, and the bleeding is aggravated as the victim swishes its injured tail, the pigs become much more aggressive and bite in earnest (Van Putten 1969). The blood itself appears to be the stimulus for play to become true aggression (Fraser 1987). Some pigs are killed outright, but more often losses occur from infection of the wounded tails. If only one or two pigs are responsible for most of the biting, they should be removed. Often, simply giving the pigs corn on the cob to chew will stop an outbreak of tail biting that has not progressed to the cannibalistic stage. On many hog farms, pigs' tails are docked at birth. This approach eliminates the target but not the vice, and ear biting may arise instead (Penny et al. 1972). Because its causes are unknown, tail biting is difficult to induce (Ewbank 1973) and, therefore, it has not been possible to study the problem objectively in the laboratory.

Sheep and Goats

Social Aggression. Feral sheep in a natural setting form separate ewe and ram flocks. The ewe flocks also include lambs and immature rams. There are seldom more than 20 adult ewes in a flock. Ram flocks are much smaller, about 6 animals, and less stable. This type of social organization is seen in Soay sheep, which are a primitive form of domestic sheep, and among mountain sheep (Grubb 1974b; Geist 1971). Feral goat herds can range in size from 1 goat to 100, but the mean size is 4 (Shackleton and Shank 1984). Goats form sexually segregated flocks. An interesting aspect of sheep and goat behavior is that the females tend to spend each night in a particular area, a night camp. Domestic sheep also "camp" in one particular area at night. Theoretically, these areas may be useful because information can be exchanged between animals even though energy must be expended traveling from food sources to the night camp (O'Brien 1988). Home ranges of male goats are larger than those of females and vary with the season. The total range is 10 to 40 hectares (O'Brien 1984a).

The formation of large commercial herds of hundreds of sheep is usually accompanied by a cacophony of baaing as the small flocks are lost within the large one and the individual sheep give separation calls. The formation of smaller or larger groups of sheep in farm situations is somewhat unnatural. Three sheep do not readily form a flock and tend to disperse; therefore, three sheep are often used in sheepdog trials as a test of the dog's ability to control the sheep.

Sheep tend to select sheep of the same breed as flock mates when randomly mixed (Arnold and Pahl 1974; Winfield and Mullaney 1973). Familiarity is very important to sheep. They quickly form associations that are slow to break down (Winfield et al. 1981). There are breed differences in the tendency to aggregate. For example, Clun Forest sheep gather in large groups, whereas Dalesbred and Jacob sheep are more dispersed and form smaller groups (Shillito Walser and Hague 1981). Similarly, there are breed differences in the tendency to form separate subgroups within large flocks. Merinos rarely form subgroups whereas Dorsets and Southdowns do. The grouping depends on the sheep's activity. Some sheep form subgroups only when grazing, whereas others form subgroups both when grazing and when camping (lying down) (Arnold et al. 1981).

Leadership in sheep is negatively correlated with the tendency to join the flock. In other words, independent sheep lead; the others follow. Within an established related flock of sheep, the oldest ewe is dominant over other ewes and is usually the leader in movements. Dominance is not related to body weight in commercial flocks of sheep of similar age. Very little overt aggression is seen among sheep, but dominance can be determined by limiting feeding space. Dominant sheep will push out subordinates (Arnold and Maller 1974; Squires and Daws 1975). Because age is so important in ovine leadership, the typical arrangement of ewes is the grandmother in the lead followed by her daughter, followed by her daughter's daughter. Each ewe will be followed by its lamb of the season (Scott 1945). Shearing may reduce the rank of a dominant ewe. Rams are dominant over ewes.

One can observe the dominance hierarchy by entering a pen of sheep. The farthest sheep will be the most dominant (Dove et al. 1974). On a large pasture, sheep will divide into flocks with individual, but overlapping, territories. Newly introduced sheep, even offspring separated since weaning, are not allowed to join the original flocks but are relegated to less productive parts of the pasture (Hunter and Davies 1963).

Dominance is much more obvious in a flock of goats than in a flock of sheep. Despite their close relationship, the two species are very different in behavior. Goats are much more aggressive and much more exploratory than sheep. In goats, the presence of horns is an important determinant of dominance as it is among horned sheep. Because horns confer dominance, most agonistic interactions are brief feints or rushes in which the dominant animal lowers its head and points its horns at the subordinate. When the animals are of equal or undetermined rank, long fights occur in which horns and heads are clashed together repeatedly. Goats horn wrestle, but sheep do not. Sheep do more shoulder pushing; once the dominant animal has been determined the ram displays behaviors toward the subordinate that appear identical to those used in courtship, that is, nudging, head low and nose up while striking with the front limb. Age and weight, but not sex, are also important in determining dominance (Scott 1946, 1948; Ross and Scott 1949; Ross and Berg 1956). Syme et al. (1974) found that a novel food will increase the level of aggression within a goat herd.

Sexual Aggression. Sexual aggression in sheep can actually interfere with breeding. Among rams, the dominant ram will usually breed more ewes than his subordinates unless he is so aggressive that his battles distract him from the ewes' estrus (Shreffler and Hohenboken 1974). Two rams may breed fewer ewes than one alone if they are often engaged in the butting contests typical of ram aggression. If three rams are present in a flock, two may fight while the third, less aggressive but evolutionarily more competent ram, impregnates the ewes.

When goats and sheep are placed in the same pasture, a strange situation can develop. The typical aggressive posture of the goat is to rear and meet its opponent in midair where their horns clash. Sheep charge with lowered head and butt one another head-on. If a goat and a sheep fight, the goat rears but is butted in the belly by the sheep, which is usually the victor.

Dogs

Social Systems

DETERMINANTS OF DOMINANCE. Size is important as a determinant of dominance, but territoriality is perhaps more important; a small terrier can attack a Doberman pinscher, with impunity, on the terrier's own territory.

When breeds of dogs are of similar size, breed temperament is important in determining who will be dominant. For example, fox terriers tend to be dominant over cocker spaniels and beagles. This was true of feeding and mating behavior in a mixed breed group; the fox terriers sired all the puppies (James 1951). Males are usually dominant over females, although there are breed differences. Female Shetland sheepdogs can be dominant over males but female fox terriers cannot dominate males (Scott and Fuller 1974). Within a litter of puppies, the dominance hierarchy does not seem to be formed until 4 weeks of age, in contrast to the early hierarchy formation in pigs. When the dominance relationship of dogs is uncertain, it can be quickly established by placing one bone with two dogs. They will very rarely share it; the dominant one will appropriate it.

Not only is it important to determine the dominance hierarchy within a group of dogs to ensure that all have adequate access to food but it is also important that owners establish dominance over their pets. This is most easily accomplished when the animal is a puppy and can be mildly punished by being lifted or restrained. Later, much more force will be necessary for the human to establish dominance. These facts are another reason for accepting only puppies rather than older dogs as house pets. Many behavioral problems arise when the dog is dominant over its owner.

Changes in the "pack" membership can result in changes in the hierarchy, especially if a dominant, or alpha, animal or person is removed. An interesting example is a dog over whom dominance was established late in its development by the owner. When the owner left the dog with his wife and children for a prolonged period, the dog quickly reestablished itself as the alpha member of the pack and expressed its newly won position by territorial marking inside the home. The urination rather than the domination, however, was the presenting problem.

Types of Aggression

SOCIAL AGGRESSION. Urban dogs are either solitary or form small groups, most often of two or three animals (Beck 1973; Fox et al. 1975; Berman and Dunbar 1983; Daniels 1983a; Lehner et al. 1983). Perhaps we have selected dogs to be less social or pack living than their lupine ancestors. Another explanation is that dogs perceive their owners as part of their packs because the majority of these dogs were free-ranging pets, not strays. Large groups of urban dogs are seen only in association with estrous bitches (Daniels 1983b). Rural dogs form slightly larger packs that often contain 2 to 5 dogs (Scott and Causey 1973). Wolf packs contain 2 to 12 members (Mech 1975).

Regardless of the size of a canine group, a hierarchy is formed. Dogs express submission as follows: the ears back and the tail lowered between the legs; a body posture that lowers with increasing fear; and in extreme passive submission, a prone position on its side as it lifts its hindleg and urinates. The posture of abject submission is that of the newborn puppy presenting the inguinal area to its mother for the licking stimulation necessary to urinate (see Fig. 1.3).

A dominant dog will have an erect tail and ears. A dominant dog typically assumes a T-position relative to the submissive dog's shoulder (Fig. 2.7). The submissive dog turns its head away, avoiding the eye contact that might elicit an attack. This turning away of the head has been misinter-

Fig. 2.7. Dominant and submissive postures in dogs. The dominant dog (*D*) forming an intimidating T-position relative to the position of the subordinate (*S*) who attempts to avoid a confrontation by turning away (Fox 1972, copyright © 1972, with permission of Coward, McCann and Geoghegan).

preted as exposing the vulnerable jugular area to the dominant dog. The submissive animal often remains stationary because running usually elicits an attack or chase (Fox 1972). Dogs seem to identify their own breeds and will choose their own littermates in a two-choice test (Hepper 1986). There is not only preference for own breed but aversion to others. For instance, the attack of fox terriers will be more aggressive toward a dog of another breed.

Dogs in packs are more dangerous to humans than solitary dogs. This kind of aggression is usually predatory; the pack, usually underfed, chases a person on foot or on a two-wheeled vehicle (Borchelt et al. 1983).

TERRITORIAL AGGRESSION. This is the most commonly observed type of aggression but not the one most commonly presented to behaviorists. Dogs barking at one another from their respective territories and dogs threatening or actually attacking either dogs or people that encroach on their territories are examples.

FEAR-INDUCED AGGRESSION. Fear-induced aggression is the type that most often directly confronts the veterinarian. The fear-biting dog will be most apt to attack when its critical distance has been invaded, which the clinician must do to examine it. Every effort should be made to reduce fear in this type of dog. The astute behaviorist should be able to judge which animal is the fear biter and should not be dominated and which is the generally aggressive dog that will be more easily handled when made submissive.

IRRITABLE AGGRESSION. The canine aggression discussed so far has been a public problem, but the veterinarian is more frequently faced with the private problems of the owner of a dog that does not tolerate moderate irritation or annoyance. If the suggested treatments of aggression do not solve the problem, euthanasia of the dog should be advised. Owners should be advised that allowing their children to play with an irritable 90-pound dog is not in the children's best interest. In most cases, it would be advisable to discourage the breeding of an irritable dog because its offspring would approximate the parents' temperaments. The inherent dangers of perpetuating this form of aggression must be carefully considered.

MATERNAL AGGRESSION. Females are not exempt from hormonal effects on behavior. Although the precise mechanisms have not been as thoroughly investigated, females with newborn young are more aggressive than at most other times. Protection of the young by the mother certainly has survival value for the species but can be dangerous to other animals and humans. Perhaps the most common example is the case of the pseudopregnant bitch. Suffering from "false pregnancy" under the influence of high progesterone levels, the bitch will not only adopt an inanimate object, such as a leash or toy, as a puppy but will sometimes viciously guard her "offspring." Removing the source of hormones by ovariohysterectomy will alleviate the problem and is advisable for both medical and behavioral reasons. The medical reason is that the bitch that tends to become recurrently pseudopregnant often is afflicted with pyometra. See Chapter 5 for a description of maternal behavior in other species. There are nonmaternal forms of female aggression. Spayed females tend to be more aggressive than intact ones, possibly because the source of progesterone has been removed.

PREDATORY AGGRESSION. Predatory behavior by dogs is often a clinical problem. Dogs that kill chickens, deer, lambs, or cats are frequently presented for treatment. The easiest approach is proper restraint of the dog. A dog on a leash or in a pen is not only prevented from killing other animals but is also no longer at risk of automobile-induced trauma. A more drastic treatment may also be used. Many families of mammals, including canids, are able to learn to avoid a food that they associate with illness (Gustavson et al. 1974). If the dog not only kills but also eats its prey, this phenomenon of taste aversion can be used to eliminate predatory behavior. For example, if the dog kills and eats chickens, it can be allowed to do so and then be injected intraperitoneally shortly afterwards with lithium chloride (LiCl) or apomorphine, or a dead chicken can be baited with an equivalent amount of LiCl in a capsule. Two or three exposures to the nausea associated with LiCl should suffice to teach the dog to avoid attacking chickens. Another method is to "socialize" the dog to the prey animal by penning the dog and prey together. At first, the prey animal might have to be caged for protection.

Guard Dogs for Predator Control. In the western United States, coyotes are the major predator on sheep; in the eastern United States, dogs are the primary predator. To prevent predation in both areas, guard dogs, not herding dogs, have been moderately successful. The breeds used are the Anatolian shepherd, Maremma, Shar Planinetz (Yugoslavian herder), the spitz, Komondor, and Great Pyrenees. Sixty to 70% of sheep producers who use these dogs felt that they were economically beneficial. The major problem is that the guard dogs sometimes (25%) injure or kill the sheep themselves, but most dogs can be trained not to chase the sheep (Coppinger et al. 1983; Green et al. 1984). Bonding sheep or goats to cattle may be a better means of reducing losses from predation (Hulet et al. 1989; Anderson et al. 1987).

Prevention of Aggression. The first step in preventing aggression is to obtain dogs of nonaggressive breeds because there are both environmental and hereditary influences on aggressive behavior, as on all behavior. Mackenzie et al. (1985) have shown that traits necessary for good, that is, aggressive, guard dogs are heritable. Fearfulness also is heritable (Scott and Fuller 1974; Goddard and Beilharz 1982–83). In fact, it was possible to develop a fearful and a normal strain of pointers within a few generations (Murphree et al. 1969). Although dogs of mixed breeds and of nearly every pure breed can be diagnosed as aggressive, some breeds are more at risk than others. German shepherds are more likely to show territorial aggression and because of their popularity as a breed are at or near the top of lists of breeds presented to behavioral clinics (Beaver 1983; Borchelt 1983; Houpt 1983). The same behavior clinics also report that cocker and springer spaniels are frequently presented, usually for either dominance or guarding (possessive) types of aggression toward the owner. If the public were to stop buying dogs of breeds that tend to be aggressive, breeders would select dogs for suitable pet temperament as well as for conformation and coat.

It is much easier to prevent the development of aggression than it is to cure it. Puppies rather than adult dogs should be obtained as pets. The owner can then establish dominance over the dog when it is easy to do so. Puppies should be acquired during the socialization period at 6 to 12 weeks (see Chap. 6). Both parents should be friendly and approachable. The puppy should be outgoing but not so assertive that it bites at hands when held or at feet when following people.

Once in its new home, the puppy should be picked up and suspended with its feet off the ground several times a day. This is effective in dominating the dog and should be continued until the dog is mature (or too heavy to lift).

To avoid aggression toward other dogs, try to socialize the puppy to other dogs as soon as it has been properly vaccinated. If the puppy has plenty of pleasant experiences with other dogs while it is at its most playful age it will be less apt to be either aggressive or fearful toward other dogs as an adult.

As soon as possible the dog should be trained to sit and to lie, to stay, and to come on command. This can be taught at home or in a special puppy kindergarten class. It is not necessary to wait until the dog is 6 months old. It can learn much earlier, especially when food rewards rather than force are used. Fifteen minutes a day doing obedience work for the dog's lifetime will promote a stronger owner-dog bond, reduce the likelihood of aggression, and provide exercise for both the owner and the dog.

AVOIDING AGGRESSION TOWARD INFANTS. To avoid aggression toward babies, the owner should prepare the dog for the arrival of the baby by holding samples of the baby's soiled clothing to smell. The dog will then be familiar with the baby's odor. The dog should not be present when the baby is brought into the house. To introduce the baby to the dog, the owner should hold the dog on a leash and give a sit stay command, then reward the dog with bits of food as the baby approaches. For the next few weeks, the owner should give the dog attention only when the baby is present. The dog will associate the baby with the attention and its absence with being ignored.

Treatment of Aggression in Dogs

GENERAL APPROACH. A general approach to aggression problems in dogs would be the following treatments in order, moving on to the next if the less potent method fails to achieve results: obedience training, specific submissiveness training, castration, or treatment with progestational agents. When aggression is severe, the veterinarian should recommend euthanasia.

There have been many studies of aggression in dogs (Borchelt 1983;

Houpt 1983; Marcella 1983; Borchelt and Voith 1985a; Wright and Nesselrote 1987; Hart and Hart 1988). All of these studies concur that males are more likely to be presented for aggression than females and that certain breeds such as spaniels predominate in the dominance aggression category, whereas the more typically aggressive dogs present as territorially aggressive. Borchelt and Voith (1985a), Tortora (1980a,b), and Hart and Hart (1985) discuss treatment of aggression.

The first thing that should be determined when treating an aggressive dog is the object of the dog's aggression. Aggression can be directed toward another dog (intraspecific aggression), toward people, or toward other animals, such as cats or livestock (interspecific). If the aggression is directed toward people, the owners may be aggressed against or strangers may be the usual target. It is important to determine the target of the dog's aggression because the dog's motivation and the physiological mechanisms underlying it may be different when it is aggressing against another dog of the same sex than when it is chasing a sheep. Different treatments would be employed.

Interspecies aggression is probably the greatest canine behavior problem (Feldmann 1974). Eleven people, including 2 infants, were killed by dogs in the United States in the years 1974–1975 (Winkler 1977); 37,896 dog bites were reported in New York City in 1972 (Harris et al. 1974). A total of 157 people were killed by dogs in the United States between 1979 and 1988 (Sacks et al. 1989). Twenty people were killed by "pit bulls"; 19 of the attacking dogs were owned by men, 7 of whom had convictions for violent crimes (Lockwood 1987). Although only 12 states and 3 cities collected data on animal bites, the number of people bitten by dogs, 164,331, between 1971 and 1972 from these areas alone, is very large (Moore et al. 1977). These data probably do not include many of the bites received. Most bites occur on or near the dog owner's property, which suggests that the dogs are protecting their territories. Delivery people and postal employees are well-known recipients of canine aggression, but children receive the most bite wounds (60%), some of them fatal (Feldmann and Carding 1973). Children tend to run, whether in play or in fear of the dog; running triggers pursuit and attack by the dog. Thus, joggers and bicyclists are also frequent targets of aggressive dogs. Stray dogs do not inflict many bites, presumably because they do not have territories to protect.

The most common kind of aggression presented to veterinarians is that directed against the owner. This type is probably not the most common aggression, but it is considered abnormal by the owners, whereas aggression against intruders or against other animals is considered "natural." Why do dogs aggress against their owners, literally biting the hand that feeds them? There are probably two main reasons: dominance and fear. Because dogs are pack animals, they have evolved tendencies to vie for rank within their social groups. For pet dogs, the social group is the family. Whether dogs really consider people as part of their pack or have separate hierarchies for people and for other dogs is unknown. All dogs do appear to be "running for higher office," especially at the time of social maturity, two to three years of age. The usual cause of dominance-motivated aggression is a dispute over some desirable resource, usually food, but sometimes space, such as on a favorite chair, for example. Challenging the dominant dog by demanding a submissive posture with the command "down" or by petting the dog on the head once too often can also result in aggression. The member of the family attacked by the dog may give some insight into the human hierarchy. In some cases the husband is aggressed against, in other cases the wife. The member of the family that is aggressed against is usually the one closest to the dog in rank. If the husband is clearly dominant, the wife may be the victim, or vice versa if she is dominant. A person who never threatens the dog probably will not be aggressed against either.

Children present a special problem. Dogs can and do kill babies. No dog should ever be left alone in a room with a baby nor should a baby be on the floor where the dog can reach it even when other people are present. A dog can kill a baby before the parent can cross the room. The dog may react like a predator to the very young baby's high-pitched cries. Aggression that occurs later, so-called jealous behavior, still must be eliminated. Toddlers are most at risk because they make direct eye contact with the dog, which the dog perceives as a threatening gesture, and are apt to pull hair, ears, and tail or grab for bones or toys. Older children can learn how to approach dogs correctly. Many owners claim that children teased their dog causing it to be aggressive toward children, but the most teasing can do is to aggravate an underlying aggressive tendency.

Although high-ranking or dominant dogs are most apt to aggress against their owners, low-ranking dogs may also be aggressive. Their aggression is less frequent and, therefore, surprises the owner. A low-ranking or submissive dog may react to a threatening stimulus, a human hand or head in close proximity, with fear-induced aggression. If the aggression causes the stimulus to be withdrawn, the dog will be encouraged to repeat the aggressive act when next threatened. It will have learned that aggression is rewarded. It is important to distinguish fear-induced from dominance aggression because the treatment is different. In particular, dominant behavior or punishment would be contraindicated for submissive fearful dogs. It is usually possible to distinguish a dominant from a submissive dog by its posture and facial expression (see Chap. 1, Fig. 1.3). It is possible for one dog to be dominant over food but fearful when cornered and approached.

Most aggression toward strangers involves territorial boundaries, the front door or the yard. Many owners encourage this type of aggression because they want a watchdog. Unfortunately, they are not teaching the dog to discriminate between people with legitimate business and burglars. The dogs are being rewarded for barking and lunging at visitors by their owner's approval and also by the flight of the victim. For example, many dogs have learned that if they bark at mail carriers, the carriers will leave. They would leave anyway, of course, but the dog has been operantly conditioned in the meantime (see Chap. 7).

Once the object of the dog's aggression is identified, the next step in treating aggression is to determine the circumstances in which aggression occurs. This is important because behavioral therapy must be tailored to the individual dog. For example, if the dog growls when it is kissed good night (a very common occurrence), either the owner should stop kissing the dog good night or the dog must learn that it will be rewarded when a human's head approaches.

The treatment depends on the severity of the aggression and the owner's motivation to remedy it. The responsible veterinarian should suggest euthanasia, especially if the dog is large, the aggression is severe, or a small child is involved. The first step in the treatment of aggression is to protect the public and the owners. Aggressive dogs should not run free, and those that bite people other than the owner should be muzzled when taken for walks. A muzzle will also protect the owner and children in the home, if it can be put on the dog. Severely dominant dogs will object to the muzzle.

PHARMACOLOGICAL TREATMENT. The most effective pharmacological treatment of aggression now available is progestin (Joby et al. 1984; Hart and Hart 1985). The progestins have serious side effects, producing mammary tumors and pyometra in intact females and diabetes mellitus in both sexes, with prolonged use. Progestins are best used for a few weeks to reduce the dog's aggression while the owner modifies the dog's behavior. A typical treatment regime would be megestrol acetate, 1 mg/lb for 2 weeks, followed by 0.5 mg/lb for 2 weeks and a final 2 weeks on 0.25 mg/lb. The serotonin agonist fluoxetin, 5–20 mg/dog, has also been used to treat aggression, but its efficacy has not been proven. Tranquilizers may increase aggression, presumably because the dog no longer inhibits its aggression. Anticonvulsants have been successful in some cases that probably were caused by psychomotor epilepsy.

SUBMISSION THROUGH OBEDIENCE. Obedience training will improve the behavior of most dogs. This is particularly true of dogs whose main problem is dominance-related aggression. A good example is a springer spaniel who aggressed against everyone in the family except the father who had responded immediately and firmly to its aggression, and one of the four children, the one who took him to obedience classes. The purpose of obedience training is twofold: (1) to teach the dog that it will be rewarded for obeying; (2) to establish that the owner is dominant. Obedience training for the dominant dog should emphasize submissive exercises such as down, stay, and sit, rather than heeling or recall. Owners should be urged to spend at least 15 minutes per day training the dog so that the dog learns to obey in the home as well as in obedience class. There are several ways to teach dogs to lie down, to sit, and so forth. Forceful methods may stimulate a dominant dog to aggress, so food rewards may be necessary for the aggressive dog as well as for one that is too easily frightened. Many dogs have had some training so that the owner can retrain it easily at home. Other owners will need the help of an obedience class or a private trainer. The veterinarian and the trainer should be in contact so that they do not work at cross purposes. An example of the difficulty that can arise if the trainer and the veterinarian are not in contact was the case of a Doberman pinscher who showed severe aggression toward strange men. The dog was taken to a group class, although the owner had been advised to work quietly at home. The dog bit the trainer and had to be taken home where it immediately scent marked in several rooms.

DESENSITIZATION. Specific behavioral modification techniques must be tailored to each case. In general, the principle is to reward the dog with small bits of very palatable food for good, that is, nonaggressive, behavior in situations that once elicited aggression. If the dog aggressed when petted, one person should hold the dog on a sit stay while the other, the "victim," approaches the dog. If the dog does not aggress, the holder rewards the dog. Next the "victim" approaches more closely. Again, the dog is rewarded for good behavior. Next, the "victim" extends a hand over the dog's head. Again, the dog is rewarded. Finally, the "victim" touches the dog's head, and the dog is rewarded for tolerating that without growling or snapping. Each step in the process should be very small so that the dog is unlikely to aggress. If it does, the holder should be able to prevent the dog from reaching the "victim." These simple exercises should be repeated 10 times or more a day, gradually re-creating the circumstances in which the dog used to aggress, that is, a rapid approach, many pats on the head, and so forth. If aggression occurred in one room and at one particular time, the exercise should first be done elsewhere, and only when the dog is consistently nonaggressive should it be tested at "the scene of the crime."

SURGICAL PROCEDURES. Castration should always be advised for aggressive male dogs (Hopkins et al. 1976). The effect of castration is twofold: (1) aggression may be reduced, particularly aggression that is sexually motivated; and (2) the dog will not be able to pass on its aggressive tendencies. For the same reasons, ovariohysterectomy should be advised for aggressive females, particularly in cases of maternal or pseudopregnancy-related aggression.

Prefrontal lobotomy has been used to treat aggression in dogs (Allen et al. 1974; Redding 1975). Although lobotomy can only be considered an experimental procedure, intraspecies aggression was reduced to the point where formerly extremely aggressive malamutes could be used in harness together. Predatory aggression was also reduced. The aggression of household pets directed at humans was not permanently attenuated, so prefrontal lobotomy is not recommended for dogs that bite people (interspecific aggression directed at humans).

REDUCING TERRITORIAL AGGRESSION. The best method of treating aggression toward people other than the owners is to have the owner properly control the dog. Because many dogs are aggressive both to their owners and to others, this can be difficult. If, as is usually the case, the dog is most aggressive to people entering the house, the dog can be taught to lie down and stay when people enter. This should be done gradually by first teaching the dog to lie in a particular place near the door and, then, when it will down stay there, adding cues, such as knocking or doorbell ringing, that have signaled a visitor. Once the dog has learned to maintain its stay during those cues, people, family members at first and then others, can come to the door and enter while the down stay is enforced. This method will work for all dogs, but easier methods will reduce aggression in some dogs. Playful dogs can be taught that everyone who enters will throw a ball for them. Dogs that are highly responsive to food rewards can be given food treats by everyone who enters. Keeping a ball or a cup of treats outside the door beside a sign that advises visitors what to do facilitates the process.

Some dogs will aggress toward visitors after their initial entrance, particularly if they walk across the room. For this reason, aggressive dogs should be kept on leashes while company is present. It is best not to isolate the dog because it will not improve; the dog can be muzzled if necessary.

Of course, any dog that is aggressive to people should not run free and should not be tied outside alone. Tying seems to increase aggression, and if the tether breaks or people get within its radius, they will be bitten.

Treatment of Aggression toward Veterinarians.

Dogs entering a veterinary clinic can be classified as follows: apprehensive and submissive (the majority of the dogs); fear biters; actively defensive (large breed dogs in particular); and friendly and outgoing (Stanford 1981).

The veterinarian faces an acute and personal problem in the dog that is aggressive upon examination. Whenever possible, of course, conditioning the dog to fear the animal hospital or the clinician should be avoided (see Learning). Placing a dog on the table often induces submission. One reason that veterinarians particularly and wisely dislike house calls is that a dog in its own territory is far more apt to be aggressive.

MUZZLING. If placing the dog on the table and manual restraint by an assistant or a competent owner is not sufficient, the veterinarian may place a gauze muzzle on the dog. Dogs sometimes are so intent on trying to remove the muzzle that they do not notice the veterinarian's actions. Other dogs, unfortunately, will be driven to frenzy. The other disadvantage is that the oral cavity and tonsils cannot be examined. Manual muzzling by holding the mouth shut has been suggested as a method of gaining temporary dominance over feisty small- to medium-sized dogs, for instance, Scottish terriers. This method takes advantage of the fact that the abductor muscles of the jaws are not as powerful as the adductors. This method is only successful if the dog makes direct eye contact with the human and resists the restraint. In a few moments the dog will stop resisting and can usually be examined with minimal restraint.

DRUGS. As restraint is escalated, various devices such as slip chain collars or nooses on poles may be used, but the development of fast-acting sedatives like xylazine, which can be given intramuscularly, has revolutionized a practitioner's ability to deal with extremely vicious large dogs. Not only will the dog be calm and manageable in a few minutes following injection, but it will not have had a traumatic and frightening wrestling match with kennel personnel. Therefore, it will probably be easier or, at least, no worse to handle on its next visit.

There are dangers in using such drugs. First, if sedation is too deep, the animal will not respond to pain, and diagnosis, especially of neurological disorders, will be difficult. Second, any powerful sedative is contraindicated in seriously ill or aged animals.

THREAT REDUCTION. What should clinicians do to reduce aggression in their patients? The most effective method seems to be threat reduction. The veterinarian should not threaten or act in a dominant manner toward the dog. This approach indicates to the fearful dog that it need not fear and to the dominant dog that it need not defend its dominance. By reviewing the actions and postures of a dominant dog one learns what to avoid. Avoid approaching the dog directly. Squat down while taking a history from the owner and address some remarks toward the dog. If possible call the dog by its name so that it will approach you. Avoid direct eye contact, especially at first. Do not place your hand on the dog's head or shoulders. Stroke it under the chin. Let the owner put the dog on the examining table, or if the dog is large, let the owner lift the forequarters of the dog. Because some dogs associate being placed on a table with unpleasant past experiences, they may be much more tractable if examined on the floor. If the initial interaction between the dog and the clinician is not unpleasant, the veterinarian may then perform a routine examination much more easily.

Cats. Cats are fairly plastic in their social structure. In a rural setting, cats have territories as large as 200 ha/female cat and 600 ha/male cat (MacDonald 1981; Wolski 1982; Turner and Bateson 1988) whereas in an urban setting the density varies from 2 to 12 cats/ha (Dards 1983; Natoli 1985). Cats are considered to be a nonsocial species because they do not live in groups as adults if they are living on natural prey (Turner and Bateson, 1988). Cats can adapt to a concentrated food source such as that found in dumps, fishing villages, and farms by living in groups, but these groups are of matrilineal female kin. Females rarely transfer from group to group, although males can. In general, the dominant tom's territory encompasses the females' (Fig 2.8). Although he will not hunt on the females' territories, he will repel any marauding male.

Social Aggression. When two cats approach each other aggressively, they walk on tiptoe, slowly lashing their tails about their hocks and turning their heads from side to side making direct eye contact: This threat may intimidate a subordinate cat so that it slinks off; evenly matched rivals will continue to approach one another (Fig. 2.9). They will walk slightly past one another before one cat will spring, trying for a grip on the nape of the opponent's neck. The attacked cat throws itself on its back, thus protecting the nape. The two adversaries will both lie on the ground belly to belly while they claw, vocalize, and bite at one another. After a few moments, one cat, usually the original attacker, will jump free. The other cat may adopt a defensive posture, attack, or run away. The victor usually pursues the vanquished (Leyhausen 1979). Cats that are aggressive to other cats are not necessarily aggressive to people and vice versa (Turner et al. 1986). These differences in personality appear to be inherited because a paternal effect has been noted (Feaver et al. 1986).

When placed together in a home, a laboratory, or on a farm, cats will form dominance hierarchies (Baron et al. 1957; Cole and Shafer 1966; Laundre 1977), but marked aggression may persist in this essentially solitary species. Cats may divide up a house: one's territory may be the first floor; the other's, the second floor. Roommates may find that the two cats belonging to one person will gang up on the single cat belonging to the other. Urination in the house, especially on beds or rugs, often occurs when strange cats are introduced.

Sexual and Territorial Aggression. Aggression directed by one sex partner toward the other is seen most clearly in cats when the male bites the nape of the female's neck even before he mounts. He may be testing, for only a cat in full estrus would submit to the bite. Immediately after copulation the female may turn on the male, hiss, and strike out with her claws. Agonistic interactions during competition for access to estrous females are the inspiration for many cartoon sequences and a source of annoyance to both pet and homeowner.

Males maintain nonoverlapping territories in the nonbreeding season but territories overlap considerably in the breeding season. Therefore, in both free-ranging and pet cats, intraspecies aggression among intact male cats is a very common problem. Many tomcats are repeatedly presented for treatment of bite wounds and abscesses resulting from fighting behavior. Castration is approximately 90% effective in eliminating roaming and fighting in adult male cats (Hart and Barrett 1973), although the disappearance of one may not be associated with a decline in the other.

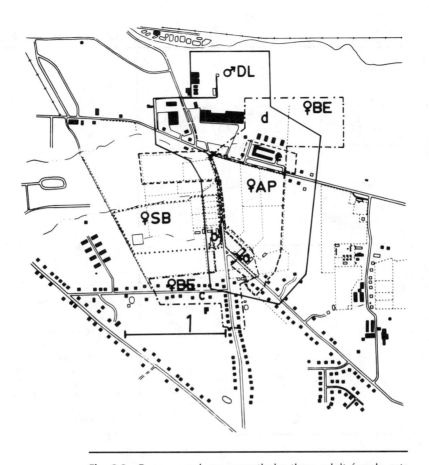

Fig. 2.8. Ranges used concurrently by three adult female cats (*AP, BE, SB*) and one adult male (*DL*) cats during 1979. *AP* and *BE* shared three of four barns at the home farm (*a*) and most of the yards and pastures immediately surrounding the barns. *SB* was the only female at her farm (*b*). *BE* used the southern section of her range (*c*) through most of 1978 and 1979 and started using the northern area (*d*) in the late summer of 1979 in a series of foraging excursions with that year's litter. One female kitten remained and eventually reproduced in this section of the range, whereas *BE* and a male kitten disappeared from this area in late 1979. The two sections of *BE*'s range were connected only by the road between *SB*'s and *AP*'s ranges; *BE* hunted frequently along the road shoulders whereas *SB* and *AP* foraged only in the adjoining pastures. *AP* and *BE* shared the area around an apartment complex (*e*) but never were noted to contact one another here. *DL*'s range included large areas of each female's range, and on most evenings *DL* visited each barn complex at least once. Note: The enclosed line at the lower end of the map represents 500 m. Immediately above it, the north-south axis is indicated. Dark quadrangles represent homes, apartments, and stores; open quadrangles define barns and other outbuildings. Narrow dotted lines indicate fence lines (copyright © Wolski 1982, with permission of Veterinary Clinics of North America: Small Animal Practice).

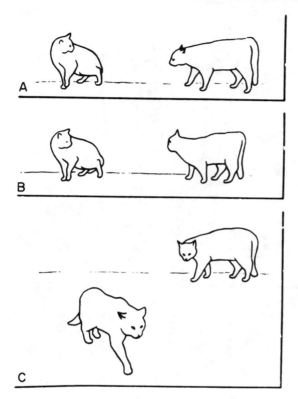

Fig. 2.9. Dominant and submissive postures in cats. The dominant cat is on the right. The submissive cat moves slowly away and avoids eye contact (Leyhausen 1975, Katzen−eine Verhaltenskunde, 6th ed., copyright ©, 1982, with permission of Paul Parey, Berlin and Hamburg).

Predatory Aggression. The tall posture of the cat engaged in territorial or sexual aggression contrasts with the stalking posture of predatory aggression. The predatory cat carries its body as close as possible to the ground. It moves toward its quarry slowly, taking advantage of any natural cover. The closer the cat gets to its prey the more slowly it advances. Almost inevitably, the cat will pause before leaping to attack. Only the tip of the tail will move as the cat lies in wait. There are usually two or three bounds from hiding to the prey. When attacking a large animal, cats try to bite the nape to sever the spinal cord (Leyhausen 1979).

   Predatory aggression is not easily elicited in cats that have not been taught to hunt. Kittens raised with a mother who killed rats in their presence killed at their first opportunity; kittens raised alone seldom did, whereas those raised in a cage with a rat never did (Kuo 1930). Apparently, kittens learn to direct various innate predatory motor patterns to the prey (see Chap. 6) their mother brings to them. She does not simply let them eat the prey; she lets the prey go and catches it again. If the kittens attempt to catch or eat the mouse, the mother will compete with them for it. In this manner, the kittens are stimulated by the hunting game and apparently learn by observation. The types of prey brought to the kittens may influence the range of prey hunted by the kittens as adults. Although the mother can influence kittens' predatory skill by bringing prey and interacting with it, adult cats without such learning experience also become competent predators, so a kitten that is not a good hunter can acquire the skills as an adult (Turner and Batson 1988).

Most cats will kill rats if fasted for two or more days, but they still prefer to eat commercial cat food rather than their prey (Adamec 1976). One feline characteristic that is distasteful to some people is that cats sometimes play with their prey before and after it is dead. They will catch a mouse, let it go and catch it again. After it is dead they will throw it up with their paws and leap upon it. The function of this behavior is obscure, although it may be appetitive or, perhaps, displacement behavior. Truly hungry cats rarely play with prey; they eat it as soon as it is dead and they have recuperated from the predatory effort.

Treatment of Aggression in Cats. Pathological causes of aggression in cats are more common than in dogs. Meningiomas, feline ischemic syndrome (deLahunta 1977) and toxoplasmosis have been associated with aggression. Although some meningiomas can be removed so that the cat's behavior returns to normal, the other conditions cause tissue damage in the limbic system that is diffuse and, therefore, untreatable. Sudden onset of severe aggression is a poor prognostic sign. Euthanasia should be recommended because the cats may climb up the owner and attack the face. The aggression is usually well directed. For example, a cat with toxoplasmosis attacked dogs but only when she had been with her kittens. Beaver (1980), Borchelt and Voith (1985a), and Hart and Hart (1985) have addressed the problem of aggression in cats.

Feline aggression toward people can be subdivided into predatory/ playful and irritable. Predatory aggression is preceded by stalking and pouncing and is usually directed toward the feet of a moving person. If the cat is young and has no other kitten with whom to play, the aggression is probably play. In that case, the bite and scratches are usually inhibited; however, if the owners have not reprimanded the cat for biting too hard in play, it may not have learned to inhibit its bite. Playful aggression should be redirected toward swinging toys. The owner may swing a toy from a string and praise the cat verbally for attacking it but punish the cat for biting at people's feet. The best way to punish a cat is to startle it. Water guns or aerosol sprays are very effective. Even a loud noise like a whistle can be used. The best punishment is one that the cat does not associate with the owner. A similar strategy is used for true predatory aggression except that neither toys nor rewards are used; the cat is sprayed for attacking.

Irritable aggression usually occurs when the cat is being stroked. The cat, particularly a male, may react by giving the nape bite that it gives during copulation. If a verbal reprimand does not suffice, the owner can flick the cat's nose with a thumb and forefinger and pet the cat more gently for shorter periods.

INTERCAT AGGRESSION. Aggression among cats in the same household is the most common feline aggressive problem. If a new adult cat is introduced, aggression is to be expected, but aggression can also occur between cats that have lived peacefully together for years. In some cases, a physical change and/or a change in odor can precipitate the aggression. For example, if one cat is hospitalized, it may be attacked when it returns either because it is weak or because it smells different. This reaction could be termed nonrecognition aggression.

To treat aggression between cats, separate the cats initially. The cats should be fed in cat carriers at opposite ends of a room. Each day the boxes are moved closer together. The cats are being rewarded with food only in the presence of the other. When the cats can eat in their boxes side by side with no hissing or growling, the original can be let free in the room while the aggressor remains caged. When neither cat hisses or growls under these conditions, it is safe to release the cats. A similar method is to feed the cats on opposite sides of a solid door and then substitute first clear plastic and then screening so that visual and olfactory clues are added gradually. Diazepam has been used and is sometimes successful, probably if fear is the cause of the aggression.

Offensive aggression usually is increased when it is treated with diazepam. Progestins may also reduce aggression. Because of the high incidence of elimination and aggression problems with cats in multicat households, potential cat owners should be advised to have one cat. It is only during the cat's playful period from 4 to 11 weeks that companions are advisable.

AGGRESSION TOWARD VETERINARIANS. Aggression of cats aimed at humans is seen most often in the clinic. Although spontaneous aggression is uncommon in cats, clinicians would be quick to add that feline patients are among the most fractious. Clawing and biting in response to handling by a stranger is probably a fear reaction. Many of the difficulties encountered in handling cats are found when handling any small animal, for example, a toy poodle. It is difficult to restrain these animals without injurying them, and it is equally hard to examine them without being within range of their teeth and claws. "Cat bags" are commercially available to assist the veterinarian in restraining cats, although many practitioners would be quick to agree that no restraint is usually the best restraint for most cats. A recommended method for holding a cat is to place it under one arm. One hand holds both forelegs and the elbow of that hand is used to hold the cat firmly between the person's arm and body. The other hand is then free to stroke the cat under the chin and behind the jaw. This action is similar to bunting between two cats and has a calming effect.

3

Biological Rhythms and Sleep

INTERNAL RHYTHMS. An animal's day is as rigidly structured as that of any student, but the animal is responding not to alarm clocks, a 5-day work week, or the traditional meal patterns but to internal signals. The patterns of behavior, especially those of activity and sleep, reflect these internal rhythms. The several types of rhythms differ in period length. The circadian rhythms with period lengths of approximately 24 hours are the best known and best studied. The activity cycles of most animals, for example, are circadian in that the periods of activity and inactivity add up to approximately 24 hours. Other types of rhythms are high-frequency, ultradian, infradian, and annual cycles.

Types of Rhythms

High-frequency Rhythms. High-frequency rhythms have periods shorter than 30 minutes, for example, heart and respiration rates. Heart rate varies inversely with body weight; so the heart rate of a cat (110–130 beats/min) is considerably higher than that of a horse (28–40 beats/min). Respiratory rate does not vary linearly with body size; so the cow breathes 10 to 30 times per minute and the pig 8 to 18 times per minute. Respiratory cycles affect cardiac rate in that cardiac rate increases during inspiration. This effect, sinus arrhythmia, is more marked in dogs than in other domestic species. The endogenous nature of biological rhythms can be best appreciated by considering the contraction rate of the embryonic heart, especially that of the chick embryo, which has not even the maternal heart rate to influence it.

Ultradian Rhythms. Ultradian rhythms are longer than 30 minutes but shorter than 20 hours; for example, the fluctuations of growth hormone output from the pituitary, which in cattle occur in cycles of 3.5 hours (Blom et al. 1976). Body temperature also varies in ultradian cycles of approximately one hour in cats (Hawking 1971b). The physiological bases for or influences upon these short cycles are unknown but are believed to be the result of oscillations of cells in central pattern generators. A most interesting ultradian behavior rhythm is that of feeding. When food is available ad libitum, nearly all species eat 9 to 12 meals a day. This pattern is seen in dogs and cats (Kanarek 1975; Mugford 1977), sheep (Chase et al. 1976), horses (Laut et al. 1985), pigs (Bigelow and Houpt 1988), and cattle (Putnam and Davis 1963).

Circadian Rhythms. A circadian rhythm is self-sustaining and maintained under conditions of constant light or dark, with a period approximately 24 hours long. In addition to gross activity, a number of cellular and endocrinological parameters vary in a circadian rhythm. Enzyme levels, for example hepatic tyrosine aminotransferase in swine (Carroll and Stanton 1974), show a circadian fluctuation. Many hormones have been demonstrated to have circadian rhythms. Corticosteroids, including both cortisol and corticosterone, increase during the day in pigs and horses, with peak levels in late morning (Whipp et al. 1970; Bottoms et al. 1972) (Fig. 3.1). Both pigs and horses are diurnal (day active animals) and show diurnal peaks in adrenocortical activity and adrenal responsiveness to ACTH during the day, whereas cats (Scott et al. 1979) show increased adrenal activity at night. In stallions and boars, testosterone levels are highest during the day (Ellendorff et al. 1975; Kirkpatrick et al. 1976). Some hormones, such as vasopressin in the cat, show circadian rhythms in the cerebrospinal fluid but not in the blood (Reppert et al. 1981).

There are age-related effects on the expression of circadian rhythms. For example, although circadian rhythms occur in cortisol concentration in adult dogs, rhythms are not observed in either puppies or old (>12 years) dogs (Palazzolo and Quadri 1987). These age-related changes may be related to the restlessness observed in some older dogs.

Not all hormones are secreted in greatest quantity during the day in diurnal animals; growth hormone, for example, decreases in output during the day in pigs (Topel et al. 1973). Circadian rhythms of heart rate, body temperature, white blood cell number, metabolic rate, liver glycogen and glucose, hepatic phospholipid, and glucose absorption from the gut have also been identified (Aschoff 1965; Evans et al. 1976). Dogs show circadian rhythms of body temperature with a period length of 23.7 hours (Kanno 1977). The various rhythms demonstrated in farm animals may not be circadian in the strictest definition of the word because they have not been demonstrated to persist under conditions of constant light or constant

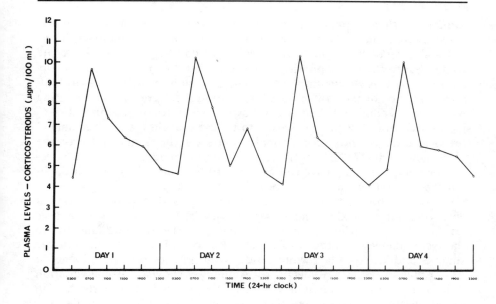

Fig. 3.1. Circadian rhythms of corticosteroid secretion in pigs. The graph is based on the mean results from six pigs. The peak of corticosteroid output occurs in late morning (Baldwin and Stephens 1973, copyright © 1973, with permission).

darkness. In fact, there do not appear to be circadian rhythms of locomotor activity in cats, dogs, and other carnivores, although there are circadian rhythms of hormone output and other functions.

Infradian Rhythms. Infradian rhythms have period lengths greater than 28 hours, but less than 2.5 days. Circatrigentian rhythms are those of approximately 30-day periods. The sexual cycles of polyestrous domestic animals show period lengths of approximately 2 to 3 weeks. The sow and cow come into heat every 21 days. Examples of species that are seasonally polyestrous include the mare, which comes into heat every 17 to 24 days in the spring, and the ewe, which comes into heat every 16 to 17 days in the fall (Farner 1961). These cycles may represent rhythms of hypothalamic, pituitary, or even ovarian activity. In rodents, such as hamsters, the estrous cycle is apparently controlled by the same biological clock as is activity, and the period of the sexual cycle is some multiple of that of the activity cycle. This may be true of burrowing solitary mammals, which must be sexually receptive during their active phase if they are to encounter mates and breed successfully (Fitzgerald and Zucker 1976).

Annual Rhythms. Annual or seasonal cycles are somewhat better understood. Horses and sheep are seasonal breeders. Horses are anestrous in the fall and winter and begin to show estrus as day length increases in late winter. Sheep show an opposite response in that they begin to be sexually active when days shorten in the fall. The evolutionary advantages to both species are obvious: the offspring of horses and sheep are born in the spring when food is abundant. Dogs now show sexual cycles of approximately 6-month duration, but there is reason to believe that they were once annual breeders also. Basenjis, for example, still show an annual fall breeding season (Scott and Fuller 1974). Domestication, abundant food, and selective breeding also may have caused swine to become polyestrous throughout the year rather than during one season.

Not all annual cycles are reproductive. Cats show annual cycles in corticosteroid, thyroxine, and epinephrine levels. Peak levels of these three hormones occur during the winter (Randall et al. 1975). More familiar are the cyclic changes in hair coat. Hair follicle activity in cats is highest in late summer and lowest in late winter, and, as a result, fur is 0.5 mm longer in winter than in summer (Ryder 1976). Adult ewes show a seasonal variation in heart rate, with a minimum in winter. Horses show seasonal rhythms in carbohydrate metabolism, but these may be related to training (Gill et al. 1974).

Feeding Rhythms. An important rhythm is that of feeding. When, instead of freely feeding, an animal has meals imposed on it, the situation for most domestic animals, the animal anticipates the meal with an increase in activity. To entrain this rhythm the meal must contain calories (Mistelberger et al. 1990). The feeding rhythm is independent of the suprachiasmatic nucleus. Rhythms of intestinal enzymes are secondary to feeding (Fisher and Gardner 1976).

Parasitic Rhythms. Still another facet of biological rhythms of importance to the clinician is the rhythm of parasites. Perhaps the best example is that of *Dirofilaria immitis,* canine heartworm. Microfilaria are most active and most likely to be found in the peripheral circulation in the evening (Hawking 1971b). The activity peak nicely coincides with that of the insect vectors, the mosquitoes, that will carry the microfilaria to a new host. The clinician may have more success in making a positive diagnosis of the presence of the parasite by examining blood taken in the evening.

External Influences. Circadian rhythms are endogenous, that is, the rhythm persists under conditions of constant light or constant darkness, but circadian rhythms are usually influenced by external factors. Some of these factors are light, temperature, barometric pressure, various drugs, and hormones. Of these factors, the most important is light.

Light. Circadian rhythms are entrained to light; that is, although under conditions of constant illumination a rhythm may have a period length of approximately 24 hours, under natural lighting the rhythm will be that of the light-dark cycle. The light must be present during a specific portion of the endogenous rhythm. Hamsters, for example, entrain to a 12-hour-light–12-hour-dark day and to a 6-hour-light–12-hour-dark day but not to a 6-hour-light–30-hour-dark day (Elliott et al. 1972).

Considerable practical advantage has been taken of the entraining function of light to bring mares into estrus early or, conversely, to avoid injuries during hierarchy formation by keeping pigs in the dark. Even the simple act of putting a cover over a parrot's cage uses the effect of light on avian activity.

Not all types of light are equally effective in entraining circadian rhythms. Green light is most effective, and red light is least effective; therefore, red light may be used when visibility is desirable but interference with an animal's circadian rhythms and dark activities is not (McGuire et al. 1973). When *Zeitgebers* (literally, "time givers") are removed, the resulting desynchronization of internal rhythms may have deleterious results. For example, thermoregulation may be impaired (Fuller et al. 1978).

Barometric Pressure. The influence of other factors on circadian rhythms has not been as well studied, but barometric pressure has been shown to influence activity patterns. Mice show higher activity levels when barometric pressure is increasing (Sprott 1967). Although the phenomenon has never been quantified, farm animals, such as horses and pigs, show high levels of activity before storms, and tail-biting episodes often occur in swine just before storms.

Drugs. Drugs can affect rhythms. Examples are caffeine and theophylline (Ehret et al. 1975) and lithium (Johnsson et al. 1980; McEachron et al. 1982; Delius et al. 1984). The action of lithium on circadian rhythms of humans as well as animals may be the basis for its amelioration of depression. The two compounds that may be useful for treatment of "jet lag" and sleep disturbances of shift workers are melatonin (see Pineal Gland) and benzodiazepines (Seidel et al. 1984; Turek and Losee-Olsen 1986).

More important from a clinical standpoint is that a given drug may have a greater effect and/or have lower toxicity at one time of day than at another. Hypoglycemic agents are more apt to precipitate hypoglycemic convulsions if administered while liver glycogen and plasma glucose are low than if administered at another point in the cycle (Johnson and Chura 1974).

BIOLOGICAL BASIS FOR ENTRAINING RHYTHMS. The rhythms discussed are manifested by changes in the animal's behavior and by changes in its physiology, but what is the biological basis for these changes? Where is the biological clock or clocks that measure time and signal the rest of the body, that is, entrain the rhythms?

Cycle Variations within Cells. At the most fundamental intracellular level there are cyclic variations in the macromolecules within the cell. Relative levels of deoxyribonucleic acid (DNA) and ribonucleic acid (RNA) show circadian rhythms (Harker 1964). Inhibitors of protein synthesis shift circadian rhythms by interfering with ribosomal activity (Jacklet 1977).

Suprachiasmatic Nucleus. There are also rhythms of cellular activity. For example, the cells of the lateral hypothalamus show a circadian rhythm of responsiveness (Terman and Terman 1970; Schmitt 1973). The rhythmic activity of the endocrine organs has been noted.

The Pacemaker. Probably at least one master clock within the body entrains the rhythms of the other organs and cells to environmental Zeitgebers, such as light. The best candidate for the integration of circadian rhythms is the suprachiasmatic nucleus of the hypothalamus. Removal of that area abolishes circadian rhythms in rodents (Ibuka and Kawamura 1975; Stetson and Watson-Whitmyre 1976). Stimuli from the optic nerve are transmitted to the suprachiasmatic nuclei, which can, in turn, transmit information to the rest of the brain.

There may be other pacemakers in addition to the supra-chiasmatic nucleus, because although circadian rhythms of activity may be abolished by lesions of that nucleus, other rhythms, such as that of body temperature, may remain (Moore-Ede et al. 1982). Normally, of course, the Zeitgeber will set both pacemakers or clocks so that the rhythms are in synchrony. One of the causes of jet lag (which may or may not affect animals, such as race horses that are transported across several time zones) is the desynchronization of rhythms.

Pineal Gland. The pineal gland is probably an important intermediary in the synchronization of circadian rhythms, for it demonstrates marked rhythms of output of several hormones and neurotransmitters (Garbarg et al. 1974). Melatonin is produced by the pineal and is present in higher quantities in plasma and cerebral spinal fluid at night (Hedlund et al. 1977) (Fig. 3.2). Melatonin has an antigonadotropic effect in long-day breeders and a progonadotrophic effect in short-day breeders. It may be the means by which the hypothalamus is apprised of day length, the link between circadian rhythms and annual sexual cycles, for melatonin would increase as dark period length increases, and the increased melatonin levels would depress gonadal activity. The practical application of this role of melatonin is that short-day breeders, such as sheep, may be brought into estrus earlier

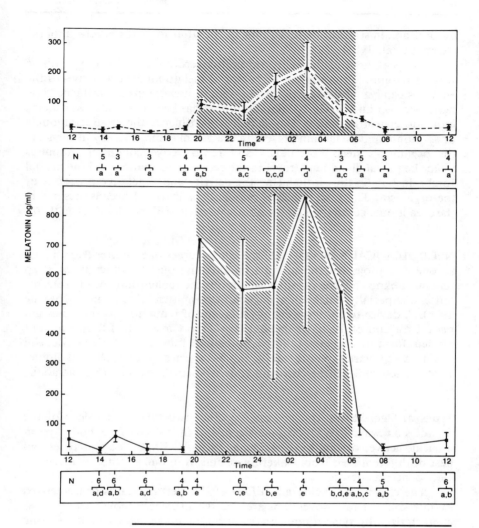

Fig. 3.2. Nyctohemoral cycle of melatonin concentration in calf plasma (*upper*) and cerebrospinal fluid (lower) at selected times of the day. The lights were on from 0600 to 2000 hours. Data points are means ± 1 standard error of samples from *N* number of calves. Letters indicate that when two means do not have any letter in common they are significantly different (Hedlund et al. 1977, American Association for the Advancement of Science. copyright © 1977 with permission of AAAS).

by oral administration of melatonin in the afternoon for several months (Arendt et al. 1983).

The peak level of the neurotransmitter serotonin is 180° out of phase with melatonin. Serotonin is a precursor of melatonin. The activities of the enzymes catalyzing the reaction (serotonin to melatonin) are influenced by light (McGeer and McGeer 1966). Serotonin has been implicated as a sleep-inducing neurotransmitter. Day length will influence the relative amounts of serotonin and melatonin present in the pineal gland, thereby influencing the organism's sleep-wakefulness and reproductive condition. Impulses from the pineal gland may travel to the reticular formation and the medial forebrain bundle thus influencing the activity and vegetative processes of the organism. The raphe nucleus may also be involved because lesions in the area influence rhythms of locomotory activity (Block and Zucker 1976).

NEUROLOGICAL BASIS OF SLEEP. Sleep occupies one-quarter (ruminants) to one-half (dogs) of the lifetime of animals, but the function of sleep remains unknown. One possible function is the replenishing of neurotransmitters, especially catecholamine neurotransmitters (Stern and Morgane 1974). A device to conserve energy, a means of remaining inconspicuous, a period for consolidation of memory, or simply a way to fill up time not needed for foraging are other hypothetical functions of sleep (Zepelin 1989). Sleep varies considerably among species (Campbell and Tobler 1984). These differences correlate roughly with body size and metabolic rate.

Types of Sleep. Sleep can be classified into two types: the "sleep of the mind," slow wave (SWS), or quiet sleep; and the "sleep of the body," paradoxical, active, or rapid eye movement sleep (REM). The two types can best be differentiated from wakefulness and from one another by means of electroencephalography.

The electroencephalogram (EEG) of the alert animal is characterized by low-voltage, fast waves that are not synchronized. Slow wave sleep is characterized by synchronous waves of high-voltage, slow activity. During paradoxical sleep, the EEG shows low-voltage, fast activity similar to that seen in the wakeful state (Fig. 3.3), but there is very little muscular activity; therefore, this type of sleep is called the "sleep of the body." The animal is more difficult to arouse than when it is in SWS. Although muscle tone is very low during paradoxical sleep, the muscles of the eyes frequently contract; hence the term rapid eye movement (REM) sleep. The low-voltage, fast activity of REM sleep does not result in many body movements because there is an inhibitory area in the medulla that, in effect, paralyzes the muscles of the body (Morrison 1983). Humans awakened from REM sleep report that they have been dreaming; the vocalizations and the muscular activity of the face and legs during canine sleep indicate that dogs may also

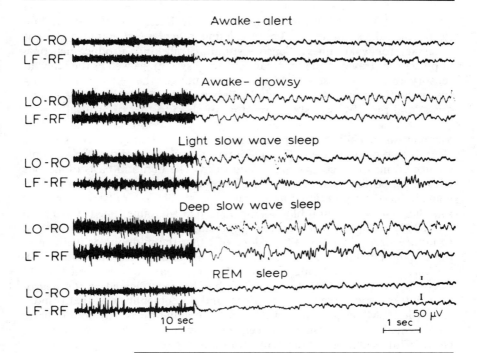

Fig. 3.3. The stages of vigilance and sleep in the cat. Polygraphic record at the speed of 30 mm per second showing the stages identified. LO-RO electroencephalographic record from the left and right occipital area; LF-RF electroencephalographic record from left and right frontal area; electromyograph (EMG) record from the muscles of the neck; electrooculogram (EOG) record of eye movement (Ursin 1968, copyright © 1968, with permission of Pergamon Press).

be dreaming. One can only speculate about the presence or content of animal dreams. REM sleep, however, appears to be the most critical or necessary component of sleep. REM sleep deprivation results in behavioral abnormalities in all species tested, and rebound or extra REM sleep occurs during recovery from deprivation (Ruckebusch 1972, 1974; Ursin et al. 1976). Rats die after 10 to 50 days of total sleep or REM sleep deprivation, and the symptoms are those of increased energy utilization (Everson et al. 1989; Kushida et al. 1989). This gives strong suppport to the energy-conserving function of sleep.

Wakefulness apparently depends upon brain areas behind the origin of the oculomotor nerves and above the lower medulla, because transection of the brain at the former location abolishes wakefulness in cats (EEG patterns of sustained SWS), while transection at the latter level has no effect on sleep-wakefulness patterns. The ascending reticular activating system is necessary for wakefulness. Sensory stimuli of all kinds produce neuronal impulses that travel to the reticular activating system, as well as the relevant sensory cortex. If these stimuli are strong enough, the reticular activating system will be stimulated to send impulses to higher central nervous system areas, and the EEG will shift from that of sleep to that of arousal.

Mechanism of Falling Asleep. The mechanism of falling asleep is not as clear as that of sleep itself. The buildup of a neurotransmitter, for example serotonin, may induce drowsiness and sleep. The humoral basis of sleep is further supported by studies in which cerebrospinal fluid of sleep-deprived goats induced sleep in rats (Pappenheimer et al. 1975). Similarly, parabiotic rats that have a common blood supply sleep at the same time (Matsumoto et al. 1972). Clinically, tryptophan, a precursor of serotonin, may be used to induce sleep. Anatomically, lesions of either the midpons or of the preoptic hypothalamus abolish SWS in cats (McGinty and Sterman 1968). Lesions of the locus ceruleus and nucleus reticularis pontis abolish REM sleep (Teitelbaum 1967). Acetylcholine has been implicated as the neurotransmitter responsible for wakefulness and norepinephrine as that responsible for REM sleep (Jouvet 1967), but the situation is, no doubt, more complicated. Acetylcholine has been shown to induce REM sleep (Sitaram et al. 1976), and more of the neurotransmitter is released from the cerebral cortex during REM sleep and wakefulness than during SWS sleep (Jasper and Tessier 1971).

PATTERNS OF SLEEP AND ACTIVITY IN DOMESTIC ANIMALS. The activity patterns of the various species described next are reported from laboratories or under specific environmental conditions. The behavior patterns may be different under different environmental conditions and, therefore, the numbers given should not be considered applicable to every animal under every condition. Allison and Cicchetti (1976) hypothesize that sleep time is inversely related to the danger of predation for a given species. Roughly speaking, predators sleep more than prey animals and large animals more than small animals. See Figure 3.4 for activity patterns of three species in the same pasture.

Dogs. Sleeping dogs lie in a characteristic posture with their hindlegs tucked up and their heads turned caudolaterally. Their eyes may be open or closed. Rapid eye movement sleep may be accompanied by leg movements, vocalizations, and either polypnea or apnea. A dog awakened abruptly from REM sleep may bite, so one should let dreaming dogs, or at least sleeping dogs, lie, or awaken them gently. Dogs show short periods of activity interspersed with periods of rest when free ranging (Beck 1973), when tethered

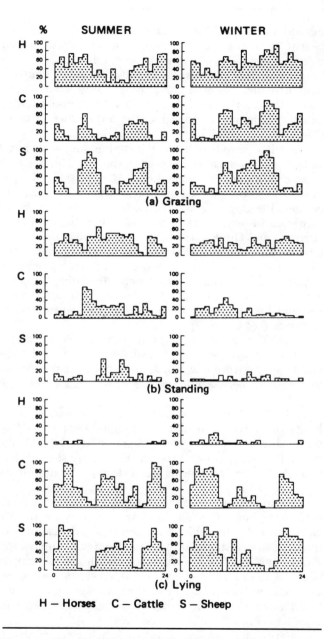

Fig. 3.4. Time budgets of domestic herbivores. Grazing, standing, and lying behavior of cattle, sheep, and horses living on the same pasture during two seasons (Arnold 1984–85, copyright © 1984, with permission of Elsevier Scientific Publishing and CSIRO).

outdoors (Delude 1986), and when caged (Hawking 1971b). Pet dogs appear to sleep at night, but their behavior may be entrained to that of their owner. Caged dogs, alternate 30-minute to 2-hour active periods with longer quiescent periods. During the day, caged dogs spend more than one-half their time sleeping (20% in light slow wave sleep, 25% in deep slow wave sleep, and 10–12% in REM sleep) and one-quarter of their time standing. The rest of the time, the dogs either sit or lie (Hite et al. 1977) (Fig. 3.5).

Similar periods of activity just after sunrise and an hour or two before sunset occur in huskies tethered to their dog houses by 8-foot leads. Under these conditions, dogs spend over 80% of their time, night and day, lying down and sit only 2% of the time, mostly while observing another dog or a person. Pet dogs are walked or let out early in the morning when their owners arise and in early evening when their owners return from work (Berman and Dunbar 1983). A peak can also occur at noon, apparently in communities where people return home for lunch (Lehner et al. 1983). These studies represent the only ones of pet dog behavior outdoors and did not include any behaviors that were not visible from a passing vehicle. The activity patterns of owned dogs free in their homes would be very interesting and might give some clue as to the causes and cures of destructive behavior (see Chap. 9).

Free-ranging urban dogs are most active in early morning and in the evening. Foraging for food, usually garbage, socializing with other dogs, and traveling from alley to street to park are their major activities and are usually interspersed with periods of rest (Beck 1973) (Fig. 3.6).

Cats. In the laboratory, cats, like dogs, show short bursts of activity (1 to 2 hours of activity distributed throughout) and are 1.4 times more active during the day than at night (Sterman et al. 1965) (Fig. 3.7). Caged cats spend 10 hours per day sleeping. Slow wave sleep occupies 39% of the day and REM sleep 8%. During REM sleep the nictitating membrane covers the eye.

Ursin (1970) has found that SWS of cats can be subdivided into light slow wave sleep (LSWS) and deep slow wave sleep (DSWS), based on electroencephalographic characteristics and ease of arousal. The usual sequence of sleep stages in the cat is from wakefulness to LSWS to DSWS to REM to either LSWS again or to wakefulness. The two major sleep epochs occur at night. Total sleep time decreases in continuous light but so does alert wakefulness. Cats in continuous light drowse more. Cats fed three times a day drowse more than those fed once a day, and fasted cats drowse even less (Ruckebusch and Gaujoux 1976).

Farm cats also spend 40% of their time asleep, most of it at night. Although active at dusk and into the evening, cats are not really nocturnal. The rest of the farm cat's time is divided into 22% resting, 14% hunting (although this will vary from cat to cat), 15% grooming, 35% traveling and

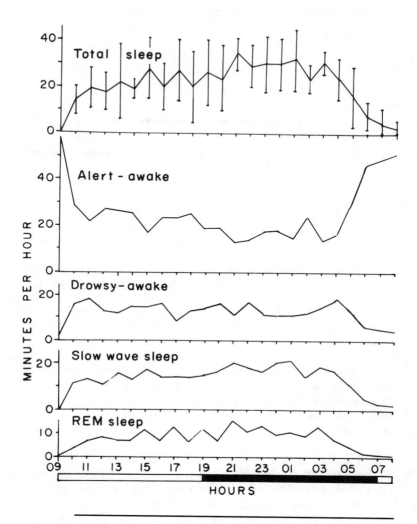

Fig. 3.5. Activity patterns of laboratory dogs. Hourly totals of the four states of vigilance and for total sleep. The ordinate designates the mean number of minutes occupied by each state for 1 hour starting with the 0900 to 1000 hour interval at the far left. Lights were on from 0700 to 1900 (light horizontal bar below). Over 60% of total sleep occurred during the 12-hour period of darkness. The dogs slept relatively little from 0500 until 1500 hours. Drowsiness is nearly equally represented in each hourly interval except during the peak of wakefulness from 0600 to 1000 hours. The vertical bars on the total sleep graph represent one standard deviation from the mean (Lucas et al. 1977, copyright ©, with permission of Brain Research Publications and Pergamon Press).

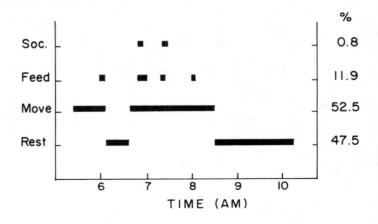

Fig. 3.6. Activity patterns of feral urban dogs. The activities of two dogs during one morning in the fall. Soc = social behavior with other dogs, including sniffing and chasing. Feed = feeding behavior, including rummaging through garbage and eating. Move = walking and running. Rest = resting or sleeping (Beck 1973, copyright © 1973, with permission of York Press).

2% feeding (Panaman 1981). Urban cats are most frequently seen and presumably most active at night between dark and dawn (Berman and Dunbar 1983). Older cats (>10 yrs) show less REM sleep and more SWS, as well as more brief episodes of wakefulness than young cats (Bowersox et al. 1984).

Pigs. Pigs are usually kept in confinement so that the 7 hours of rooting (for example, food searching) noted on pasture (Hafez and Signoret 1969) fall to 2 hours per day of eating in a pen. Pigs spend more time resting than any other domestic animal (Haugse et al. 1965). They are recumbent 19 hours a day. They drowse 5 hours a day. Slow wave sleep occupies 6 hours a day and REM 1.75 hours in 33 periods. Only 1 to 3 hours per day are spent in other activities, such as drinking, walking, playing, or fighting (Fraser 1975a). Domestic pigs are diurnal, and most activity takes place during the day (Morrison et al. 1968). Although motor activity and food intake increase during the day in pigs, these rhythms disappear in constant light, indicating that they are not circadian. Rhythms of body temperature depend on feeding and do not occur in pigs fed ad libitum (Ingram and Dauncy 1985). Pigs can entrain to 9 hours of light and 9 hours of dark as well as to 12:12 cycles (Ingram et al. 1985). Feral pigs and wild boars are more nocturnal during the summer months, probably to avoid predation.

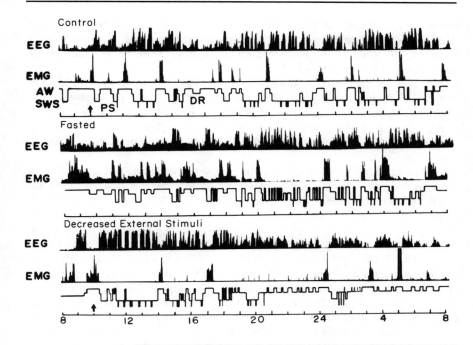

Fig. 3.7. Distribution of sleep episodes in the cat. Integrated electrocorticogram (EEG) and electromyogram (EMG) and diagram showing wakefulness (*AW*), drowsiness (*DR*), slow waves (*SWS*), and paradoxical sleep (*PS*). Arrows indicate feeding; slow wave sleep episodes are fragmented on 3d day of fasting. Concentration of sleep during daytime with reduced external stimuli (Ruckebusch and Gaujoux 1976, copyright © 1976, with permission of Elsevier Scientific Publishing).

Even wild pigs are not very active. Trapping and retrapping indicated that sows were usually found within 0.3 km of the point at which they were originally trapped; boars were within 2.0 km (Martin 1975). The sex difference may be in range rather than in activity.

Pigs are characterized by extreme muscle relaxation during sleep (see Fig. 3.10). It is difficult to evaluate muscle tonus in a 400-pound sow, but when a sleeping piglet is picked up it is as relaxed as a rag doll.

Circadian rhythms are disrupted by changes in physical, social, or reproductive conditions. For example, putting a pig into a group, or tethering it, after it had been housed individually, as well as surgery or estrus, will disrupt circadian rhythms of cortisol for one to four days (Becker et al. 1985).

Horses. Horses may be able to drowse and even to engage in slow wave sleep while standing by means of the unique stay apparatus of the equine hindlegs, but they lie down (Hale and Huggins 1980) when engaging in REM sleep. The horse in REM sleep lies either in lateral recumbency or in sternal recumbency with its muzzle touching the ground so that its head is supported during the atonia accompanying that phase. A healthy horse seldom remains lying when it is approached, probably because a standing horse is better able to flee or to defend itself. It is interesting that a dominant stallion lies down first (Ruckebusch 1975), that is, before subordinate horses lie down.

During the day the horse is awake 88% of the time, and most of this time the animal is alert. Even at night the horse is awake 71% of the time, but it drowses for 19% of the night (Ruckebusch 1972). Stabled horses are recumbent 2 hours a day in four to five periods. Ponies are recumbent 5 hours a day and donkeys even more (Littlejohn and Munro 1972). Slow wave sleep occupies 2 hours a day and REM sleep occurs in nine periods, each five minutes long. Horses, unlike ruminants, show tachycardia, leg movements, and an increase in respiratory rate during REM sleep (Ruckebusch 1972). Management practices can affect equine sleep patterns. Previously stabled horses sleep less on pasture; they don't lie down during the first night, and total sleep time remains low for a month. If horses are tied short in a straight stall so that they cannot lie down, they do not REM sleep. The horses compensate by sleeping while free during the day. Care must be taken not to deprive horses of sleep inadvertently. This is most apt to occur when horses are transported long distances and must be tied in straight stalls. Diet also affects length of sleep in horses, as it does in ruminants. Lying increases when the protein content of grasses increases in the spring (Duncan 1985); a similar trend, an increase in lying, occurs when oats are substituted for hay; fasting has the same effects (Dallaire and Ruckebusch 1974a,b).

Horses feed, lie, stand, and travel, but the main activity is feeding, either eating hay when that is available free choice or grazing. Grazing time varies from nearly 50% to 80% of the 24 hours and occurs both day and night. The time of day when horses graze varies with the presence of biting insects. More time is spent grazing during the day in the winter when forage is scarcer (more steps between bites of food) and when biting insects do not drive the horses to refuge in snow, water, or barren areas (Duncan and Cowtan 1980; Keiper and Berger 1982–83). Horses graze 15 minutes less for every extra hour of sunlight per day in the spring. Lactating mares graze more than barren or pregnant mares, reflecting the greater energy demands of milk production. Table 3.1 lists the percentage of time spent grazing by different populations of free ranging or pastured horses and ponies. The large amount of time occupied by oral behavior indicates the reasons for the appearance of oral vices, such as wood chewing and cribbing in stalled horses on low roughage diets.

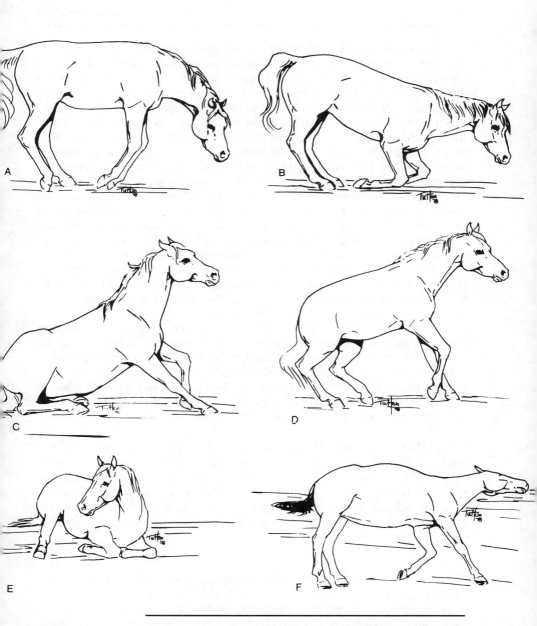

Fig. 3.8. The postures of the horse when lying down and getting up. (*A, B*) Lying down; (*C, D*) getting up; (*E*) sternal recumbency, the horse not lying symmetrically but with the lateral surface of one foreleg on the ground; (*F*) lateral recumbency, the upper foreleg anterior to the lower one. The horse exhibits REM sleep in *F* or in *E* with the muzzle touching the ground.

Ponies on pasture lie down 7% of the night, 2% in lateral recumbency. Lying in lateral recumbency occurs only in the hours just after midnight. In stalls, the same ponies lie down 12% of the time, probably in response to the drier stall environment. Horses may lie in lateral recumbency during the day, usually after bad weather has kept them from lying down. The amount of traveling a horse does depends on two things: the availability of nutrients and its social status. Young bachelor stallions travel more than harem stallions or mares, but otherwise the distance that must be traveled to procure water or enough forage determines the amount of movement (Berger 1986). Horses isolated from other horses walk considerably more than those in sight of other horses. This type of walking, like that of the bachelor stallion, is presumably a search for companions. Camargue horses grazing on a large pasture walk 7%–10% of the day and so do horses in a grassless corral (Duncan 1980; Houpt and Houpt 1988). Rapid traveling, that is trotting and cantering, occupy a very small fraction, less than 1%, of the time budgets of adult horses studied in a variety of environments. Horses do not spend as much time traveling at night as they do during the day. Ponies on lush pasture walk 3% of the night whereas stalled horses and ponies walk less than 1% of the night time.

Standing behavior occurs when horses are not engaged in acquiring food, socializing, or sleeping deeply. Standing increases when feeding decreases (see Table 3.2 for time budgets of confined horses and ponies on a variety of diets). Standing in pastured horses is also influenced by weather conditions. Horses stand rather than lie when it rains and stand 20 minutes more per day for every degree (Celsius) drop in environmental temperature in a relatively mild Mediterranean climate (Duncan 1985).

Cattle. Cattle are essentially diurnal (day active). Their major activities are grazing, ruminating, and resting. Lying occupies nearly one-half the cow's day; when deprived of the opportunity to lie down, she will compensate by lying for longer periods when she is free to do so. This compensatory behavior indicates that rest is necessary. In fact, when deprived of both rest and food, cattle lie down rather than eat when given the opportunity to do one or the other (Metz 1985).

Grazing. Most grazing takes place during the day (Atkeson et al. 1942). Cattle on pasture spend anywhere from 5 to 8 hours grazing.

Grazing time is inversely proportional to the quality of the pasture. Cattle on moderately good pasture spend 5 hours actually gathering food with their prehensile tongues and 2 hours walking (Johnstone-Wallace and Kennedy 1944). As herbage becomes more sparse, more walking between mouthfuls is necessary. The distance covered by a grazing cow varies from 0.6 to 9 km per day, depending on the size of the pasture or range and the abundance of forage. The distance traveled by cattle or sheep can be

Table 3.1. Time budgets of stabled horses and ponies

Time/Equid	Environment	Diet	Feeding	Standing	Moving	Drinking	Lying	Reference
					Percentage of Time			
Day/Horse	Metabolism cage	Limited hay	50	45	0	1	4	Willard et al. (1977)
		Concentrates	32	62	0	1	5	Willard et al. (1977)
Day/Pony	Box stall	Ad lib hay	76	19	3	2	1	Sweeting et al. (1985)
Day/Horse	Corral	Limited hay and grain	43	27	6	–	0	Houpt and Houpt (1988)
Night/Pony	Box stall	Limited hay and grain	15	17	1	–	13	Houpt et al. (1986)
Night/Horse	Box stall	Limited hay and grain	27	67	0.3	–	6	Shaw et al. (1988)
24 hr/Pony	Pen	Ad lib grain	17	–	–	2	–	Laut et al. (1985)
24 hr/Pony	Pen	Ad lib pellets	31	–	–	–	–	Ralston et al. (1979)
Day/Horse	Small pen		68	18	12	1.5	0	Boyd (1988)
Day/Przewalski	Large pen		44	45	8.5	1.5	0	Boyd (1988)

Table 3.2. Percentage time grazing by various populations of free-ranging horses

0600	0800	1000	1200	1400	1600	1800	2000	2200	2400	0200	0400	Reference	Season	Population
					Time of Day									
98	95	33	85	70	80	80	81	–	–	–	–	Salter and Hudson (1979)	Winter	W. Alberta
100	65	45	90	70	80	92	80	–	–	–	–	Salter and Hudson (1979)	Summer	W. Alberta
80	77	83	83	75	72	75	–	–	–	–	–	Keiper (1981)	Summer	Assateague Is. MD–VA
–	–	–	–	–	55–64[a]	–	63	53	53	40	70	Keiper and Keenan (1980)	Summer	Assateague Is. MD–VA
–	–	–	–	–	51–60[a]	–	–	–	–	–	–	Duncan (1980)	Winter	Camargue, France
–	–	–	–	–	25–50[a]	–	–	–	–	–	–	Duncan (1980)	Summer	Camargue, France
80	80	75	85	80	85	70	75	70	–	–	–	Berger (1977)	Summer	Grand Canyon, CO
70	70	60	55	50	60	65	65	60	–	–	–	Rubenstein (1981)	Winter	Shackleford Is. SC
–	–	–	–	–	–	–	–	–	–	–	–	Rubenstein (1981)	Summer	Shackleford Is. SC
–	40	–	31	–	38	41	41	68	60	60	60	Boyd et al. (1988)	Summer	Front Royal, VA

[a]Average percentage of time spent grazing over total time period.

measured by a rangemeter, a device like an odometer (Cresswell 1959; Anderson and Urquhart 1986).

Grazing usually occurs in bouts and is engaged in by the entire herd; social facilitation is strong in cattle. There are two major grazing bouts: one just after sunrise and the other during late afternoon until sunset (Hughes and Reid 1951). Midmorning and midafternoon are resting and idling times (Walker 1962). By 1 hour after sunset, most cattle are lying down, although they will usually rise to graze during the night (Gary et al. 1970). Night grazing may increase during warm weather (Seath and Miller 1946). Cattle drink two to four times a day during summer on the range but only once or even every other day during the winter (Cory 1927).

Housing Conditions. Cattle live not only on the range but also in varying degrees of confinement. Dairy cattle are milked at least twice a day, and their grazing habits are organized around the milking schedule. The most intense grazing activity follows each milking. There are two bouts after the afternoon milking and one before the morning milking. A brief bout of rumination follows each grazing bout. A total of 5.5 hours during daylight is spent grazing, and an equal time is spent ruminating. In contrast, cattle in a loose-housing situation spend only half as much time eating and ruminating as cattle on pasture. They spend 6 to 7 hours loafing, that is, standing neither grazing nor ruminating, and 12 hours resting. Time spent walking decreases and the frequency of urinations increases with the size of the idling area available to the cattle. See Figure 3.9 for activity patterns of stanchioned cattle.

Cattle on feedlots are in a highly unnatural environment, as reflected in their activity patterns. Grazing bouts are replaced by 9 to 14 feeding periods, 70% to 80% of which occur during daylight hours. If hay and/or silage are fed, a total of 5 hours a day is spent eating, but the time decreases as the percentage of concentrates in the diet increases or if the roughage is ground (Putnam and Davis 1963). Standing increases at the cost of lying during rain or snow (Gonyou and Stricklin 1984).

In hot climates moving into the shade is another activity that appears to be a response to light rather than to temperature per se. Cattle should, of course, have access to shade, and shading behavior should be considered when management plans are made. Cattle defecate 7 to 15 times a day and urinate 5 to 13 times. The frequency of both excretory activities decreases in hot weather (Corbett 1953). Rumination time also decreases under these conditions. See Table 3.3 for activity patterns of cattle in different environments.

The presence or absence of true sleep in ruminants has been controversial (Balch 1955; Bell 1960; Merrick and Scharp 1971), but the extensive studies of Ruckebusch (1972, 1974, 1975) and Ruckebusch et al. (1974) indicate that cattle show both REM sleep and SWS. Rapid eye movement sleep occurs in 11 periods so that the total of 45 minutes of REM sleep

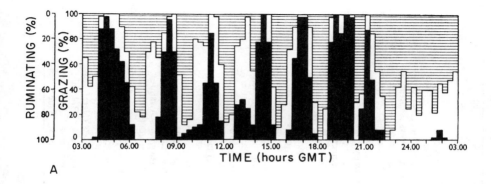

A

Fig. 3.9. (*A*) Activity patterns of Hereford steers in a paddock. Black areas indicate grazing; cross-hatched areas, ruminating; and blank areas, idling (Hughes and Reid 1951). (*B*) Activity cycles of stanchioned cattle. Indicated for each cow are the periods of standing, rumination, eating (solid), drinking (vertical lines), and social interaction (dots). The x-arrow indicates the period of feeding and cleaning (horizontal line), milking (dot), and ear movement rest (squares) (Hedlund and Rolls 1977, copyright © 1977, with permission of *J. Dairy Sci.*).

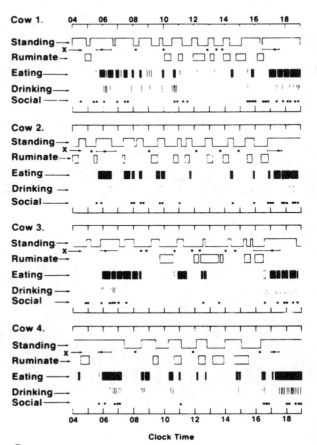

B

Table 3.3. Activity patterns of cattle in different environments

Author	Grazing (hr)	Number of Grazing Bouts	Ruminating (hr)	Lying (hr)	Walking (hr)	Standing (hr)	Idling (hr)	Type of Cattle
Cattle on pasture								
Atkeson et al. (1942)	5.5–7.5	6 (2 at night)	—	13	—	4	—	Dairy cows
Bailey et al. (1974)	5.5–10	—	—	—	—	—	—	Beef steers (Hereford)
Castle and Halley (1953)	6.5	—	5.5	9.25	—	—	8.25	Dairy cows (shorthorn)
Corbett (1953)	8	5–7 (1 at night)	5.5	9.25	—	—	3.50	Dairy cows (shorthorn)
Culley (1938)[a]	7–9	2	—	—	—	—	—	Beef cattle
Dalton et al. (1967)	7.25–7.5	4–5	4	—	—	—	2	Dairy calves
Gary et al. (1970)	6	3	—	8.25	—	—	9	Beef cows (Charolais)
Hancock (1954)	10–12	6 (1 at night)	8	—	—	—	4	Dairy cows
Hardison et al. (1956)	9	4 (1 at night)	8.5	9	—	15	6	Dairy cows (Holstein)
Harker et al. (1954)	7–8	—	4.5	5	0.25	3.25	—	Zebu cattle
Hein (1935)	9(8–11)	2	8	2	—	—	—	Steers
Herbel and Nelson (1966a)	9–10	4	8	2	2–3	2	—	Beef cows (Hereford, Santa Gertrudis)
Holder (1960)	11.50	—	8.50	—	—	4	—	Dairy cows
Hughes and Reid (1951)	8	5	8	—	—	—	9	Beef steers (Hereford)
Johnstone-Wallace and Kennedy (1944)	7–8	5	7	12	—	—	—	Beef cattle (Hereford)
Kropp et al. (1973)	11.50	5	7	—	1.25	—	5	Beef and dairy heifers
Lampkin et al. (1958)	7	2	7	—	—	—	—	Zebu and grade steers
Larsen (1963)	—	4	—	—	—	—	—	Dairy cows (brown Swiss)
Lofgreen et al. (1957)	6–8	4–8	—	9–11	—	—	7–12	Steers
Moorefiled and Hopkins (1951)	10–10.5	—	—	9–11	—	2–3.5	—	Beef cattle (Hereford)
O'Donnell and Walton (1969)	9.5–12	3	—	10–14	—	1.25–4	—	Nonlactating cattle
Seath and Miller (1946)	8–9.5	4	—	—	—	—	—	Dairy cows
Sneva (1970)[a]	10	2	—	—	1.5	—	—	Beef cattle
Walker (1962)	—	3	—	—	1.5	—	—	Beef cattle
Wardrop (1953)	7(5.5–8)	5	6.25(4.5–9.5)	5	1[a]	—	—	(Dairy cows Ayrshire)
Zemo and Klemmedson (1970)	9	6	7	5	1[a]	6	—	Beef steers (Hereford)
Cattle in Confinement								
Friend and Polan (1974)	3–5	—	—	—	11	—	—	Dairy cows[b] (Holstein)
Lewis and Johnson (1954)	4–5	—	—	—	8–11	—	—	Dairy cows[c] (brown Swiss)
Putnam and Davis (1963)	3.5–5.25	9–12	—	—	—	—	—	Steers[d]
Schake and Riggs (1969)	3.5	4	7.5	9.5	6.5	14	1.25	Beef cows[d] (Hereford)
Schmisseur et al. (1966)	3–4	—	—	—	12.25	—	6–7	Dairy cows[c] (Holstein)
Turner (1961)	5	10	—	10.5	—	8.5	—	Dairy cows[d] (Ayrshire)
Webb et al. (1963)	6.25	18	—	—	—	—	—	Dairy cows[e] (Guernsey)

[a]Daylight observation only.
[b]Free stall.
[c]Loose housing.
[d]Feedlot.
[e]Cowshed

3.5 hours of SWS is divided into many short naps. Cattle in REM sleep usually are lying down with their heads resting on the ground and turned back into the flank. Cattle sniff the ground before lying down and on rising lick and scratch themselves. Cows in slings are sleep deprived as are cows that have not yet adjusted to stanchioning or newly mixed groups of cattle. The stress of sleep deprivation should be considered by clinicians and stock managers.

When kept in a corral at night, cattle, or Zebu cattle at least, tend to sleep in areas that remain constant for each individual from night to night. The resting places do not appear to depend on dominance (Reinhardt et al. 1978a). Most characteristic of ruminants are the extensive periods of drowsiness usually associated with rumination. Cattle are in a drowsy state 7.5 hours per day, divided into 25 periods that precede and follow sleep. Rumination and sleep are inversely related so that sleep time decreases with rumen development (see Chap. 6) and decreases as the percentage of roughage in the diet increases (Balch 1955).

Social changes can disrupt activity rhythms in ruminants as well as in pigs. For example, calves in a stable group show definite diurnal activity patterns; those in continually changing groups do not (Kondo et al. 1983–84).

The normal bovine day depends on the diet and on the housing conditions, and, in general, consists of alternating periods of eating and ruminating interspersed with resting or loafing and short periods of sleep (Fig. 3.4). Activity patterns in cattle have been studied by other investigators in addition to those listed in Table 3.4 (Hancock 1950; Harker et al. 1956; Sheppard et al. 1957; Fontenot and Blaser 1965; Ray and Roubicek 1971).

Table 3.4. Activity patterns of sheep

Author	Grazing (hr)	Number of grazing bouts	Ruminating (hr)	Standing (hr)	Lying (hr)	Walking (hr)
Doran (1943)[b]	7	—	5.5[a]	—	—	2
England (1954)	9–12	5	9–10.5	2.5	3.5	0.5
Hughes and Reid (1951)	9	2	—	3.5	11.25	0.25
Squires (1974)	4–5.5	2	—	—	—	—

[a]Includes idling and resting.
[b]Observed for 14.5 hr/day (daylight).

Sheep. Until recently, sheep were seldom kept confined, so therefore, most studies of their activities have dealt with range or pasture conditions (Table 3.5). Sheep on the range spend 50% of the daylight hours grazing (Cory 1927), of which 7 hours are spent grazing and 2 hours traveling (Doran 1943). On the range, sheep travel 6 to 14 km a day, but they travel only 0.8 km a day on pasture (Cresswell 1960). On pasture, sheep spend 9 to 10 hours grazing in 4 periods, and they spend an equal amount of time

Table 3.5. Mean values of comparative data of sleep-wakefulness states and attitudes in four species of farm animals (three subjects of each species)

Species and Time Period	Duration and Percentage					
	Wakefulness		Sleep		Attitude	
	AW	DR	SWS	PS	Standing	Recumbent
Horse						
24-hr period	19 hr 13 min 80.8%	1 hr 55 min 8.0%	2 hr 05 min 8.7%	47 min 3.3%	22 hr 01 min 91.8%	1 hr 59 min 8.2%
Nighttime (10 hr)	5 hr 14 min 52.4%	1 hr 54 min 19.0%	2 hr 05 min 20.8%	47 min 7.8%	8 hr 01 min 80.1%	1 hr 59 min 19.9%
Cow						
24-hr period	12 hr 33 min 52.3%	7 hr 29 min 31.2%	3 hr 13 min 13.3%	45 min 3.1%	9 hr 50 min 40.9%	14 hr 10 min 59.1%
Nighttime (12 hr)	1 hr 55 min 16.0%	6 hr 14 min 51.9%	3 hr 06 min 25.8%	45 min 6.3%	1 hr 30 min 12.5%	10 hr 30 min 87.5%
Sheep						
24-hr period	15 hr 57 min 66.5%	4 hr 12 min 17.5%	3 hr 17 min 13.6%	34 min 2.4%	16 hr 50 min 70.1%	7 hr 10 min 29.9%
Nighttime (12 hr)	5 hr 59 min 49.8%	2 hr 45 min 22.9%	2 hr 43 min 22.5%	34 min 4.8%	7 hr 10 min 59.7%	4 hr 50 min 40.3%
Pig						
24-hr period	11 hr 07 min 46.3%	5 hr 04 min 21.1%	6 hr 04 min 25.3%	1 hr 45 min 7.3%	5 hr 10 min 21.5%	18 hr 50 min 78.5%
Nighttime (12 hr)	4 hr 23 min 36.5%	2 hr 30 min 20.8%	3 hr 52 min 32.2%	1 hr 15 min 10.5%	1 hr 20 min 11.1%	10 hr 40 min 88.9%

Source: Ruckebusch (1972).

ruminating in 15 bouts. Sheep allowed to graze only during daylight hours also spent 9 hours grazing (Berggren-Thomas and Hohenboken 1986), similar to the time spent by sheep with 24 hours to graze. As has already been noted in cattle, more time is spent grazing on a poor pasture (up to 12 hours a day) and twice as much distance is traveled as on a good pasture. Sheep are awake for 16 hours a day. They drowse 4.5 hours a day, far less than cattle. Slow wave sleep occupies 3.5 hours a day, and REM sleep occurs in 7 periods for a total of 43 minutes (Ruckebusch 1972). Sleep increases in sheep fed a low-roughage diet (Morag 1967). While sleeping, sheep expend 10% less energy than while awake (Toutain et al. 1977), so sleep deprivation would be expected to increase energy expenditure. Sheep will stand up 8 to 11 times during the night, usually to urinate or defecate (England 1954).

The ocular muscle relaxation and other physiological parameters that accompany the various states of vigilance in farm animals are shown in Figure 3.10A. The percentages of the night spent in sleep, wakefulness, and various attitudes are shown in Figure 3.10B. Comparative data for sleep and wakefulness in farm animals are given in Table 3.5.

Goats. Very little work has been done on activity patterns of goats. Adults spend 41%–47% of the time foraging, and kids spend 59%–65%, depending on the stocking rate (Provenza and Malechek 1986). Goats graze less and travel more when it is raining or when flies are abundant. These changes are more pronounced in shorn goats (Brindley et al. 1989).

CLINICAL PROBLEMS ASSOCIATED WITH DISORGANIZATION OF RHYTHMS AND SLEEP.

Human mental illness is often associated with disorganization of diurnal rhythms (Nikitopoulou and Crammer 1976) and with sleep disturbance (Sweetwood et al. 1976). This possibility has not been investigated in veterinary medicine, but owners of destructive dogs, in particular, complain that the animals are restless at night. In one case, a cat that lacked the medullary inhibition of movement during REM sleep was very active at that time (Hendricks et al. 1981).

Common Problems

Hyperactivity. Hyperactivity is probably the behavior problem most frequently diagnosed by owners and least frequently confirmed by the clinician. Most dogs that owners perceive as hyperactive are simply unruly. The owners have usually unconsciously rewarded the dog for hyperactivity by ignoring the dog when it is quiet and paying attention to it when it is rambunctious. For these dogs, even negative attention is preferable to no attention. The culprits usually are young dogs of working breeds. Although they may actually require more exercise than smaller breeds, it is more likely that they attract negative attention because they cause more disruption when they are active than a little dog would. The running, jumping,

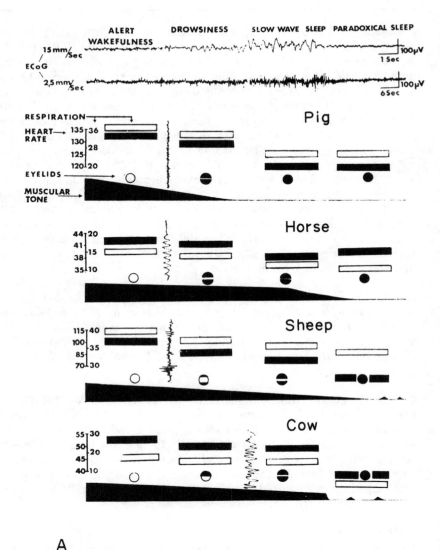

Fig. 3.10. (A) Physiological parameters of sleep in various species. The different electrocorticogram patterns (ECoG) for each species are shown at speeds of 15 mm per second: theta rhythm (horse), delta rhythm (ruminating cow), spindles (sheep), and alpha rhythm (pig) (Ruckebusch 1972). (B) Mean comparative data of sleep wakefulness states and attitudes during the night. The inner circle shows the relative duration of the ECoG (electrocorticogram) pattern (REM in black and the outer circle the relative duration of the attitudes) (Ruckebusch 1972, copyright © 1972, with permission of Bailliere Tindall).

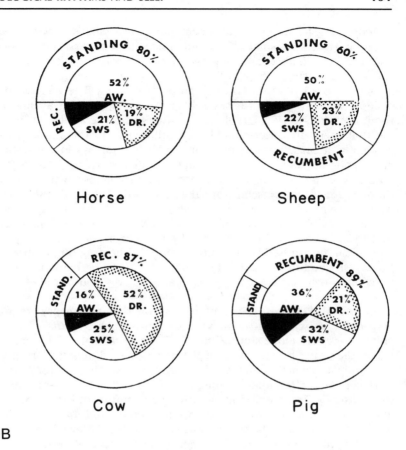

Horse

Sheep

Cow

Pig

B

and mouthing behavior may cause the owner to isolate the dog for long periods, which, of course, leads it to be more excited when it sees people again.

If the dog appears normal in the absence of the owner, then recommending more exercise, a canine companion (two or more owners could leave their dogs together so that extra dogs would not have to be acquired), and obedience training usually suffice to improve the dog's behavior. A specific behavioral exercise, "quiet training," is also recommended. The owner should pet and praise the dog whenever it is lying quietly. More formally, the dog can be given food rewards for lying quietly. If it becomes hyperactive, the owner should either ignore it or, if destruction of property is likely, isolate the dog, but only until the dog is heard to be resting quietly. If the dog is truly hyperactive, dextroamphetamine 0.2–1.3 mg/kg or methylphenidate (Ritalin) 2–4 mg/kg may be prescribed.

Narcolepsy. The number of clinical problems associated with sleep and circadian rhythms is small when compared with those associated, for example, with dominance and aggression. A narcolepsy syndrome in dogs (Mitler et al. 1976), cats (Knecht et al. 1973), and horses (Foutz et al. 1980) is characterized by attacks of inappropriate sleep. The affected dog will collapse and fall asleep for several seconds or minutes at a time. Play or food, especially very palatable food, often elicits the attacks. It is interesting that there are species and breed differences in the stimuli that elicit cataleptic attacks. Food is most apt to elicit catalepsy in dogs but not in narcoleptic horses, who are most apt to collapse when petted or saddled. Young dogs and Labradors are most apt to collapse when playing. Although Doberman pinschers are the breed most affected, the disease appears to be more severe in small breeds of dogs. The best diagnostic test is to space small bits of food a foot or so apart and time how long it takes the dog to consume all the food and the number of times it collapses. The animal can be aroused with auditory stimuli. The drug imipramine reduces the severity and incidence of the attacks. In the Doberman pinscher breed, narcolepsy is inherited as an autosomal recessive (Foutz et al. 1980).

Nocturnal Wakefulness. Much more common than narcolepsy are problems with dogs and cats that demand attention at night. In most cases, these animals are left alone during the day. The usual pattern is that the dog wakes the owner because it has a genuine need to eliminate. The dog learns after only a few nights that it can waken the owner and go for a walk or at least get some attention. Although sedation may be necessary in extreme cases, usually firmly enforcing a down stay command, if the dog demands attention, combined with an increase in exercise and attention during the hours the owner is at home and awake will solve the problem.

The problem in cats occurs either in young cats that are seeking play or in cats subjected to a change in social or physical environment. More play in the evening will help to reduce nighttime, usually dawn, play periods in kittens. Free choice food often is helpful, especially if the cat is crying for breakfast. Sometimes punishing the cat with a water pistol, for example, suppresses the behavior, but cats are often able to avoid a stream of water and still disturb the owner.

Management Practices. The veterinarian should be aware of the activity and sleep patterns of animals so that abnormality can be detected. A horse that is lying down at night is probably sleeping; an adult horse that lies down during the day (especially a cold, cloudy day) is abnormal and should be carefully observed. Cows do not sleep much but are difficult to arouse when they are sleeping. Cats and dogs sleep a great deal, so somnolence is no cause for alarm. In fact, one sign of cardiac disease in dogs is interference with sleep. A dog in congestive heart failure may sleep fitfully and rise often because recumbency further compromises its cardiac function.

Management practices should interfere as little as is economically feasible with normal grazing patterns and activity cycles. The interruption of these activity cycles and loss of sleep may play an important role in the etiology of the stress diseases associated with livestock transport and mixing of strange animals. It is often difficult to decide objectively what treatment of animals is inhumane, but normal behavior patterns may be used as a guide. For example, forcing a horse to stand for long periods is probably not cruel, but forcing a pig to stand is probably cruel, because pigs are usually recumbent 80% of the time.

4

Sexual Behavior

A goodly portion of the large animal practitioner's work has traditionally involved sexual disorders of both organic and behavioral etiology, and small animal practitioners may find that both kennels and "backyard" breeders often turn to them for consultation on reproductive disorders.

Mating systems have evolved within the framework of the morphological and physiological parameters of the individual species under continual ecological pressures. Although the nutritional, climatic, and health stresses to which domestic animals are exposed have been drastically reduced and "good breeders" have been heavily selected for, one still finds remnants of this long-term evolutionary selection hindering the goal of high reproduction rates. Thus, one may continue to see seasonal breeding in housed stock subjected to artificial control of photoperiodicity; poor fertility in some artificial insemination (AI) programs despite the absence of organic disease; and mate selection preferences that do not coincide with one's ideas of desirable matings as problems that hinder breeding and production schedules. In discussing the normal and pathological reproductive patterns of the domestic animals, some reference will be made to the wild species from which the domestic varieties were derived or to which they are related to help explain the origin of designated normal or problem behavior.

PHYSIOLOGICAL BASES OF SEXUAL BEHAVIOR. Adult male and female sexual behavior depends on a variety of physiological, environmental, or psychological factors, for its expression: (1) the genetic sex of the animal; (2) its perinatal organizational action of hormone; (3) its past social and sexual experience; (4) its adult activational action of hormonal and anatomical status; (5) the attractiveness of the potential mate; and (6) the external environment. Figure 4.1 illustrates the factors that affect sexual behavior, using the stallion as an example.

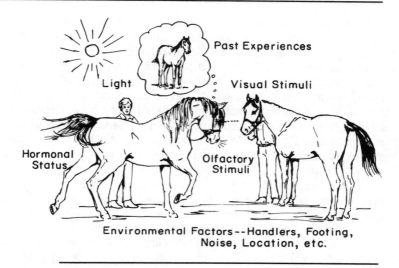

Fig. 4.1. Factors that affect sexual behavior.

GENETIC SEX. The sex of the animal is determined at the moment of conception, and the chromosomal sex will determine whether the indifferent fetal gonad develops into an ovary or a testis. Nevertheless, the potential for masculine and feminine behavior remains in both sexes. Studies on laboratory animals have revealed that the brain, and therefore behavior, is usually female unless the fetus is exposed to androgen during development. Similarly, the external genitalia will be female without androgenic stimulation.

ORGANIZATIONAL PERINATAL HORMONAL INFLUENCES. The role of sex hormones during ontogeny has been studied extensively in laboratory rodents and also in dogs. Male puppies have been castrated at birth and their behavior compared to that of intact dogs and dogs castrated as adults. Anatomically, these dogs were altered in that their penes were very small. Despite this anatomical change, they still urinated in the masculine posture and were no different in their response to exogenous testosterone from dogs castrated as adults (Beach 1974). They were attracted to estrous females and mounted them, but their underdeveloped external genitalia prohibited normal intromission. Both sexes have the genetic potential for male or female behavior, but neonatal androgens defeminize males, that is, make them less likely to show female sexual behavior. Therefore, the male puppies castrated at birth show receptive behavior to other males when treated with estrogen as adults.

Female puppies treated with androgens in utero and neonatally were markedly altered anatomically. They had no external vagina and did have small phalluses. They urinated in the masculine posture half of the time. As adults they were ovariectomized and treated with either female (estrogen and progesterone) or male (testosterone) hormones. The estrogen treatments revealed that the dogs had been "defeminized" because they would not stand or show any other signs of sexual receptivity. They were not even attracted to the male as is a normal intact estrous female or ovariectomized estrogen-treated female. When treated with testosterone, the experimental dogs were obviously masculinized because they were attracted to and would mount other female dogs that were in estrus.

These studies indicate the powerful effects of perinatal androgens on the anatomy and behavior of animals of either genetic sex (Beach and Kuehn 1970; Beach et al. 1972; Beach et al. 1977). Sheep and cattle are also defeminized by their neonatal androgens. The freemartin cow (born twin to a bull) is an example of a naturally occurring manipulation of the perinatal hormonal environment; as will be described later in this chapter, these females exhibit masculinized behavior. Pigs appear to be unique in that they are defeminized not during the pre- or neonatal period, but at puberty (Ford and Christenson 1981).

ACTIVATIONAL: ADULT HORMONAL INFLUENCES. The most important feature of the hormonal basis of sexual behavior is that hormones have a permissive role. That is, an animal requires a certain level of hormones for normal sexual behavior, but a higher level of hormones will not increase libido or receptivity. Hormonal treatment will not cure a deficiency of sexual behavior unless there is a deficiency of that hormone.

Male sexual behavior depends on the activational action of testosterone; female receptivity depends on estrogen, but the ewe, doe, and bitch require progesterone in addition. This is demonstrated by the fact that the first ovulation of the ewe is never accompanied by estrus — no progesterone has been secreted yet. When estrus is produced in the ovariectomized female, progesterone must be administered before estrogen in the ewe and doe but afterwards in the bitch.

SOCIAL AND SEXUAL EXPERIENCE. It is much more common to observe animals that have been influenced by lack of adequate sexual experience than animals influenced by abnormal hormone levels. This situation can arise not only when experimental animals are raised in isolation but also when rams or boars are raised in all-male groups or when puppies are weaned too early. Total lack of experience, homosexual experience only, or too many or too unpleasant sexual experiences, as when young stallions are overused for breeding, can all lead to sexual abnormalities.

Lack of Socialization. The concept of critical periods of development will be discussed in Chapter 6, but it should be emphasized here that the most obvious effect of lack of socialization to conspecifics is on sexual behavior. Dogs raised without physical contact with other dogs from the age of three weeks showed normal libido toward estrous bitches, but were very poor at orientation; they would mount improperly and seldom achieved intromission. This is probably an effect of lack of mounting experience because mounting forms a large part of male puppy play (Beach 1968). Similarly, boars raised in isolation from the time they were three weeks showed very little sexual behavior (Hemsworth et al. 1978).

Within all male groups, individuals may direct their sexual attentions to other males or be subordinate to other males. The dominant males may continue to mount males even in the presence of an estrous female; the subordinate animals often have little or no mounting experience. Some of these inappropriate responses will cease with time, but breeding efficiency is affected (at least temporarily).

Negative Sexual Experience. Unpleasant experiences during mating will have a deleterious effect on future sexual behavior, especially if the animal is young and has not had many (pleasant) experiences. The negative associations can be the result of overt aggression on the part of the sexual partner, rough treatment by a handler, or an injury sustained during mating.

The effects of experience on sexual behavior have not been as thoroughly studied in females as in males. The effect may be less, in part because the female plays a less active role in mating in most management situations. Sows raised in isolation showed normal sexual behavior and when in heat, were attracted to males (Signoret 1970). More research should be done on the effects of experience and age at weaning on female sexual behavior. Cats, in particular, may not be adequately socialized under normal rearing techniques and may reject toms, at least initially.

ADULT HORMONAL AND ANATOMICAL STATUS. The complex relationships of the central nervous system, gonadal hormones, and behavior will be discussed in more detail later in this chapter, but, in general, ovarian hormones result in an attraction to males and receptivity to male mounting. In some species (cats and pigs), the complete pattern of estrous behavior can be elicited by estrogen alone. In others (dogs and sheep) (Beach and Merari 1968, 1970; Bermant et al. 1969; Signoret 1975), progesterone must also be administered. In ungulates, the behavioral action of estrogen is facilitated by a rapid preovulatory fall in progesterone, whereas in dogs a rise in progesterone is important. Progesterone is administered before estrogen in the ewe and after estrogen in the bitch to induce estrous behavior.

Ovariectomy and Castration. Ovariectomy usually abolishes estrous behavior in females. Castration (orchiectomy) generally abolishes sexual behavior in males, but there are many exceptions. The more experienced the male, the longer sexual behavior, both arousal and copulation, will persist after castration. Prepubertal castration is, therefore, more effective than postpubertal castration in eliminating sexual behavior. There are species differences in the effectiveness of prepubertal castration. Cats are markedly affected whereas dogs are not (Rosenblatt and Aronson 1958a,b; Hart 1968; Le Boeuf 1970).

Anatomical Factors. As mentioned, anatomical factors are important because a small penis precludes successful intromission. Intact afferent pathways from the penis are also necessary; experimental or traumatic neural damage to the penis results in misorientation in tomcats (Aronson and Cooper 1977) and failure to ejaculate in bulls (Beckett et al. 1978). Similarly, desensitization of the vagina inhibits ovulation in the cat, an induced ovulator (Diakow 1971).

The species differences in the structure of the penis are also important. For anatomical reasons, castration reduces the copulatory ability of male cats and horses more than that of male ruminants and dogs. The muscular penis of the horse is more dependent on erection for successful intromission than the fibroelastic penis of the ruminant. Similarly, the penile spines of the tomcat, which atrophy in the castrated male, are important for successful intromission and ejaculation.

ATTRACTIVENESS OF POTENTIAL MATE. Attractiveness, or lack of it, of the sexual partner is not often considered, but higher mammals are influenced by this factor as well as by their hormonal levels. The attractivity of an estrous bitch's urine depends on her hormonal state. If the donor of the urine is treated with estrogen, her urine becomes more attractive to males; if treated with testosterone, her urine is less attractive. Marking behavior of males is apparently an attempt to mask the attractiveness of bitch urine (Dunbar and Buehler 1980; Dunbar and Carmichael 1981).

Females may show individual preferences for one male over another. All females do not prefer the same male, indicating that the differences in attractiveness of males are based on the female's innate preferences and on past experience rather than on some physiological characteristic of the male, such as pheromone release. Male preferences can be based on physiological factors. Rams prefer unmated ewes, which will be discussed later. The action of ovarian hormones not only renders the female receptive but also increases her attractiveness, presumably by pheromonal release. Care must be taken when comparing the results of experiments on animals artificially brought into estrus with those involving females in natural estrus, because the latter may be more attractive than the former.

Like females, males also show individual mate preferences that are apparently psychological or idiosyncratic. Some of these preferences may have an evolutionary basis. Males may prefer females that are similar, but not too similar, to themselves, so that inbreeding will not occur and yet the same gene pool will be propagated (Bateson 1978). Evidence for such behavior influencing the sexual behavior of sheep (Hayman 1964; Lees and Weatherhead 1970), cats (Liberg 1983), and horses (Keiper and Houpt 1984) has been found. The mechanisms and ramifications of mate selection are topics of considerable current interest to researchers; extensive reviews of this work (primarily nondomestic species) may be found in Wilson (1975) and Krebs and Davies (1978).

EXTERNAL ENVIRONMENT. The external environment is very important for optimal sexual behavior. It is obvious that extremely inclement weather will inhibit sexual behavior, but more subtle environmental factors, such as too many human spectators or a slippery floor, may also inhibit it. Not all additions to the environment are detrimental. In some cases, the presence of another male may stimulate sexual behavior, as has been well documented in cattle and goats (Blockey 1981b; Price et al. 1984C). Males generally are more influenced by the environment than females, so the female is usually brought to the male. Nevertheless, environmental factors in female sexual behavior deserve study. Time of day, for example, is important in cows that show more signs of estrus at night. It is becoming clear that olfactory stimuli from the males, probably androgen-derived pheromones, affect sexual cycles in female ruminants.

CENTRAL NERVOUS SYSTEM AND CONTROL OF SEXUAL BEHAVIOR
Females
Hypothalamic Factors. In the female, gonadotropin releasing factor (GnRH) in the hypothalamus stimulates the release of follicle-stimulating hormone (FSH) and luteinizing hormone (LH) from the anterior pituitary. Follicle-stimulating hormone induces follicular development, and FSH and LH together stimulate estrogen and progesterone production by the ovary. For most of the estrous cycle, estrogen and progesterone maintain low levels of LH and FSH through a negative feedback action on the hypothalamic pituitary axis.

Near proestrus, however, the situation is reversed; rising estrogen levels have a positive feedback on LH secretion, resulting in the LH surge that causes ovulation. This preovulatory rise in estrogen is responsible for the hormonally based components of estrous behavior. Cells scattered in the hypothalamus between the preoptic area and mammillary body near the basal midline take up estrogen and concentrate it in their nuclei. The central nervous system affected by estrogen shows a refractory period during which biochemical changes presumably take place that will result in estrous behavior (Michael 1973). For example, radioactive estrogen is taken up by cells in the hypothalamus within hours of administration to an ovariectomized cat, but estrous behavior does not appear for three days.

Evidence for two separate areas of the hypothalamus, one mediating gonadotropin release and the other sexual behavior, has been obtained in sheep and in cats in which lesions of the anterior hypothalamus abolished sexual behavior even after treatment with exogenous estrogen (Clegg et al. 1958; Michael 1973) (Fig. 4.2). If ventromedial hypothalamic lesions are made in female cats they still crouch and solicit the male but will not permit intromission (Leedy and Hart 1985).

Cortex and Midbrain. Other areas of the brain also are involved in female sexual behavior. Both the midbrain and the cortex appear to be involved in rodents (Clemens 1971), but in the cat decortication has no effect on hormonally induced sexual receptivity (Bard 1936). The cerebral cortex is probably necessary for the more active aspects of female sexual behavior, such as seeking the male and courtship, but apparently it is not necessary for the passive aspects, such as standing for the male and lordosis.

Cyclical Ovulation. In spontaneous ovulators (bitch, ewe, mare, and sow), LH surge and, consequently, ovulation take place cyclically; but in induced ovulators, such as cats, external stimuli from the vagina either by natural coitus or artificial manipulation are necessary to trigger the LH surge. The relationship between hormone levels and sexual behavior in the bitch is shown in Figure 4.3. The female's active solicitation of the male is called *proceptive behavior.*

The seasonal nature of reproductive behavior is also due to central neural variation in responsiveness to gonadal hormones. The same ewes had to be injected with a larger amount of estrogen to induce estrous behavior in the spring than was needed in the fall.

Olfactory Influences

PHEROMONES. Odors can have important effects on reproduction; which has been demonstrated in a variety of rodents, mice, in particular (Bronson and Whitten 1968). In domestic animals, odor is probably not as important, nevertheless, the age at first puberty is lower in gilts exposed to a strange boar (continuous cohabitation with a boar will not produce the effect). Olfactory bulbectomy eliminates the effect, which indicates that the odor is the important stimulus (Kirkwood et al. 1981). Ewes show a similar response, called the ram effect, but odor may not be essential. Postpubertal cows show another response to odor; they will come into estrus sooner if exposed to the odor of estrous cow urine or mucus (Izard and Vandenbergh 1982). The best-studied farm animal pheromone is the boar odor that stim-

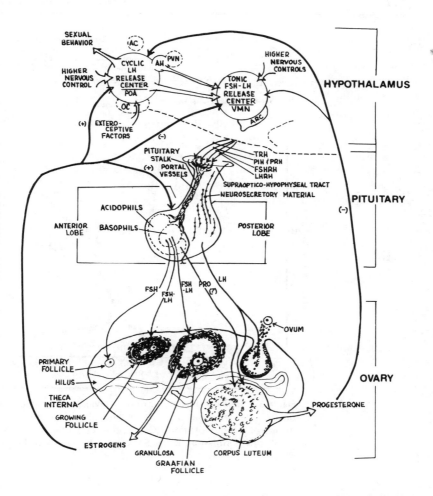

Fig. 4.2. Interrelationships among the hypothalamus, anterior pituitary, and ovaries of a female animal. The diagram of the ovary represents successive stages in follicle development and corpus luteum formation as they occur during a normal estrous cycle: *OC* = optic chiasm; *AC* = anterior commissure; *AH* = anterior hypothalamus; *ARC* = arcuate nucleus; *POA* = preoptic area; *PVN* = paraventricular nucleus; *VMN* = ventromedial nucleus; *FSH* = follicle-stimulating hormone; *LH* = luteinizing hormone; *PRO* = prolactin-releasing hormone; *FSHRH* = follicle-stimulating hormone-releasing hormone; *LHRH* = luteinizing hormone-releasing hormone (Hansel and McEntee 1977, copyright © 1977, with permission of Cornell University Press). *FSHRH* and *LHRH* are now believed to be one hormone, gonadotropin-releasing factor (GnRH).

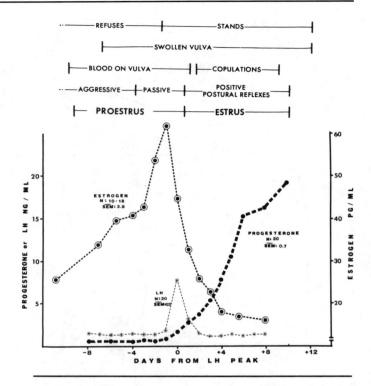

Fig. 4.3.. Relationship of hormone levels to behavior in the bitch. Times of onset of behavior and physical signs represent means of a range of times (Concannon et al. 1975, copyright © 1975, with permission of the Society for the Study of Reproduction).

ulates the estrous sow to assume the immobile, rigid posture that can be used for estrus detection.

Males

Hypothalamic Factors. The hypothalamic pituitary axis is also involved in male sexual behavior. Follicle-stimulating hormone is released in response to hypothalamic-releasing factor. In the male, FSH stimulates spermatogenesis, and LH stimulates testosterone release. Testosterone, in turn, acts upon the anterior hypothalamus-preoptic area in conjunction with appropriate stimuli from an estrous female to produce male sexual behavior. Inhibin, a putative testicular factor produced in the spermatic tubules, acts as a negative feedback on FSH release (Setchell 1978). Lesions on the preoptic area abolish male sexual behavior and scent marking, but not aggression, in dogs (Hart 1974b) and cats (Hart et al. 1973). Medial preoptic lesions combined with bulbectomy not only abolish male sexual behavior in male cats, but increase their female sexual behavior (rolling, soliciting and crouching with pelvic elevation) (Hart et al. 1973; Hart and Leedy 1983). The medial preoptic area is also important for normal sexual behavior in male goats (Hart 1986), indicating that the same brain area is responsible for male sexual behavior in a wide variety of species.

Considerable evidence is accumulating that not testosterone itself, but a metabolite of testosterone, estradiol, acts on the central nervous system to produce male sexual behavior. In general, estrogen and testosterone have similar actions in stimulating male sexual behavior in castrated animals, but an androgen that cannot be metabolized to estradiol, dihydrotestosterone, does not (Cerny 1977; Clarke and Scaramuzzi 1978). When freemartins are treated with estrogens and androgens, both hormones increase aggressiveness, mounting, and sniffing of the vulva of other cows, but only testosterone stimulates the flehmen response (Greene et al. 1978). Testosterone itself, rather than a metabolite, may be responsible for this response (see Fig. 4.4). Dihydrotestosterone does stimulate enlargement of the penis and of the sex glands, so when it is administered in combination with estradiol to castrated pigs, the full sequence of male sexual behavior including intromission occurs (Parrott and Booth 1984).

Neocortex and Amygdala. As in females, areas other than the hypothalamus are involved in male sexual behavior. Lesions in the sensorimotor area of the neocortex suppress copulatory performance but not arousal in cats (Beach et al. 1955; Beach et al. 1956). There are also areas of the brain that are inhibitory and areas that are excitatory. Lesions of the amygdala or pyriform cortex produce hypersexuality in male cats. The lesioned cats would mount a furry toy, another male cat, or one another in tandem (Green et al. 1957). Similar lesions in dogs did not produce hypersexuality (Clemens and Christensen 1975).

Stimulation of the anterior hypothalamus by testosterone or its metabolites may facilitate appearance of male sexual behavior, but it is the appearance of a female that triggers the behavior. The stallion, for example, will exhibit the flehmen response and begin the courtship rituals of nibbling the crest and rump of the mare. At the same time hypothalamic stimulation of the parasympathetic nervous system results in secretion of the accessory sex glands. The combination of penile stimulation after intromission and sympathetic stimulation leads to ejaculation.

Olfaction. Olfaction is, no doubt, important for identification of the estrous female; but elimination of the sense of smell by olfactory bulbectomy does not impair the sexual performance of cats or rams, which indicates that olfactory stimulation is not essential and that the male can identify the receptive female by visual or auditory means. The lack of resistance to his

Fig. 4.4. Effects of testosterone (*T*), estradiol (*E₂*), estrone (*E₁*), and dihydrotestosterone (*DHT*) on agonistic behavior, vulvar interest, the flehmen response, and mounting (Greene et al. 1978), copyright © 1978, with permission of Academic Press.

mounting attempts may be the most important information to the male (Lindsay 1965; Hart and Haugen 1971; Aronson and Cooper 1974).

Summary. Hormones are important for normal sexual behavior, but the central nervous system can function relatively independent of them as indicated by the persistence of copulatory behavior in castrated male cats (Rosenblatt and Aronson 1958a), dogs (Beach 1970a), and rams (Clegg et al. 1969) that had considerable precastration experience. Even prepubertal castration may not eliminate sexual behavior; one-third of prepubertally castrated bulls mounted cows (Folman and Volcani 1966). A similar percentage of prepubertally castrated geldings showed sexual behavior (Line et al.

1985). Apparently once a behavior is "learned," external stimuli are sufficient to stimulate male sexual behavior.

SEXUAL BEHAVIOR OF DOMESTIC ANIMALS
Cattle

The Cow. The cow is a nonseasonal, continuously cycling breeder but shows peak fertility from May to July and low fertility from December to February. Puberty occurs anywhere from 4 to 24 months of age, usually at 6 to 18 months. The estrous cycle is 18 to 24 days long (mean = 21 days), although it is somewhat shorter in heifers and in the Zebu breed.

ONSET OF ESTRUS. Onset of estrus occurs more often in the evening and ceases in the morning. Actual sexual receptivity lasts 13 to 14 hours. The estrous cow shows a general increase in motor activity and a decreased food intake (Hurnik et al. 1975) (Fig. 4.5A). Investigative behavior of and by the estrous cow, such as flehmen, sniffing, rubbing, and licking, increases as does premounting behavior, such as standing behind the cow and resting the chin on the back of the estrous cow. She will bellow a great deal, switch her tail, raise or deviate it to one side, and urinate frequently. Estrogen levels peak when the cow stands to be mounted (Esselmont et al. 1980; Glencross et al. 1981). The cow that mounts is usually preovulatory. The mounting cows are also usually dominant over the mounted cows, which represents an interaction between hormonally mediated behavior and social influences. Homosexual mounting in cows may have been selected for in dairy cattle because as long ago as medieval times bulls were not routinely kept with cows; so only those cows that the owner noticed to be mounting were taken to a bull and bred (Baker and Seidel 1985). Beef cows mount much less frequently than dairy cows, thereby confirming this hypothesis.

Aggressive behavior also increases markedly (Hurnik et al. 1975). Cows mount cows both before and after the period in which bulls would mount (Kilgour et al. 1977) (Fig. 4.5B). Mounting behavior by cows is not observed in cows free ranging with bulls (Schloeth 1961). The intensity of estrus can be measured only subjectively using the degree of restlessness and the conspicuousness of mounting, but intraindividual and intrabreed differences may be noted. For example, the brown Swiss breed shows the least marked estrous activity of the dairy breeds, and it has been reported that black cattle may show stronger signs of estrus than red, roan, or white cattle (Alexander et al. 1974). The physiological bases of these observations are not known but are probably under genetic control.

DETECTION OF ESTRUS. Detection of estrus becomes more and more critical in the dairy industry as artificial insemination becomes more widespread (70% of Holstein calves are conceived via AI). The physical signs of estrus, such as a copious vaginal secretion and vulvar relaxation, may be weak or

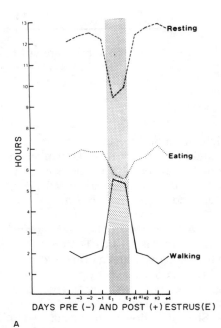

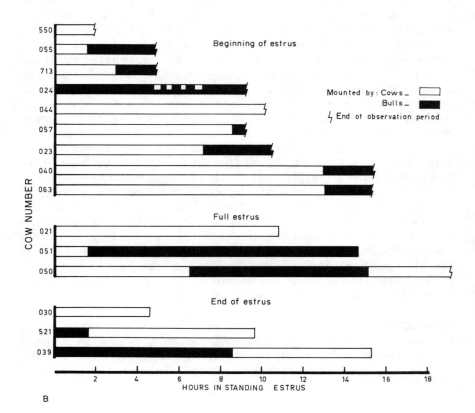

Fig. 4.5A. (*A*) Average composition of daily activities of cattle during estrous and nonestrous stages of the reproductive cycle (Hurnik et al. 1975, copyright © 1975, with permission of Department of Animal and Poultry Science, University of Guelph, Ontario and Elsevier Science Publishers). (*B*) Mounting of cows at different stages of estrus by cows and bulls (Kilgour et al. 1977).

absent. Traditionally, the bull is the best detector of estrus in the cow, with human detectors a poor to fair second. An estimated 19% of heats may be missed because behavioral signs are absent (silent heats) even though ovulation occurs, and an additional 15% are labeled as estrous periods even though ovulation does not occur (false heats). Some silent heats may have been behaviorally evident but were not observed by the herders. Pregnant cows may also show signs of estrus (Fraser 1968).

Observation of mounting behavior helps the farmer to detect heat, but it may not occur around milking time or when the cows are usually observed. A commercial estrus detection device, which consists of a plastic vial containing a red dye, is glued to the dorsal tail base of a cow suspected to be in estrus. As she is mounted by other cows in the herd, the dye is gradually expressed into a viewing chamber; a full chamber is supposedly correlated with a sufficient number of mountings to indicate a full estrus (Williamson et al. 1972). Several cows may be in heat at once, and one may not be mounted; so even the mechanical heat detector can give misleading false negatives (Mylrea and Beilharz 1964).

A vasectomized bull or a teaser bull with a surgically deviated penis may also be used to detect heat. The danger of keeping an unpredictably aggressive bull in close contact with people, however, has discouraged this practice. A freemartin heifer may be used satisfactorily. Fetal exposure to the androgens produced by the male fetus masculinizes the nervous system of freemartins; they are infertile and, thus, of little value as dairy animals. Freemartins are even more likely than the other cows in the herd to mount an estrous cow, and treatment with injectable androgens may heighten this behavior even further. Dogs have been trained to detect estrous cows, an interesting innovation presumably based on canine olfactory acuity (Kiddy et al. 1978). The dogs were 80% to 90% correct in detecting estrous cows.

BEHAVIORAL ABERRATIONS. Silent heats have already been mentioned. Only a conspecific male or female will be able to detect this type of estrus. Heifers are especially subject to this phenomenon.

Nymphomania is more common in high-production dairy cows than in beef breeds. The cow shows intense estrous behavior either persistently or at frequent, irregular intervals. Milk production drops noticeably because these cows are often the best producers in the herd. Most commonly, the affected cow is 4 to 6 years old and has calved two to three times. Unlike a cow in normal estrus, however, the nymphomaniac does not stand for other cows. She actively seeks other cows and mounts them. She paws and bellows like a bull, and with time also becomes more malelike in voice and body conformation. Nymphomania is usually associated with follicular cysts, and treatment is sometimes successful using a source of LH, such as pituitary extract or chorionic gonadotropin (Roberts 1971).

The Bull. Bovine male courtship behavior is becoming an unobserved rarity in modern dairy farming with the advent and spread of AI. Bulls are now selected for (among other things) their willingness and ability to mount dummy cows or steers and to ejaculate into an artificial vagina. One might wonder how this willingness to leave behavioral interactions with the opposite sex up to random genetic drift will eventually affect the species. The courtship sequence is actually a series of reciprocal interactions between male and female. The female behavior has already been described.

COURTSHIP BEHAVIOR. Starting late in the cow's proestrus, the bull will begin to graze beside the cow, guarding her from any other cattle. His attempts to mount will be repulsed by the cow. During proestrus, most females are attractive to the male and attracted to him but not yet receptive. The bull may attempt to drive the cow away from the herd. Bovids generally do not show territorial behavior, and the cow is not driven to a specific mating arena.

Periodically, the bull will smell and lick the cow's vulva, and often exhibit the flehmen response (Fig. 4.4). Some, but not all, dairy bulls exhibit flehmen to estrous urine. They exhibit flehman more often to estrous urine than to mucus or to nonestrous urine (Houpt et al. 1989) (Fig. 4.6). Flehmen is followed by an increase in LH (Lunstra et al. 1989).

As estrus approaches, guarding becomes more marked as both other

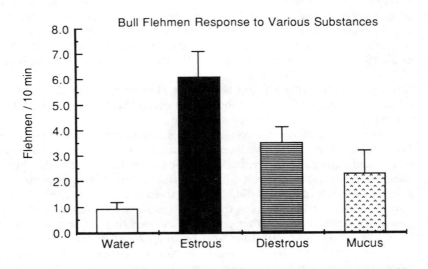

Fig. 4.6. The mean (± SEM) response of bulls to estrous and nonestrous urine, mucus, and water (Houpt et al. 1989, copyright © 1989, with permission of Butterworth Publishers).

bulls and nonestrous cows are kept away. When the cow is in full estrus, the bull will have a partial erection while guarding her, and accessory gland fluid or precoital discharge will drip from the penis (Kerruish 1955). The bull frequently nudges the female's flanks and either maintains head-to-head contact, or because of mutual genital sniffing and licking stands in a reverse parallel (bigeminal) position with the cow. This position is common to the courtship sequence of all ungulates. The bull may rest his head across the cow's back while they stand in a T-position (Fraser 1968). He makes several mounting attempts with a partial erection before the female will stand for him. When she is ready, the cow remains immobile, and the bull mounts immediately. He fixes his forelegs just cranial to the pelvis of the female as he straddles her. Ejaculation occurs within seconds of intromission and is noted by a marked, generalized muscular contraction. The bull's rear legs may be brought off the ground during this spasm (Fig. 4.7). Dismount and retraction of the penis follow rapidly. When bulls are used for

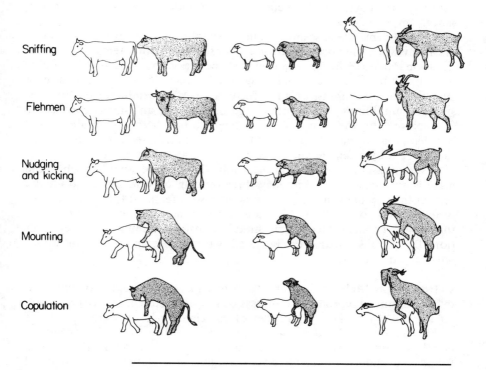

Fig. 4.7. Sexual patterns in cattle, sheep, and goats (Y. Rouger, unpublished data, Hafez 1974, copyright © 1974, with permission of Lea and Febiger).

hand breeding or for AI, lack of the stimulatory effects of the prolonged courtship may result in poor semen quality or poor reproductive behavior.

SENSORY STIMULI. Experimental manipulations have shown that a variety of sensory stimuli are needed to elicit male sexual behavior. Tape recordings of cows will sexually stimulate a bull (De Vuyst et al. 1964), but these sounds are probably of a nonspecific arousal nature. Bulls used for AI apparently become conditioned to the sounds in the collection arena, which may help stimulate these animals.

Olfactory cues seem to be used by the bull during early estrus to monitor the stage of the cow's reproductive cycle, but no studies have been able to show sexual excitation caused by olfactory stimuli alone. Sexually experienced animals with olfactory blocks still display normal sexual behavior, but this is undoubtedly due to the role of learning in sexual performance. It seems unlikely that free-ranging ungulates would be able to identify estrous females without the benefit of sexual pheromones. But, with modern husbandry techniques of confinement of males with females during estrus and with the use of AI, olfactory cues have been reduced to secondary importance.

Visual deprivation seems to hinder sexual response generally, but most markedly when the blinded individual is presented with a female in a novel situation (Hale 1966). The inverted U-shape seems to be the visual stimulus most likely to stimulate sexual behavior, whether this is in the form of a standing cow, a dummy, or an unfortunate person bending over. Without special conditioning, a wild bovid will only mount an estrous conspecific female; one of the aims of domestication has been to obtain "easy" breeders, that is, bulls that respond to a less specific series of stimuli than that offered by a female and do so with little sexual foreplay.

Sexual responsiveness is influenced by the range of stimuli to which the male is exposed during ontogeny. Free-ranging or wild bovids gradually restrict their sexual behavior to interactions with females. Bulls not exposed to females do not establish this more narrow range of sexual stimuli; hence the ability of some rather bizarre stimuli to instigate mounting and ejaculation. Dairy bulls are also rarely raised with females and so can be stimulated by other bulls.

MALNUTRITION. Malnutrition rarely affects sexual performance, as evidenced by the continued reproductive success of starving cattle in parts of Asia and Africa. Likewise, there are hordes of emaciated cats in southern Europe and hordes of emaciated people in other parts of the world. This phenomenon has been experimentally demonstrated by Wierzbowski (1978) who found that underfed bulls were actually quicker to copulate than well-fed bulls.

BEHAVIORAL ABERRATIONS. Impotence, the loss of libido, must be approached as a clinical problem with a rather large differential diagnosis in mind. The problem may be secondary to almost any other organic disease. Bulls used for AI seem especially susceptible to musculoskeletal diseases that render them lame and unable or unwilling to mount. Obesity may be considered a pathological condition that caused many problems in the early development of the AI program. Obese bulls were difficult or impossible to arouse. Now the diets of bulls are more closely controlled, but obesity should still be considered when assessing loss of libido. Balanoposthitis or injury to the penis are specific conditions of the genital urinary tract that might be confused with a libido problem (Roberts 1971). Testicular atrophy involving both the Leydig cells and the seminiferous tubules may occur and be recognized by the reduced size and soft consistency of the testis. This is one case in which androgen administration might be expected to lead to a return of normal sexual behavior. A sperm count is recommended, however, because a hypospermatic animal will still be an unsuitable breeder.

Management practices may influence male libido. A large bull may be unwilling to mount a teaser steer or a cow in an icy corral or on wet concrete floors. One or two slips may be sufficient to condition him to ignore the stimulus of the teaser. Distractions of other animals or human observers should be kept to a minimum; this may be more of a problem in some bulls than others. By 2.5 years of age, most bulls are dominant over cows, but before this a dominant cow may prevent a young bull from mounting. Restraint of the cow may be of value if another, older bull cannot be substituted. Young bulls also lack experience, and their tentative approach while courting may stimulate aggression in a cow. If AI is not possible, and a proposed breeding is between animals far enough apart that one must be transported to the other, it is advisable to bring the cow to the bull. Males seem more sensitive to environmental factors; a new or different surrounding may trigger a reluctance to mount. Fortunately, it is also usually easier to truck a cow than a large, aggressive bull.

For reasons that must be classified as "psychological," a bull trained to mount another male and ejaculate into an artificial vagina may eventually lose interest and refuse to mount, especially if he has been used too frequently. This inhibition may be overcome by changing the sexual stimulus to a new steer or bull. The stimulation of sexual behavior caused by a new female is called the *Coolidge effect*. A new teaser, preferably a cow, and even better an estrous cow, or a moderate change in the environment or location of the mating arena may be sufficient, but cows are not allowed at artificial insemination centers—to reduce the incidence of disease. Other bulls are more attractive teasers than steers. Although psychological in its origin, the Coolidge effect has a physiological limit. The maximum number of ejaculations will be the same as that obtained by electroejaculation.

Restraining the bull from the teaser, or allowing him to watch another bull being collected also may arouse a previously disinterested male (Kerruish 1955; Blockey 1981b; Mader and Price 1984). A caution regarding the latter technique must be given: a submissive or young bull being allowed to observe a strongly dominant animal may be further conditioned against servicing through intimidation. In most ungulate species copulations are restricted to a small percentage of dominant males in the population. The lack of sexual activity in the remainder of the male population has been termed "psychological castration," although the effect is usually and quickly reversed if the dominant individual is eliminated.

Individual bulls show marked differences in their levels of sexual behavior as measured by such parameters as mounts/ejaculation, the number of ejaculations per unit time, the latency to ejaculation, the number of ejaculations required for satiation, and the length of time to recover from postsatiation refractoriness. Although it is impossible to separate completely the genetic from the learned components of these differences, evidence is strong that the differences between individuals as measured by such parameters are strongly genetic in origin. Work with monozygotic twins and triplets and comparisons between sires and sons show a high degree of similarity between these groups of related animals (Bane 1954; Hafez and Bouissou 1975) (Fig. 4.8). European breeds seem more likely to mount an inappropriate sexual object (a male or anestrous female) than the Zebu breed, which may indicate a lower threshold in the European breeds. It seems unwise, therefore, to continue breeding a male that shows limited interest in mounting females because doing so will only perpetuate the defect. Early experience and management, in addition to genetic factors, are important determinants of sexual behavior. For instance, Zebu bulls raised in small groups had a faster reaction time (time to the first mount) than those raised on the range in large herds (Macfarlane 1974).

Blockey (1981a,b,c) has developed a procedure for testing libido in potential beef cattle sires in which several bulls are tested for 20 minutes with heifers restrained in stanchions. The relative performance in this short test is well correlated with sexual performance in a 19-day pasture breeding test. Because dominance hierarchies are not strong in bulls under 2 years of age they do not fight in the group breeding situation, but they can interfere with one another (Garcia et al. 1986). In a pasture situation, older dominant bulls will mate more than younger, low-ranking bulls, which can be a problem because the older bulls are less fertile. Therefore, within the same pasture, bulls of the same age should be used for breeding.

Problems can arise with short-term tests of libido. A high environmental temperature can reduce libido, and the effect may not be the same in all the bulls. So one that normally has high libido may be more affected than one with low libido (Christensen et al. 1982). In addition, bulls are more responsive to pre- than to postovulatory cows (Garcia et al. 1986).

Masturbation, although commonly noted among bulls, causes no known reproductive problem. In general, masturbation is frowned upon on an anthropomorphic basis. Sperm quality or counts do not seem correlated with the frequency of this "vice." The bull performs pelvic thrusts, with his

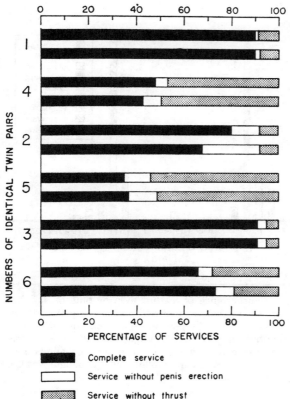

Fig. 4.8. Ejaculatory behavior in identical twin dairy bulls showing percentage distribution of complete copulations, mounts without penis erections, and mounts without thrust (Hafez and Bouissou 1975, copyright © 1975, with permission of W. B. Saunders Co.).

back arched, with a partially erect penis. Thus the penis moves in and out of the preputial sheath until ejaculation occurs. All bulls masturbate, especially at times of inactivity (Houpt and Woolney 1989). As will be seen in the following sections, all domestic animals have been observed to masturbate; so it would be difficult to label this an abnormal behavior. See Figure 4.9 for patterns of masturbation in dairy bulls.

Although most problems of sexual behavior in food-producing animals are those of insufficiency, manifestation of sexual behavior in steers is also a problem. Approximately 2% of feedlot steers are buller steers that are mounted by other steers. This occurs more often in steers that are implanted with stilbestrol or estrogen, but the hormone level may actually be lower in the buller's plasma than it is in that of normal steers. There is a large component of dominance-related aggression in the buller syndrome. Penile erection and intromission rarely occur. More aggressive animals are the bullers, and the rate of mounting increases dramatically when new steers are added (Klemm et al. 1983). The syndrome is most frequently seen

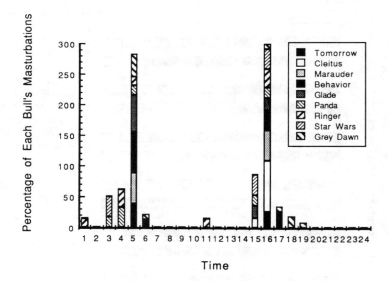

Fig. 4.9. The occurrence of masturbation by time of day. The masturbations observed are plotted by the hour of the day in which they occurred as the percentage of each bull's masturbations. Some bulls were observed longer than others, so it was necessary to use percentages rather than absolute numbers of masturbations (from Houpt and Wollney 1989, copyright © 1989, Elsevier Science Publishers).

when groups of animals are mixed, especially in crowded conditions. The economic losses resulting from the syndrome are due to the increased activity of the mounting steers and the harassment of the buller steer. None of the animals will gain weight as they should (Irwin et al. 1979). The usual means of treating bulling behavior is to remove the steers involved, but electrified wire placed above the pens so that a steer that reared to mount would be shocked can be used to reduce the incidence of the behavior (Kenny and Tarrant 1987).

Sheep and Goats

The Ewe and the Doe

ESTROUS CYCLE. Both the ewe and doe are short-light breeders, that is, they usually cycle in the fall of the year as the light phase of the photoperiod decreases. The pathway of information to the hypothalamic-hypophyseal tract about the photoperiod is uncertain, but the pineal gland has been suggested as a possible source of this information, and its product, melatonin, has been used to hasten the onset of the breeding season (see Chap. 3). The ewe is polyestrous and will cycle several times during one breeding season if not bred; the average cycle for the ewe is 16 days (14–20). The actual period of estrous receptivity is 30 to 36 hours in the ewe. Signoret (1967) has shown that breed differences in the duration of estrus in sheep exist even in estrogen-induced estrus in ovariectomized ewes. The duration of estrus in lambs and in the first estrus of the season is less than the normal estrus. Puberty depends on the time of birth. Females born early enough in the year will cycle their first fall season so that puberty may occur as early as 4 months of age.

Other environmental factors influence the estrous cycle of sheep. Domestic sheep raised in the tropics and subtropics are nonseasonal breeders (Hulet et al. 1975); as with many other life forms in these areas, reproductive activity occurs all year. Breeds originating at higher latitudes and higher altitudes, such as the border Leicester, Scottish blackface, and Welsh mountain breeds, show a shorter breeding season than their more tropical or lowland cousins (Hafez 1952). Thus, environmental input seems to temper genetic constraints.

Olfactory cues are not necessary, but they may be sufficient because the the synchronization of the first estrus of the breeding season in sheep can occur in the absence of direct contact or visual cues (Alexander et al. 1974). Synchronization of estrus is a desirable property of the ewe cycle for management purposes, because breeding may be accomplished in a short time and the lamb crop will be born synchronously and tend to reach market size simultaneously.

THE RAM EFFECT. Most sheep breeding is still done naturally, with the rams pastured with the ewes all year or introduced in the late summer. The introduction of the ram tends to synchronize estrus in a high proportion of the ewes 15 to 17 days later (Thompson and Schinckel 1952; Hulet et al. 1975). Continuous exposure to a ram tends to increase the incidence of estrus in the normally anestrous period (Riches and Watson 1954). The presence of a ram also shortens the duration of estrus, but this depends on direct physical contact with the ram and probably mounting of the ewe by the ram rather than on pheromones (Parsons and Hunter 1967).

Estradiol-treated wethers can stimulate estrus in anestrous ewes, indicating that testosterone must be aromatized to produce the ram effect. The pheromone is present in the wool and wax of the ram (Knight and Lynch 1980). Notice that in sheep the odor of wool rather than of urine, which is so important in rodents, appears to be involved in both sexual and maternal recognition. Farm animals do not appear to be completely dependent on odor; for example, the ram effect can occur even in ewes without the sense of smell (Cohen-Tannoudji et al. 1986).

Synchronization may also be accomplished by using progesterone compounds, and this procedure is used more as the practice of AI in sheep increases. Estrus occurs 48 hours after withdrawal of progesterone (Tomkins and Bryant 1974). Synchronization is more successful using these compounds if a male is added to the flock following the treatment.

Feral sheep display an even shorter breeding season than most of the domestic breeds, and it is likely that both predation pressures and the necessity of foraging over large areas for limited resources established estrous synchrony as an evolutionarily stable strategy in the ancestral forms. In feral sheep, the adult rams form a flock separate from that of the ewes and lambs and only join the females during the breeding season. Flocking tendencies are strong in sheep, and a ewe lambing after the remainder of the flock would be unable to keep up with the movements of the flock.

COURTSHIP BEHAVIOR. The ewe may actively seek out the male and sniff the male's body and genitals and then thrust her head against his flanks. The ewe may call frequently with nonspecific bleats. Increased general motor activity is common, as is tail wagging (Bryant and Tompkins 1973). Standing occurs when the female is receptive, and she will look over her shoulder at the ram as he investigates and nudges her. Like cows, most ewes are in standing estrus at night.

The active role of the ewe in seeking the ram was demonstrated in an experiment in which two-thirds of a flock of ewes was inseminated even though the rams were tethered (Lindsay and Robinson 1961b; Lindsay 1966a). The ewes apparently use olfactory cues to locate the ram, because anosmic ewes were unable to find tethered rams (Fletcher and Lindsay 1968). Ewes can be attracted to a male in the next field or to an infertile male, so ram-seeking does not guarantee pregnancy. Ram-seeking behavior seems correlated with estrogen levels (Lindsay and Fletcher 1972) but does not occur without the visual and olfactory stimuli of the male.

Competition involving agonistic behavior has been observed between estrous ewes over access to a ram. Older ewes are generally more successful in gaining access to the ram; otherwise experience does not seem to be an important factor in ewe sexual behavior. Ewes seem to prefer rams of their own breed (Lees and Weatherhead 1970).

The Ram. The reproductive capacity of the ram is not seasonally limited like that of the ewe. Estrous ewes may be satisfactorily fertilized at any season, although semen quality may decline somewhat during the spring (Pepelko and Clegg 1965a,b). Thus, the breeding season in domestic breeds is primarily determined by the environmental input to the female's hypothalamic-hypophyseal tract. Although the ewe may seek out the ram, courtship is more elaborate in the male than in the female (Figs. 4.10, 4.11).

COURTSHIP BEHAVIOR. Like the female, the male spends a great deal of time sniffing the other's genitalia and urine; the flehmen response may be noted. As stated previously, the flehmen response has no visual communicative

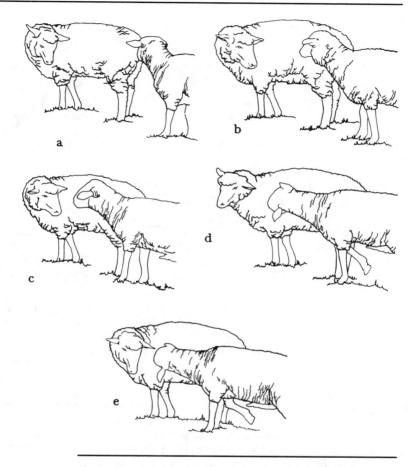

Fig. 4.10. Nudging sequence in sheep courtship behavior (Banks 1964, copyright © 1964, with permission of E. J. Brill).

properties but may be a method of introducing material into the vomerona-sal organ. Rams exhibit flehmen less frequently when ewes are in estrus, apparently because estrous ewes urinate less (Bland and Jubilan 1987). Urination appears to be a sign that the ewe is not in estrous. The male may also lick the female's genitalia, a form of tactile stimulation that may also be part of the testing procedure. Bulbectomized rams in a range situation show some difficulty in identifying estrous females; they are nonselective in the ewes they approach and test but are able to identify estrous ewes pre-sumably via visual and auditory cues (Lindsay 1965). In fact, confined rams may not use any sensory cue to detect estrus; they may rely entirely on the willingness of the ewe to stand. When given a choice of restrained

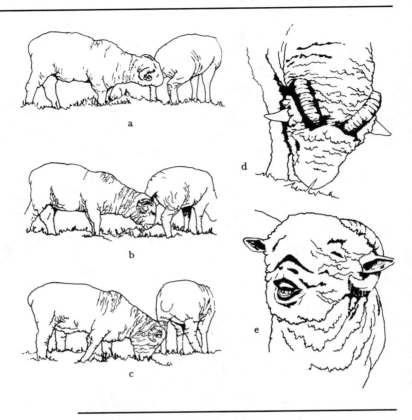

Fig. 4.11. Response of the ram to urine of an estrous ewe. (*A*) Urination by ewe; (*B-D*) ram nosing urine on the ground; (*E*) the flehmen response (Banks 1964, copyright © 1964, with permission of E. J. Brill).

ovariectomized ewes, rams sniffed, nudged, mounted, and copulated with ewes with equal frequency whether or not they had received estrogen (all had received progesterone) (Signoret 1975). Ewes do differ in their sexual attractiveness to rams. This attractveness is stable from estrus to estrus and depends, at least in part, on wool. Wooly ewes are preferred to shorn ones (Tilbrook 1987).

A ritualized kicking with a foreleg is performed as the ram orients himself behind the ewe; the leg is raised and lowered in a stiff-legged striking manner. Tongue flicking accompanies the nudging. Nudging is also stereotyped: the head is tilted and lowered while the shoulder is brought into contact with or oriented toward the flank of the ewe; simultaneously the ram utters low-pitched vocalizations or courting grunts. Six nudgings of the ewe per minute are observed at the peak of the ram's sexual interest (Thiery and Signoret 1978). Several abortive mounts may be made with pelvic thrusting, but without intromission. When the tip of the glans penis contacts the vulvar mucosa, a strong pelvic thrust accomplishes intromission and ejaculation occurs immediately. After dismount the ram may sniff the ewe's vulva, appear depressed, and stand near the female with his head down; he may urinate. A ram will copulate with a ewe approximately six times during the receptive period (Hulet et al. 1962).

The latency to ejaculation increases with the number of observed ejaculations during an observation period but is subject to large individual variation. The number of ejaculations that occur when a ram is presented with a large number of estrous ewes is similar to the number that occur in response to an electroejaculator, indicating that physical capability rather than libido is limiting. Males show a reluctance to remate a recently mated female, whether or not the test ram was the male who inseminated her originally.

If rams are mated to sexual exhaustion with one ewe (usually three to six matings), rapid recovery may be obtained by introducing a new ewe; whereas a much poorer recovery rate is obtained by removing and reintroducing the original ewe (Beamer et al. 1969). This effect of novelty has already been described as the Coolidge effect. Another method of demonstrating this phenomenon is to present four receptive ewes to a ram. He mates three times as much as he does when only one ewe is in estrus (Alexander et al. 1974). There is some evidence that habituation to a sexual partner may last several weeks after a mating as measured by the degree of recovery in the sex drive following satiation. These effects make evolutionary sense for the species and are fortunate for the rancher. A ram is capable of something less than 20 ejaculations daily. If many of these copulations were repeat breedings to especially persistent ewes, the fertility rate for the herd would probably be low. This breeding efficiency allows a rancher to introduce only a few rams to service a flock. In wild or feral sheep, a dominant male can spread his genes over a much larger proportion of the population than if the mechanisms for reduced repeat breedings by either himself or subordinate males did not exist.

THE EFFECT OF DOMINANCE. When more than one ram is being used to service a flock, dominance effects may influence the percentage of the flock each ram inseminates. Mature rams may dominate other mature rams and will almost always dominate yearling rams. Dominance orders, as determined by the number of bunts each animal delivers when the animals are confined together, also exist and may be more marked between yearling rams. Rank in a food competition hierarchy corresponds to rank in the sexual hierarchy (Fowler and Jenkins 1976). In a confined ewe flock, a dominant ram may copulate 12 to 15 times daily over the breeding season, whereas the subordinate ram(s) may average only 2 to 5 copulations daily (Hulet 1966). The subordinate rams may service ewes but mostly at times when pregnancy is not likely to result, that is, long after ovulation. In a feral population or a multiram pasture breeding situation, a number of rams may copulate with an ewe. The ram who copulates between the 9th to 15th hour of estrus is most likely to sire the lambs because that is when the ewe is most fertile (Jewell et al. 1986). In less confined conditions, the monopolization of mating by the dominant ram may be restricted only to the harem of ewes he is able to hold (Mattner et al. 1967).

Hulet (1966) suggests using uniform age groups and allowing adequate space when groups of rams are to be used together in a mass-mating system. Three rams per 100 ewes or 20 to 50 ewes per ram has been suggested for a range-breeding situation. A pair of rams should never be used to breed a group of ewes, because agonistic activity may take precedence over mating behavior. Lindsay and Robinson (1961a,b) showed that three rams served 50% more ewes than a single ram over a 4-day period. It is unfortunate that dominance or sexual aggressiveness does not necessarily correlate with fertility; the combination of dominance with infertility in a flock ram could be disastrous to a rancher.

In pen breeding, a teaser, or vasectomized, ram is used to mark ewes by means of a harness and crayon attached to his brisket. A ewe that is in estrus will be mounted and, therefore, marked (Lindsay 1966b). The marked ewes are then added to the breeding ram's pen. Problems can arise if the breeding ram is submissive to rams in nearby pens. He may be inhibited and not mount. Another problem is seen when young and mature ewes are added simultaneously. Rams prefer or are monopolized by the older ewes. For most effective pen breeding, ewes should be checked for heat at 12-hour intervals, and the breeding rams should be free of disturbance from rams or ewes in adjacent pens (Kilgour and Winfield 1977).

BEHAVIORAL ABERRATIONS. Sexual behaviors, notably mounting, are common components of play behavior in lambs and may be seen as early as 1 week of age. Rams are often reared in large monosexual groups or individually, and Zenchak and Anderson (1980) have shown that the isolated males show some aberrations in behavior when first exposed to estrous females, but most are eventually able to mate. The previously isolated rams were more proficient than group-raised rams (Bryant 1975). Some of the rams raised in monosexual groups are sexually inhibited or show homosexual preferences and do so for long periods or permanently after maturity. In one study, after 9 days of exposure to estrous ewes, some rams (17%) still had not copulated (Hulet et al. 1964). The frequency of mounting in an all-male group cannot be used to predict sexual behavior in a mixed-sex group (Price et al. 1988).

Zenchak and Anderson (1980) argue that some individuals in the male groups learn to associate courtship and mounting behavioral activities with agonistic behavior. When presented with a conflict, they may have used courtship and mounting instead of fighting to assert and maintain a social position. The repeated association of sexual and agonistic behaviors in maintaining a social rank may make it difficult for the ram to respond to the cues provided by the estrous ewe. Another explanation is that the inhibited rams learn to associate sexual behaviors with the odors of rams only. It seems reasonable, therefore, to avoid sexual inhibition or the development of homosexual preferences to suggest not raising rams in monosexual groups. Mixed groups are probably ideal, with isolated rearing less preferable but acceptable.

The effects of the presence of the ram on the reproductive status of the ewe has been discussed above. The ram is influenced in his turn by the presence of the ewe. Testosterone levels, testis size, and levels of aggressive behavior are all higher in rams in pens next to ewes than in isolated rams (Illius et al. 1976). Rams, like ewes, prefer mates of their own or their dam's breed (Key and MacIver 1977).

The effects of castration in prepubertal rams are extremely variable (Banks 1964). Sexual activity may decline or be completely eliminated. Postpubertal castration causes decreased sexual activity in rams (Alexander et al. 1974), but erection and intromission may persist for years postoperatively, with only a marked decrease in frequency. The effects of postpubertal castration are similar to those seen in the stallion, bull, dog, and cat.

SEXUAL BEHAVIOR OF FREE-LIVING SHEEP. The sexual behavior of the primitive and free-living Soay sheep may give some clue about the optimal way to raise breeding rams. Ram lambs remain with their dams from late spring when they are born until the following fall when they begin to chase ewes. They will be rejected by most adult ewes until they are larger. Meanwhile, the adult rams have been living in separate all-male flocks on separate feeding sites. Before the ewes are in estrus, the rams begin to move toward the ewe flocks. They will chase away strange rams but remain on friendly terms with their own flock mates. As the season progresses, however, they will begin to fight with one another. The fights consist of head clashings in these horned sheep and nudging behavior similar to that seen in courtship. All this takes place before the ewes are receptive, although young rams are harassing the ewes already. When the ewes come into heat, the sexual contests between the males have been settled, and the males can turn their attention to breeding. The winner of the most contests between males breeds the most ewes, but he does so sequentially by tending each ewe for a day, or at least half a day, grazing with her and performing the courtship activities described above for domestic sheep (Grubb 1974a). The rams graze and rest less while courting and fighting and may lose weight. These observations indicate that the best way to rear breeding rams would be to leave them with their dams for as long as possible, certainly until puberty, and then put them in an all-male group. Whether or not this would be economically feasible is unknown.

Goats. Collias (1956) noted the close relationship between agonistic and sexual behaviors in male goats, and the comments for rams are probably applicable to this species also. Goats, however, have not yet been subject to the same degree of management as sheep.

The Doe. Does in estrus show an increase in tail wagging, vocalization, urination and mounting of other females.

The Buck. In general, buck behavior is similar to ram behavior. The buck holds his tail straight up during courtship. A component of the mating sequence unique to the goat is the urination by bucks on their own forelegs and beards during courtship. This behavior is termed enuration. Occasionally mouthing of the penis also occurs. Although various functions have been attributed to enuration, including increasing the intensity of the buck's odor and advertising his nutritional fitness, the behavior occurs most frequently in a situation of sexual frustration when the buck is restrained from mating (Price et al. 1986).

Before mounting, the buck usually stands directly behind the doe and sniffs her. Following copulation the buck may lick his penis. Fraser (1964) has noted that seasonal or permanent impotence is associated with misalignment of the buck to the doe. He stands at an angle to the doe rather than directly behind her.

A detailed study of feral goats indicated that does permitted large, horned bucks to mount, but small, hornless goats were virtually excluded from breeding. When pursuing a doe the buck gobbled, a sound produced by moaning and rapidly thrusting his tongue in and out. Courtship activity ceased when the buck exhibited the flehmen response to a nonestrous doe's urine. Although large males were usually able to chase off rivals, occasionally many males would mount a female in quick succession without regard to their position in the dominance hierarchy (Shank 1972). Postpubertal castration causes a decline in sexual behavior in goats (Hart and Jones 1975).

Clinical Problems

THE DOE. Silent Heat. More silent heats occur in the spring (Chemineau 1986).

THE BUCK. Not all goats that mount homosexually in an all-male group will fail to mount heterosexually; those that do fail to mount heterosexually probably have a strong preference for one partner who happens to be male. The bucks that are mounted by other males are less apt than the mounting bucks to restrict their sexual activity to males. White bucks appear to be mounted more often than colored ones, which may be a function of the calm disposition of white Saanen goats rather than of color preference by the mounters (Price and Smith 1984).

Horses

The Mare

ESTROUS CYCLE. Mares show the opposite seasonality in estrous pattern to that of sheep and goats; that is, they are long-light breeders and cycle in the spring. Foaling season is late winter and early spring, and the 11–11.5 month gestation period dictates a rapid recycling and rebreeding if foalings are to coincide with spring grasses. Sexual receptivity may be seen at various times during the year, but ovulation normally occurs only in the spring. Most free-ranging horses are probably bred on the first, or at latest the second, heat after foaling (Tyler 1972). If breeding does not take place, estrus recurs approximately every 3 weeks. The average length of receptivity is 5 to 6 days. Breed differences in the length of estrus in different localities have been noted but are of uncertain significance due to the probable interplay of genetic and environmental factors, methods of testing for the presence of estrus, and statistical analysis (Waring et al. 1975).

Modern management usually isolates the stallion from the mares. Teaser animals are used to determine the receptivity of the mare during the breeding season; behavioral signs are often unclear because of the gradual onset of estrus. The teasers are usually stallions that are introduced to a mare over some form of barrier; the mare is restrained and sometimes hobbled. A nonreceptive mare will react to the teaser's advances by squealing, striking, kicking, and moving away. During full estrus, the mare indicates receptivity by her immobility and by permitting the teaser to nibble her rump and withers. The mare exhibits a characteristic breeding expression, in which her ears are turned outward and her lips are held loosely (see Fig. 1.11). She adopts a basewide squatting stance, urinates frequently, and rhythmically exposes the clitoris with a series of labial muscle contractions known as winking (Back et al. 1974; Asa et al. 1979) (Fig. 4.12). Rectal palpation may be helpful to determine whether follicular development coincides with the behavioral signs of estrus. The optimal breeding time is usually the 2nd or 3rd day of estrus.

Mares will show the same behaviors when a recording of a courting stallion's vocalizations is played and/or their genitalia are manipulated (Veeckman and Odberg 1978), which makes it possible to detect behavioral estrous in the absence of a stallion. Exposure to male odor, a procedure that is effective in sows, elicits the signs of estrus in only half of the mares tested.

COURTING BEHAVIOR. Free-ranging mares will often seek a stallion during estrus and display the normal behavioral signs of estrus before him (Tyler 1972; Feist and McCullough 1975). It is unusual for free-ranging 2-year-old mares to breed (1%) and even uncommon for 3 year olds to breed (13%) (Tyler 1972). Stallions generally do not exhibit much interest in the sexual displays of young mares. Yearling mares may show very exaggerated signs

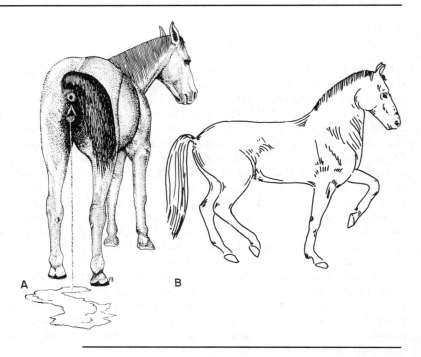

Fig. 4.12. (*A*) The posture of the estrous mare. The clitoris is everted and the tail deviated. (*B*) The prance, or piaffe, of the courting stallion (Houpt 1977a).

of estrus and may attract males from bachelor herds. It has not been established whether the yearlings are in true or psychic estrus because it has been most often observed in free-ranging horses. It is rare, but not unknown, for two-year-old feral mares to deliver a foal, which indicates that they can conceive as yearlings. It is hypothesized that the exaggerated signs of estrus in young mares may be a means of attracting stallions from a distance (i.e. unrelated stallions). Incest is unusual in free-ranging bands; mares either leave the band when sexually mature or are "stolen" by a stallion forming a new harem; or the herd stallion is replaced by a young stallion. Those that remain in their sire's herd have a much lower foaling rate than those that join an unrelated male's herd. Those that join a half-brother herd have an intermediate rate (Keiper and Houpt 1984). Stallions base their avoidance of incest on familiarity rather than on kin recognition (Berger and Cunningham 1987).

FOAL HEAT. Foal heat following parturition is quite predictable and begins from 5 to 18 days, usually 9 days, after foaling. Regular estrous cycles usually continue thereafter at approximately 21 to 22 day intervals, although a few mares appear to have a lactational diestrus owing to maintenance of the corpus luteum following the foaling heat. Other mares are cycling but do not show estrous behavior in the presence of a stallion. Instead, the mares protect their foals. Habituation to the stallion might result in normal estrous behavior. Many mares are bred on the foal heat because it is predictable and easily detected and planned for, although the endometrial epithelium is rarely restored or complete at foaling heat and usually does not accomplish restoration until day 13 to 25 following the foaling heat. The 30-day heat following foaling is a safer time to breed; conception rate is higher.

BEHAVIORAL ABERRATIONS. Most of the sexual problems of mares occur at each end of the breeding season. Mares are seasonal breeders that come into estrus as daylight increases. Feral horses in the United States have a restricted breeding season. The foals are born in early summer and the mares are bred on the foal heat. No doubt, if all domestic mares were pasture bred during the late spring and summer, they would be increasing in number at the same rate as feral horses. Mares normally are anestrus from October to February. However, mares can and do show estrus and conceive all year-round, especially if artificial light is used to simulate long days and the mares are well fed and well sheltered. Despite our ability to manipulate the equine reproductive cycle, seasonality is expressed in the large numbers of abnormal heats observed in mares.

It is at the transitional periods, the beginning and end of the breeding season, that such sexual abnormalities as *prolonged estrus* and *split estrus* are usually seen. The length of estrus can be prolonged from 5 to 7 days to 90 days. Other mares may show split estrus. Split estrus may consist of 1 to 2 days of "shallow" estrus in which the mare does not react to the stallion with vigorous squatting, tail deviation, and urination but tolerates him and occasionally exhibits signs of estrus. A few days later, she may show strong signs of sexual receptivity. Both prolonged estrus and split estrus may be accompanied by active ovaries in which follicles are present but do not mature. In the fall, follicles may remain on the ovaries rather than regress, and this condition may be accompanied by complete anestrus or abnormal estrus. Mares that have either prolonged estrus or split estrus may begin to have normal cycles as the breeding season progresses and daylight length increases.

One of the most common reproductive problems of mares is *anestrus*. Mares may be either physiologically or behaviorally anestrous. The former includes mares with persistent corpora lutea and those with inactive ovaries. Older mares may have inactive ovaries due to degenerative changes in the endometrium and/or degenerative changes in the ovary itself. There may be senile changes in the ovaries, including the formation of germinal inclusion cysts. The age at onset of these senile changes varies from the midteens to well into the twenties.

A developmental cause of anestrus in mares is hypoplasia of the gonads and the reproductive tract. This condition, which is relatively rare, is similar to Turner's syndrome in humans because it is associated with an XO rather than the normal XX pair of sex chromosomes.

The solitary mare may actually be anestrous because she has not been exposed to the pheromones that stimulate the central nervous system to initiate estrus.

Although most pregnant mares do not show estrous behavior (Asa et al. 1983), a few may. Hayes and Ginther (1989) have shown that mares that show estrus while pregnant are carrying female foals. This is most likely to occur between day 35 and 40 of pregnancy when accessory follicles are formed. The behavior of the pregnant mare can be distinguished from that of the nonpregnant estrous mare by the rapid tail lashing or wringing of the former rather than the deviated tail of the latter. The pregnant mare will not usually stand for the stallion; she is not truly receptive. Her urine will be clear rather than cloudy as is that of the estrous mare.

The common sexual behavior problems of mares are (1) silent heat; (2) psychic estrus; and (3) excessive sexual behavior.

Behavioral anestrus is also known as *silent heat*. Palpation of the ovaries may reveal normal follicles. Ovulation will take place normally. There is physiological, but not behavioral, estrus because the mare will not accept the stallion. One reason for silent heat may be related to the fact that mares do show mate preferences. These preferences should be considered in teasing a mare in silent heat. More than one stallion should be used. Environmental factors can influence the behavior of the mare just as they influence the behavior of the stallion. A mare may not show estrus if she has just been trailered to a strange stud farm or been handled by a strange person. If the mare does not show signs of estrus even in a familiar environment and with exposure to several stallions, she may still conceive if she is artificially inseminated, or tranquilized, restrained, and forcibly mated by the stallion.

Not all sexual abnormalities of mares are caused by a deficiency of sexual behavior; some result from an excess of sexual behavior. Some mares, for example, show estrous behavior without the normal physiological correlates of estrus. This abnormality is known as *psychic heat*. It may occur when any horse is brought into the environment of a solitary mare. In that case it is usually a relatively short-lived phenomenon. Much more serious is psychic heat of performing mares. A mare that stops frequently to urinate and is attracted to stallions, geldings, or mares will not be a good competitive trail horse, or three-day-event performer. Despite her posture and behavior, the mare in psychic heat may not tolerate mounting by a stallion. Progestins, such as altrenogest 0.02 ml/kg orally or progesterone 0.4 mg/kg im daily, have been used successfully to treat mares that show severe psychic heat. The progestins are believed to act on those neurons in the brain, probably in the hypothalamus, that control sexual behavior. The independence of psychic heat from endogenous hormonal control is demonstrated by the failure of ovariectomy to relieve psychic heat in some mares. If psychic estrus persists in an ovariectomized mare, dexamethasone may be used in an attempt to suppress endogenous adrenal steroids that might be inducing estrous behavior.

Two pathological conditions of the ovaries can cause *excessive sexual behavior* in the mare. The abnormalities are granulosa cell tumors and persistent follicles. These conditions should be differentiated because one, granulosa cell tumor, should be treated by ovariectomy; the other, persistent follicles, usually resolves itself. The excessive estrouslike behavior of mares with persistent follicles is sometimes called nymphomania. The cysts will regress in time and do not need to be manipulated as follicular cysts of cows do. The persistent follicles may also be treated with progesterone, luteinizing hormone, or an increase in artificial day length to 16 hours or more. Mares suffering from granulosa cell tumor can show signs of constant estrus, but they are often more aggressive than mares in normal estrus (Fretz 1977). Some tumors are virilizing, in which case the mares will attempt to mount other mares and will attack other horses of either sex. Other mares with granulosa cell tumors may show anestrus or a mixture of changing signs as described.

PREGNANCY. Pregnancy may be terminated by forced copulation during the first 6 months (Berger 1983). Whether the loss of the foal is due to the odor of the strange male or to infection introduced by the stallion or to general social stress is unknown.

The Stallion. Stallions exhibit libido throughout the year but show peak sexual behavior in the spring. Seasonal changes are also seen in sperm number and testosterone levels (Pickett et al. 1975). Sexual behavior may be seen in 2- to 3-month-old colts, with full penile erection during resting, play fighting, or mutual grooming (Tyler 1972); but the age at the first successful copulation varied from 15 months to 3 years in the same study. Tyler also noted that stallions actively prevented young males from mounting females. Feral stallions rarely acquire a harem until they are 5 or 6 years old. The years between the time they leave their natal herd at age 2 and acquire their own mares are spent in bachelor herds.

COURTSHIP BEHAVIOR. Courtship behavior will vary with the management practices involved. The following description assumes that the mare and stallion have free access to one another. Intromission may only be achieved with a full erection, and this is correlated with the degree of sexual excitement. Thus, an adequate period of sexual foreplay is essential (Waring et al. 1975). Males may tend a female for several days before she is fully receptive. Nipping and nuzzling begins at the mare's head and proceeds gradually along the body of the mare to the perineal area. As the mare shows receptivity, nickers replace the louder vocalizations. During this testing phase he exhibits the flehmen response. As sexual excitement increases, the male calls with neighs and roars. The female allows the male to lick her around the rear legs and back. Full erection usually develops over several minutes in the mature stallion. Several mounts are usually made before intromission and ejaculation. During copulation the stallion rests his sternum on the mare's croup and may reach forward to bite her neck. Ejaculation occurs around 15 seconds after intromission and after approximately 7 thrusts (Tischner et al. 1974), and intromission lasts less than 45 seconds.

Postcopulatory tending was not noted by either Feist and McCullough (1975) or Tyler (1972) in their study herds in Wyoming and England, respectively. The male may sniff the mare's genital area and exhibit the flehmen response, but the pair soon separates. Under test conditions, sexual satiation occurs after 1 to 10 ejaculations (Average = 2.9) (Bielanski and Wierzbowski 1961). Although stallions are usually limited to a few hand breedings per week, a 6-year-old Belgian stallion bred 20 mares in 9 days with an 85% conception rate. Prostaglandin had been used to synchronize estrus so that 8 mares were in heat and were bred by the stallion on 1 day (Bristol 1982).

SENSORY STIMULI. Visual and sensory stimuli are probably vital to the display of sexual behavior, but their importance may be modified through learning. Tyler (1972) feels that the visual stimuli of the mare's posture with raised tail are important for attraction and penile erection. This reaction may be generalized so that a dummy in the general shape of a mare will be mounted by a sexually experienced stallion. Inexperienced stallions will not mount the dummy. Experienced males will also mount a mare or dummy while blindfolded. The stallion is undoubtedly stimulated by olfactory information, but the stimulation may precede the copulation by several minutes to hours. Typically, a stallion will stop chasing an estrous mare to sniff and exhibit flehmen at the small volume of urine she has expelled. Odor stimulation of the vomeronasal organ may lead to an increase in LH and testosterone and consequently libido, so his behavior is synchronized with that of the receptive mare. Experienced males do not show any inhibition to mounting a mare or dummy when olfactory input is blocked, but inexperienced males will mount a dummy only when it has been sprinkled with urine from an estrous female.

Driving, herding, or snaking with a distinctive head-down position, is a behavior usually elicited by the presence of another stallion. Piaffelike prancing is a display to other stallions. Masturbation is normal behavior in a stallion. The stallion flips the erect penis against the abdominal wall. Ejaculation rarely occurs. Stallions masturbate four times a day spending

38 minutes with an erect penis sometimes, but not always, accompanied by masturbation (Tischner 1982). This behavior usually occurs in the resting stallion, even one at pasture with mares available. Masturbation may occur in association with recumbency (Wilcox et al. 1991).

BEHAVIORAL ABERRATIONS. Ten to 25% of stallions presented for breeding soundness examination have some behavioral problem. Those stallions most at risk are young and/or novice breeders, frequently bred stallions, those in a new environment, and those in transition from racing to breeding (McDonnell 1986). Young stallions appear to be particularly affected by exercise; as little as 30 minutes per day of lunging can decrease libido (Dinger and Noiles 1986).

Some common problems of sexual behavior in the male are

1. Stallions that show sexual interest in mares but will not mount, or mount but do not ejaculate

2. Stallions that have low or no libido, that is, do not show interest in a sexually receptive mare

3. Stallions that will mount mares only when another, specific horse is present

4. Stallions that injure, or "savage," mares or handlers

5. Males that self-mutilate

6. Stallions that mount, but do not intromit

7. Geldings that behave like stallions

Too much serious sexual experience too early is very detrimental to normal libido. Many stallions overused as youngsters are presented with sexual behavior problems as adults. The most common problems are impotence or low libido. Other stallions that are overworked as studs may bite the mares viciously or be uncontrollable by their attendants. It is not always clear whether the young stallions have had traumatic or unpleasant experiences. They may remember being kicked by a mare, or they may simply remember that they were exhausted. The resultant loss of libido can persist indefinitely unless treated (Pickett et al. 1977).

Figure 4.13 illustrates a case of a stallion that shows different behaviors to different mares. He mounted without erection and bit a Morgan mare with whom he had been housed as a 2 year old. His sexual behavior was normal towards another mare with whom he had had no previous contact. He had been noted to display snapping (see Chaps. 1 and 6) to the Morgan mare as a youngster and presumably was subordinate to her. As an adult he showed aggressive behavior, perhaps because of an approach-avoidance conflict. He was sexually stimulated but was afraid.

Stallions that are used as teasers, that is, to detect estrus in mares, but are not used for breeding, may eventually show a loss of libido. In addition, they may show stereotyped behavior, such as stall weaving. It has not

Fig. 4.13. A stallion showing different behaviors to different mares. He is showing aggression toward a Morgan mate by biting her after mounting without an erection; he had been housed with her as a 2-year-old. (He displayed snapping [see Chaps. 1 and 6] to the Morgan mare as a youngster and presently is subordinate to her. As an adult he showed aggressive behavior, perhaps because of an approach-avoidance conflict. He was sexually stimulated but was afraid.) Five minutes later his sexual behavior was normal towards another mare with whom he had had no contact as a juvenile.

been determined how often a teaser stallion should be allowed to copulate to prevent these abnormal behaviors, and there are, no doubt, individual differences in response to use as a teaser.

Stallions can learn to inhibit sexual behavior as easily as they learn to express it. Stallions may be fitted with stallion rings, devices placed on the penis that cause discomfort if erection occurs. They are used on stallions that are in training or in other circumstances where sexual behavior would be inappropriate. Many stallions apparently can be fitted with these devices and learn not to respond sexually when wearing them and yet respond normally when the rings are removed. Other stallions have learned too well and are impotent even when the ring is removed. Stallion rings and belly brushes are also used to prevent masturbation, which as noted above should not be considered abnormal behavior nor a cause of infertility. As noted, ejaculation rarely occurs, so the behavior is unlikely to lead to a drop in fertility, but attempts to punish masturbation do cause libido problems.

The treatment of any behavior problem must begin with elimination, or at least identification, of any *physical impairment*. The two most common physical problems associated with breeding are genital injury or limb injury. Any lameness or limb injury will inhibit the stallion's ability to mount so that he may exhibit penile erection and interest in a mare but will not mount. An older stallion with navicular disease or chronic arthritis is a typical example of the effect of organic limb disease on sexual behavior. When pain is the cause of libido problems, nonsteroidal anti-inflammatory agents such as phenylbutazone and flunixin meglumine may be of value. Another factor that may be responsible for failure to mount is improper flooring. If the flooring is slippery, for instance, the stallion will be reluctant to mount, especially if he has fallen when trying to mount a mare on other occasions. He is much more likely to mount if taken to a different environment with grass or tanbark beneath his hooves. A stallion with a mild locomotor problem may mount but be reluctant to ejaculate in cold weather. In warm weather, when he is pain-free, he will be normal.

Injury to the genitalia can cause breeding difficulties. Naturally, a stallion will avoid intromission if his penis is painful. Stallions may be reluctant to copulate long after the injury is apparently healed because they may not have learned that copulation will no longer be painful.

Stallions with physical impairment of the legs or back should mount secure mounts, that is, sturdy mares. The flooring should be adequate and the mare should be the correct size so that the sternum of the stallion rests on her croup. Anatomical fit is important because even normal stallions may lose their balance and slide off a mare that is too small or too large. Physically impaired stallions can be trained to a secure dummy mount if collection of semen and artificial insemination are permitted for that breed. At first an estrous mare may have to be held next to the dummy, but stallions, like bulls, can be conditioned to ejaculate in the absence of the normal stimulus of the mare.

The total *breeding environment* must be considered because another cause of injury to the stallion can be a low roof or an overhang. A stallion rearing on his hindlegs to mount a mare is considerably taller than when he is standing on all four feet. If a stallion strikes his head while mounting, he may not only sustain serious injury but may also be inhibited from mounting on subsequent occasions.

Perhaps the most important factors in the breeding environment are handlers. Handlers must be familiar with the breeding routine and experienced in controlling horses. Some handlers are better able to calm stallions than others with equal experience, and the calmer the stallion, the less likely are accidents. Such simple arrangements as placing all attendants on the same side of the mare and stallion can facilitate communication between them and prevent difficulties. Most breeding injuries and accidents occur when an inexperienced or highly nervous mare or nonreceptive mare is bred without adequate restraint or judgment. Because errors in detection of estrus can be made and because the situation is unnatural, mares should be hobbled before hand breeding. Distractions in the form of superfluous people or animals should be avoided.

There is no single treatment for all abnormal sexual behavior in stallions. Patience and time are necessary with almost all cases. It is advisable to take advantage of the stallion's seasonal breeding pattern and institute behavioral therapy during the spring and summer. Advantage should also be taken of the stimulatory effect of the presence of other stallions. It was noted that wild stallions are most likely to copulate when other stallions are present; the same appears to be true of domestic stallions. Some stallions have been known to breed mares only if another horse, even another mare, were present. The stallion may regard the second horse as a competitor — another stallion — or there may be another, unknown reason.

The antianxiety drug, diazepam (0.05 mg/kg slowly IV) has been used successfully to overcome impotence caused by pain associated with breeding and for the loss of libido shown in a novel environment (McDonnell et al. 1985, 1986). Imipramine (500 mg IV) has been used to treat stallions that will mount and intromit, but not ejaculate (McDonnell et al. 1987).

Finally, some stallions have definite mate preferences, and they should be allowed to exercise these preferences while recovering from loss of libido or impotence. One should tease the stallion with several mares and use the mare to which he is most responsive for further treatment. A quiet mare is necessary for a stallion that has been injured by another mare. A stallion that will not mount a mare may ejaculate into an artificial vagina. He may gain confidence and overcome his fear by this process and can later be induced to mount a mare.

Vicious behavior toward the mare and the attendants, like most other abnormal sexual behavior, is most apt to occur when stallions are used for breeding outside the normal breeding season. Therefore, stallions may be unmanageable in January, but well mannered by May.

Overuse and *rough handling* are often the cause of stallion misbehavior during breeding. They may bite the mare or be generally intractable. Attempts to improve the horse's breeding manners should wait until normal libido and copulation have been reestablished. Punishment of a horse with sexual abnormalities will retard its progress. If the stallion's viciousness is not attenuated as his libido improves, various physical devices, such as a muzzle and breeding bridle for him and a withers protector for the mare, may be used.

Self-mutilation is a very common behavior problem. Although it occurs in horses of both sexes, it is much more common in stallions. The behavior consists of biting at or actually biting the flanks or, more rarely, the chest. The horse usually squeals and kicks out at the same time. The signs mimic those of acute colic, but can be differentiated because self-mutilation does not progress to rolling or depression and is chronic. The cause of the behavior is unknown, but because it usually responds to a change in the social environment it is probably caused by sociosexual deprivation. Most breeding stallions do lead similarly deprived lives in that they are kept in stalls in isolation from other horses, particularly from mares, but do not self-mutilate.

The question arises as to whether stallions that self-mutilate should be used for breeding. Castration usually, but not always, stops self-mutilation. Preventing the behavior by using cradles and side poles does not remove the cause; the stallion will continue to vocalize and kick so that, although he can no longer injure himself, he can still injure a bystander. Sometimes providing a stall companion such as a donkey will reduce the incidence of self-mutilation. Allowing the stallion to live on pasture with a mare will eliminate the problem in most cases. The chances that the stallion will be injured by the mare are less than the chances that he will injure himself or someone else by self-mutilating. Opiate antagonists will prevent self-mutilation, (Dodman et al. 1988). Unfortunately, naloxone, the antagonist now available, is metabolized very quickly by horses. See cribbing, Chapter 8, for a discussion of the involvement of endogenous opiates in equine "vices."

The *effects of castration* can cause varying behavior. A horse that exhibits stallionlike behavior could be either a cryptorchid from whom the undescended testicle was not removed at castration or a gelding in which sexual behavior persists. A negligible plasma testosterone will distinguish the gelding from the cryptorchid stallion (Ganjam and Kenney 1975). Sexual behavior persists in over half the geldings (Line et al. 1985). The sexual behavior may be as innocuous as exhibiting flehmen or as extreme as mounting and intromission. The sexual behavior itself could cause pregnant mares to abort, but the usual problem is aggression directed toward other geldings by the one who is acting like a harem stallion. Another unwelcome stallionlike behavior is attacking foals, particularly newborn foals. Management can be used to prevent these problems. A gelding that acts like a stallion should be pastured only with other geldings and should not have access to foals.

Geldings may also self-mutilate. These are usually geldings that are displaying other stallion behaviors, but unlike intact males they self-mutilate in the presence of mares. Stall confinement or pasturing without visual contact with mares usually reduces the incidence of self-mutilation. If not, progestins may be used.

Pigs

The Sow. Like the cow, the sow is a nonseasonal breeder. Once regular cycling commences, the sow will cycle every 18 to 24 days (mean = 21) until bred. Puberty occurs at 5 to 8 months.

ESTROUS CYCLE. Like females of other species, the sow shows an increase in activity as estrus approaches (Altmann 1941). Urination is frequent, as is calling to the male with a soft, rhythmic grunt. An estrous female approaches the boar and sniffs him around the head and genitals. Estrous sows attempt to mount other estrous females. The increased motor activity eventually takes the form of searching behavior, which seems vital initially to uniting an estrous sow with a boar (Signoret et al. 1975). Olfactory stimuli alone will instigate this searching; anesthetized boars readily attract estrous sows. Olfactory bulbectomy drastically impairs the ability of the sow to discriminate between males and females (Alexander et al. 1974). Signoret and Mauleon (1962) have reported that bulbectomy also eliminates sexual behavior and prevents normal ovulation and estrus. This result has not been confirmed by Meese and Baldwin (1975b) who found that bulbectomized females mated, conceived, and reared their litters although there were deficits in maternal recognition (see Chap. 5).

Searching behavior appears to be under endogenous control and requires estrogen during behavioral ontogeny for full development. Gilts reared in isolation still show this behavior upon reaching estrus (Signoret et al. 1975). The immobility response of the fully receptive sow, however, seems to require both auditory and olfactory stimuli. The specific auditory stimulus is the courting song of the boar (Signoret et al. 1975). The immobility reaction, like searching behavior, does not seem susceptible to learned modifications. The olfactory stimuli to which the sow responds are pheromones present in both the saliva and preputial secretions of boars. The chemicals involved are metabolites of androgens and have been identified. An example is 5α-androst(16-ene)3-one (Melrose et al. 1971). These compounds have been used experimentally (Reed et al. 1974) and are available commercially (Boarmate from Antec International) to elicit the immobility reactions.

Olfactory and auditory stimuli from adult boars, supplied by an aerosol spray and a tape recorder, will increase the proceptivity of gilts toward young boars (Hughes et al. 1985). The presence of a boar has a slight positive effect on the rate of conception and number of piglets per litter (Hemsworth et al. 1978).

Estrus is less apt to be detected in sows with less than 1 m² of pen space and those living in pairs. It is also easier to detect estrus if the sows are housed across an aisle from a boar rather than in adjoining pens. Because the sow has had close olfactory contact with the boar when the attendants were not present, she may have shown the immobility response when no person was there to notice (Hemsworth et al. 1978, 1984, 1986a,b).

BEHAVIORAL ABERRATIONS. Perhaps because of the somewhat rigid genetic control of sexual behaviors, aberrations are unusual. Breed differences in the length of estrus are seen (Signoret 1970), but these are minor and unimportant clinically. Although not quantified, some sows seem to show decided mate preferences and display strong aversions to specific males (Signoret et al. 1975).

Failure to reproduce in confinement is the most important problem with confined gilts. Confinement and the social environment appear to play a role in inhibiting estrus in young gilts (Ford and Teague 1978). The presence of a boar leads to the occurrence of estrus at an earlier age and in more gilts (Thompson and Savage 1978). Puberty is accelerated in gilts older than 160 days exposed to a strange male for 20 or more minutes per day. Other nonspecific factors are social and nutritional; the latter indicates that puberty occurs when a set point of body weight or body fat is reached. Puberty is delayed in regrouped or crowded pigs (Clark et al. 1985; Barnett et al. 1986; Hemsworth et al. 1986b). The stress of trailering can also stimulate the onset of estrus. It is interesting that chronic stress of overcrowding delays puberty, but the acute stress of transport accelerates it.

The Boar

COURTSHIP BEHAVIOR. Once contact with an estrous female has been made, the boar will pursue the female, attempting to nose her sides, flanks, and vulva (Fig. 4.14). Unique to the pig is the boar's "courting song," which is used during this phase of courtship—a series of soft, gutteral grunts, about 6 to 8 per second (Signoret et al. 1975). Tactile stimulation of the female continues and increases in intensity as the boar's sexual excitement increases. The boar usually emits urine rhythmically; pheromones in the urine may further increase the female's willingness to stand. Several mounting attempts may be made until the female becomes immobile, after which mounting and intromission follow rapidly. The boar's ejaculatory time approaches that of the dog, although no copulatory lock occurs in swine. Ejaculation occurs within 3 to 20 minutes, with an average of 4 to 5 minutes (Signoret et al. 1975). Consort behavior continues for a short time following copulation. Burger (1952) observed that boars mated an estrous female

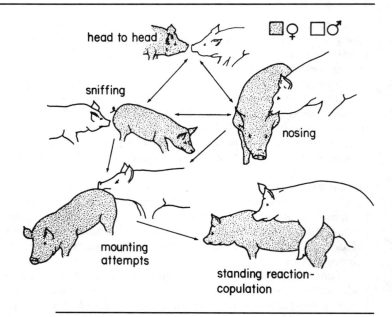

Fig. 4.14. The courtship sequence of pigs (Signoret et al. 1975, copyright © 1975, with permission of W. B. Saunders Co.).

around 10 times over a 2-to-3-day estrous period. Although domestic pigs are not considered seasonal breeders, libido and testosterone levels increase earlier in prepubertal boars if the day is artificially lengthened to 15 hours (Hoagland and Diekman 1982). Boars tend to have greater libido in separate mating pens than in their home stalls (Hemsworth et al. 1989).

OLFACTORY STIMULI. Olfactory cues seem unimportant in stimulating a boar to mount a female or a dummy. Olfactory bulbectomy, for example, does not prevent normal sexual behavior (Booth and Baldwin 1980). Some, but not all boars can distinguish between estrous and anestrous females from a distance, that is, on the basis of olfactory information (McGlone and Morrow 1987). The initial contact with the sow, however, triggers a behavioral response from her, reflecting the degree of her sexual receptivity. It is this tendency toward immobility that seems to arouse the boar. Like the female response, this reaction is under fairly strict genetic control and not subject to much learned modification. Thus, what is noted by human observers as homosexual mounting or aberrant mounting of artificial stimuli is explained as normal response to immobile objects of approximately the correct size and shape. Signoret et al. (1975) mention the occurrence of stable homosexual relationships in all-male groups. Although some breeds are easier to train for semen collection than others (e.g., Yorkshires are easier to train than Durocs [Alexander et al. 1974]), training a young boar to mount a dummy will usually be successful on the first few attempts.

EARLY SOCIALIZATION. In the boar, as in most species, the early social environment is important to later sexual behavior. Boars raised in isolation from 3 weeks of age copulated less often with estrous sows than boars raised in groups. Boars with visual and olfactory contact with other boars were not nearly as inhibited as the isolates, indicating that contact with other pigs, even male pigs, is important (Hemsworth et al. 1978). Later, contact with females is also important because boars with visual and olfactory contact with sows copulated more often and ejaculated longer than boars kept in isolation or with visual and olfactory contact with other boars (Hemsworth et al. 1977a,b). This is probably due to a pheromone that has a definite physiological effect; testosterone and corticosteroid levels are higher in boars that are in contact with sows (Liptrap and Raeside 1978).

BEHAVIORAL ABERRATIONS. As with the other domestic animals, differences in the level of "sex drive" appear to be larger between individuals than between breeds. Low libido, however, has been associated with a high plane of nutrition; and, at least in the United Kingdom, is seen more frequently in landrace than in boars of the Large White breed (Anonymous 1975). Libido may be impaired through mismanagement of a young boar. A young male turned in with a group of gilts may be frustrated by excessive curiosity or bullied by the gilts. This incompetence or fear may become conditioned and a permanent problem. Supervision of early matings is recommended. A quiet sow or one recently serviced by a mature boar should be used for the first mating (Anonymous 1975). Another common problem is aggression by the boar toward humans. This problem is usually resolved by culling the boar, which removes the danger and prevents an aggressive animal from reproducing.

Dogs

The Bitch

ESTROUS CYCLE. The domestic dog, unlike most of its canid relatives, is a nonseasonal breeder. The length of each estrous cycle is extremely variable from individual to individual, and sometimes from one heat to the next in the same bitch. From one to four cycles yearly may be seen, usually two. The basenji is an exception; one seasonal breeding per year is seen in the early fall (Fuller 1956). Basenji-cocker spaniel crosses show both mono- and polycyclic activity, which indicates genetic control of this aspect of the reproductive cycle. The onset of puberty also varies widely among individuals. No strict correlation of age at puberty may be made with either body size or conformation (Fox and Bekoff 1975), but generally the smaller breeds reach puberty earlier than the larger breeds. It would be very rare for a St. Bernard to cycle at 6 months, for example, but not unusual for a miniature poodle to do so. Thus, puberty onset for all dogs ranges from 6 to 15 months with 7 to 10 months being the usual for the "average" dog.

COURTSHIP BEHAVIOR. The correlation of hormonal levels and sexual behavior in the dog is illustrated in Figure 4.3. The first proestrus and estrus of a bitch's life is shorter than subsequent ones and the levels of LH and estradiol are lower (Chakraborty et al. 1980). She is less attractive to the male and less proceptive (Ghosh et al. 1984).

Courtship behavior is marked by play behavior in the proestrous part of the cycle, but this play behavior decreases during estrus. The female will run with the male, approach him using the typical play bow of puppies, and even whimper submissively. She will sniff and lick the male's body and genitalia. Urination becomes more frequent as estrus approaches and the posture used will frequently be the squat-lift (see Fig. 1.4). She may stand momentarily during proestrus but turns before the male can mount, often with a bark or growl (Christie and Bell 1972). Attraction of males and proceptivity appear in proestrus, but receptivity occurs later, during estrus (Beach et al. 1982).

During estrus the female stands more quietly to allow male investigation, and eventually, intromission toward the end of estrus. When the male touches her vulva she will flex her body laterally (Hart 1970); while he is thrusting she will move her perineum from side to side and ventrally, a motion that increases the probability of intromission. After the lock or copulatory tie has been established, she may roll or twist and turn. (The copulatory lock will be discussed in the following section.) Contractions of the constrictor vestibuli muscles and the anus occur as an afterreaction.

If the male does not mount, she will "present" her hindquarters to him and even back into him and deviate her tail. An older and more experienced bitch may mount a young male and execute a few to many pelvic thrusts (Fox and Bekoff 1975).

Social behaviors, other than sexual ones also are influenced by the reproductive condition of the bitch. Dominance relationships between females may shift, especially during metestrus. Males may defer to females in food competitions not because they are chivalrous males but because they are more motivated to mount than to eat.

Courtship in free-ranging dogs is possible. Stray bitches avoid their male littermates but can be bred by a persistent brother. Estrous females attract two to seven males; they will show less proceptive behavior in the presence of many males and also less active rejection of nonpreferred males.

BEHAVIORAL ABERRATIONS. Owners unfamiliar with canine courtship may be upset because the bitch appears to tease the male by soliciting and then threatens him if he mounts, but this is normal. Females may refuse males for any of several reasons. A bitch may display dominance over a male by not allowing the male to *stand over* her (a normal canine signal of dominance) or to approach from behind. Dominance relationships are learned but can be established rapidly. Thus, dominance relationships probably help inhibit mother-son matings and may prevent certain sib-sib matings (Fox and Bekoff 1975) but may also develop quickly if an aggressive bitch is placed with a more submissive dog. Le Boeuf (1967) and Beach and Le Boeuf (1967) demonstrated definite female preferences for certain males. Refusal can range from avoiding a particular male to actively chasing and biting him. Not all the females rejected a male to the same degree, and Beach (1970b) showed that dominant males were not necessarily chosen as preferred mates. Beach also found that sexual preference could not be correlated with social affinity outside the mating period.

The Dog. Sexual behaviors may appear in 5-week-old male pups, and mounting behavior becomes an important part of the male's social repertoire as it matures. As with many other mammals, mounting is used as a sign of dominance; a submissive animal will stand for a more dominant male, but standing over is not tolerated by the dominant animal. Fox and Bekoff (1975) point out that most dogs are sexually mature physiologically long before they copulate for the first time. Perhaps the lack of dominance in young dogs inhibits early mating. Social contact is vital in the ontogeny of normal sexual behavior. Dogs raised in social isolation showed abnormal mounting orientation that persisted for longer than it did in dogs with similarly limited sexual experience but more social experience (Beach 1968).

COURTSHIP BEHAVIOR. Male dogs are attracted to bitches (Beach and Gilmore 1949). Urine of the estrous bitch appears to be more attractive to the dog than vaginal secretions (Doty and Dunbar 1974; Dunbar 1978), but a component of the vaginal secretions, methyl p-hydroxybenzoate, has been shown to induce male sexual behavior when applied to the vulva of an anestrous bitch (Goodwin et al. 1979). The pheromone may be considered a *releaser* of sexual behavior in the male. Mammalian behavior is not as stereotyped as that of fish and birds, so although sexual arousal may occur in all male dogs exposed to the pheromone, the expression of that arousal may vary considerably; therefore, male courtship behavior is extremely variable. Males may show extreme interest or indifference to females, although mating may occur successfully in either case. Play behavior may be marked or absent. The male sniffs the female's head and vulva; he may lick her ears. Although canids do not show the classic flehmen response of ungulates, it is possible that the *tongueing* response seen during this olfactory investigation transports pheromones to the vomeronasal organ in a manner similar to that postulated for ungulates. The length of the play activity and olfactory investigation probably varies with the past experience of the male and female, perceived degrees of dominance within the pair, stage of estrus, and sexual satiation of either partner. A bitch can be forced to accept copulation by a strong and aggressive male who chases her until she is exhausted and holds her with his teeth by the neck or with his paw on her back. Only half of the stray male dogs are able to copulate, and those under 1 year of age rarely copulate (Ghosh et al. 1984).

The male mounts in response to female immobility; he grasps her with his forelegs just cranial to her pelvis. He thrusts with his pelvis, and when intromission has been achieved, the rate of thrusting increases. Engorgement of the bulbus glandis and contraction of the vaginal muscles following intromission result in the copulatory lock or tie, a phenomenon most closely associated with canids, but not restricted to them. The male will usually dismount and turn around so that male and female are facing opposite directions while ejaculation occurs (Figs. 4.15, 4.16). The lock may last 10 to 30 minutes (mean = 14 min., Hart 1968), after which the bulbus decreases in size and the pair separates. Following copulation, recovery from sexual refractoriness may be rapid. Fox and Bekoff (1975) report records of up to five copulations by a male dog in one day.

BEHAVIORAL ABERRATIONS. Male *impotence* or loss of libido can be the result of organic disease; most commonly, musculoskeletal disease such as hip dysplasia, arthritis, or trauma-induced pain in the hindquarters. Balano-

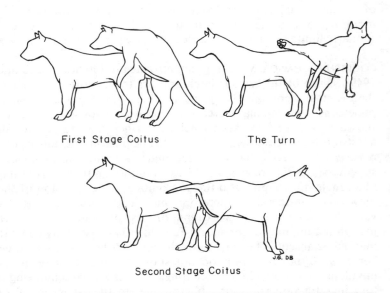

First Stage Coitus   The Turn

Second Stage Coitus

Fig. 4.15. Coital positions of the dog (Grandage 1972, copyright © 1972, with permission of *Vet. Rec*).

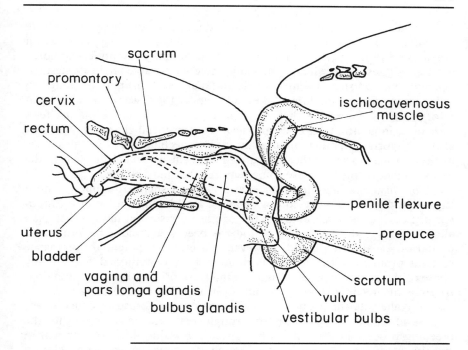

sacrum

promontory

cervix

rectum

ischiocavernosus
muscle

penile flexure

prepuce

uterus

bladder

vagina and
pars longa glandis

bulbus glandis

scrotum

vulva

vestibular bulbs

Fig. 4.16. Relationship of the male and female genitals during the copulatory lock of the dog (Grandage 1972, copyright © 1972, with the permission of *Vet. Rec*).

posthitis is generally a mild disease in dogs and unlikely to affect sexual performance, although a severe form could conceivably do so.

The *lack of sufficient social contacts* as a puppy may inhibit successful copulation. This becomes a very real problem, not only an experimental one, when a pup is purchased through one of the mail order "puppy mills." Pups are weaned as early as possible (4–5 weeks) and shipped shortly thereafter. The pup, if it survives transport, is kept isolated in the new owner's home, protecting it from infectious disease. When the owners finally do try to use their sheltered pet for breeding they have great difficulty persuading the dog to perform or find it impossible to do so. Their dog has been essentially isolated from social contacts with conspecifics. Not only have the motor patterns of sexual behavior not been perfected, but they have never been placed in a proper social context. Some dogs may overcome this void in socialization, but their sexual and other behaviors may never be normal in direction or quantity. An example of poor libido is a pointer who was kept with his sister. Both the sister and his owners reprimanded the dog for sexual interest in his sister and when presented with an unrelated estrous bitch he had no libido and seemed frightened.

Females may rebuff a male, even a dominant one, because of *mate preference.* The traits that the female may be using in making her selection are not known. In species with bright breeding plumage, or antlers or horns of various sizes, or in which a male may hold a desirable territory, nest, or den site, mechanisms of female choice (sexual selection) may be imagined to be based on these qualities. Le Boeuf's work (1967) was done with laboratory beagles, noted for their physical uniformity, and the mating tests were done in neutral territories; physical and territorial characteristics may also be important in other circumstances. The mechanisms of female choice in these beagles seem rather obscure. The form of sexual display or the temperament of the male are some features that may be used.

Timidity, especially in poodles and German shepherds, may be both learned and genetic. Affected dogs may be inhibited to the point of impotence. Leaving the male and female together for several days rather than allowing only a short breeding period has been suggested. This prolonged period of socialization could, however, exacerbate a potential dominance reversal problem if the female is dominant. A conditioned fear of any phase of the breeding program may be inadvertently instilled in a stud dog. Analysis of breeding techniques and reversal of the conditioning may successfully alleviate the problem. Thus, a young dog forced to court a very aggressive female may associate the rough treatment he received with the breeding process in general, with a specific breeding location, or with a specific color or type of female. Young dogs may make some clumsy mounting attempts and may otherwise prolong courtship but should successfully mount a bitch within the first few exposures to a female. If the dog is a persistently timid breeder, however, he should be dropped from the breeding program. Artificial insemination is recommended if the timidity is suspected to be learned or the result of the particular dominance relationship in the attempted mating. Most dogs will breed with most other dogs; therefore, mating problems with behavioral etiologies are unusual and require serious consideration.

Like other domestic animals, male dogs are sensitive to *environmental disturbances* in the breeding process. If one member of a breeding pair must be transported to the other, the female should be brought to the male. Noise and other disruptions in the breeding area should be minimized. The rather curious insistence of some breeders on helping a male dog mount and copulate might actually cause more difficulties than it is thought to prevent. Besides the physical disruption of the observer-helper, the breeder will likely dominate the dog and, thus, somewhat inhibit the male's sexual performance. A timid dog may require the owner's presence if his dominance over a female is doubtful; but, as already mentioned, the continued use of a dog this timid would be unwise. Slippery floors, such as waxed linoleum, may prevent mounting, but should be a problem easily avoided.

Masturbation is not an unusual problem in house dogs. Semen quality or value as a breeder is not affected, but the habit becomes embarrassing or annoying for the owner. Masturbation using inanimate objects is probably seen in most puppies but will become an insignificant behavior in the normally socialized adult. It is not unusual, however, for owners to deliberately or unconsciously reward this behavior by lavishing attention on the pup while he is masturbating (Tortora 1977). A great deal of physical contact between owner and pet seems to be a common aspect of the pet-owner relationship in these cases. As the dog matures, this habit may be transferred from inanimate objects to people's arms and legs as a signal of dominance and continue to be rewarded with attention. The owner may still find the display "cute" or only mildly annoying but especially troublesome when the dog mounts a guest's leg. To resolve the problem, the owners should teach the dog submissive behavior by counterconditioning the dog to down stay when it attempts to mount and consistently punish mounting. Castration may eliminate or decrease the problem. Owners reporting homosexual behavior in their dogs should also be informed that mounting is a sign of dominance.

CASTRATION. Prepubertal castration greatly reduces sexual interests. Because mounting is an integral part of the dog's behavioral ontogeny and is used in agonistic interactions all sexual behavior will probably not be eliminated. Hopkins et al. (1976) studied the postoperative effects of castration on 42 postpubertal dogs in normal home situations (Fig. 4.17). They found that

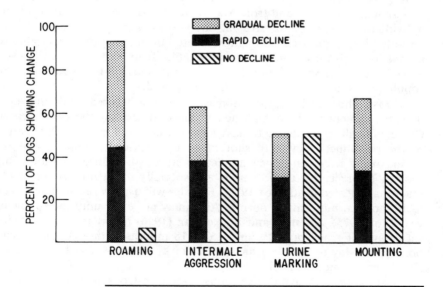

Fig. 4.17. Percentage of dogs experiencing rapid decline, gradual decline, or no change in four behaviors after castration (Hopkins et al. 1976, copyright © 1976, with permission of the School of Veterinary Medicine, University of California at Davis, and *J. Am Vet. Med. Assoc.*).

90% of the dogs castrated to control roaming showed a rapid or gradual decline in this behavior. Intermale aggression was reduced noticeably in only 60% of the dogs and urine marking in the home only 50%. Sixty-seven percent of the dogs showed a decrease in mounting behavior following surgery; mounting of people was reduced in seven of eight dogs castrated specifically for that problem, but reduced in only one of four dogs castrated because of mounting of other dogs. The authors do not report any changes in the owner's handling or attitude toward their dogs following surgery, however, which might have been responsible for some of the behavioral changes noted. Age at castration was not correlated with the noted effects. Although in no cases did the occurrence of an objectionable behavior return to its preoperative level, some behaviors appeared intermittently for long periods following castration. Hart (1968) reported that castrated dogs may retain sexual mounting behavior, with intromission, lock, and ejaculation for several years postcastration or indefinitely, although the frequency of the behaviors decreased.

Cats

The Queen

ESTROUS CYCLE. The queen is seasonally polyestrous and most cats will cycle at least twice yearly if not bred. Although population peaks occur from mid-January to March and from May to June in the Northern Hemisphere, individual cats may be in estrus at any season (Fig. 4.18). The nadir for reproductive output of a population is late fall, making the availability of kittens as Christmas presents very unreliable. If unbred, the cat will cycle every 3 weeks for several months. Actual estrus lasts 9 to 10 days without copulation, and around 4 days if the cat is bred.

Most felids, including the domestic cat, are induced ovulators, and, thus, breeding may be accomplished whenever the female shows receptivity. Owners unwilling or unable to have a pet cat neutered may use this feature of the reproductive cycle to shorten estrus; artificial stimulation of the vagina using a cotton-tipped applicator stick will induce ovulation and shorten the receptive time. This is an especially useful procedure in terminating repeated heats (Fox 1975). Females will usually reach puberty at 6 to 10 months, but females born in April may not cycle until the following year (Fox 1975). Eckstein and Zuckerman (1956) report that free-ranging cats may not reach puberty until 15 to 18 months, although "barn" cats born from May to July in Ithaca, New York, routinely give birth to their first litters 12 months later.

Cats appear to avoid incestuous mating; estrous females will travel farther from home if the closest male is related (Liberg 1983).

CATNIP. Nepetalactone, a volatile terpenoid found in the catnip plant (*Nepeta cataria*) (Waller et al. 1969) elicits behaviors similar to sexual behav-

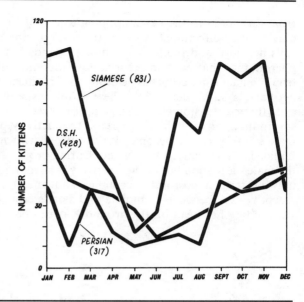

Fig. 4.18. Seasonal distribution of litter births. DSH = domestic shorthair. Numbers in parentheses are numbers of kittens. This study was performed in Australia, so the seasons are the reverse of those in the northern hemisphere (Prescott 1973, copyright © 1973, with permission of *Austr. Vet. J.*).

iors, and debate continues about whether catnip initiates a release of sexual behaviors or acts as a nonspecific pleasure inducer (Todd 1963; Palen and Goddard 1966; Jackson and Reed 1969; Hatch 1972; Leyhausen 1973; Hill et al. 1976; Hart 1978). As Hatch points out, however, estrous behavior is similar to, but not identical with, catnip-induced behavior. Catnip does not cause vulvar presentation, vocalization, or foot treading, and cats in estrus do not head shake as do cats exposed to catnip. Male and female cats respond identically. Todd (1963) found that the body-rolling and head-rubbing behaviors characteristic of both the estrous and catnip-induced states could be induced in males and females by an extract of tomcat urine. As Hart (1978) concludes, nepetalactone may be mimicking one of the compounds in male urine to which an estrous female may be especially primed but to which most or all cats are sensitive. The response to catnip depends on the main olfactory system, not on the vomeronasal organ (Hart and Leedy 1985).

COURTSHIP BEHAVIOR. An estrous female will call and purr. She is restless and shows increased general motor activity. If she is a house cat she may run from one room to the next, stopping to call at each door or window. She may be very affectionate toward the owners. She urinates frequently. She rubs her head and flanks on furniture; glands in these areas may produce pheromones that contain information announcing the presence of an estrous female. She crouches, elevates her perineal region, and treads with her back legs (Whalen 1963). These actions usually will be accompanied by a rhythmic extension and retraction of the claws of the front feet. Rolling, squirming, and stretching are seen. This activity occurs whether or not a male is present, but becomes synchronized with male behavior when the female is interacting with the male. During proestrus she will roll and solicit the male's attention but act aggressively if he mounts (Michael 1973), which may be termed postural acceptance and affective rejection. When fully receptive, she becomes immobile and stands crouched with lumbar lordosis, holding her head on the ground between her forelegs (Fig. 4.19). She

Fig. 4.19. Sexual behavior of the female cat. (*A, B*) The typical posture adopted by the queen in full estrus—note leg flexion, lordosis, and deflection of the tail; (*C*) tom holding queen with neck grip during intromission; (*D*) postcoital rolling by the queen (Scott 1970, copyright © 1970 and 1987, with permission of Lea and Febiger).

deviates her tail to one side and allows the male to mount. The latter sequence may be stimulated in an estrous female by scratching her over the dorsal tail base.

An estrous female may show darting behavior in the presence of several tomcats. She will repeatedly run a short distance from the toms; this may be her means of assessing the relative strength of the males as they chase her and try to displace one another.

BEHAVIORAL ABERRATIONS. Mounting behavior is seen in female cats only rarely and usually in colony situations. Two estrous cats may try to squirm underneath one another when presenting to an inaccessible male; one may mount the other and perform malelike pelvic thrusts. Leyhausen (1979) reports definite examples of female mate preferences in colony cats. He does not offer a clue about the mechanisms of this selection but does observe that the dominant members of the colony were sometimes bypassed as mates.

Although pair bonding does not occur in the domestic cat, short-term consort behavior does occur normally. A male and female may remain together for several hours or days, mating many times. This behavior is typical also of larger cats; a mating pair of lions, for example, leaves the pride for several days to a week, often fasting the entire time.

Ovariohysterectomy, or spaying (ovariohysterectomy, or less commonly, ovariectomy or tubal ligation) is commonly performed on the domestic house cat. Spaying eliminates sexual behavior but may affect other behaviors. Care must be taken not to ascribe changes in behavior to surgery, when the changes might be caused by maturational changes in the cat or changes in the cat's environment. A common behavior problem following spaying is an increase in aggression between cats in a multicat household resulting either from the lowered progesterone levels of the spayed cat and/or the change in her odor that precipitated attacks by the other cat. Maternal behavior has also been observed in newly spayed cats, perhaps triggered by the fall in ovarian hormone levels similar to the fall that occurs at parturition and in the presence of neonatal kittens.

The Tom

COURTSHIP BEHAVIOR. The male probably locates an estrous female via olfactory cues deposited as pheromones in the urine and by some sebaceous gland secretions. A male placed with a female in a mating arena will spend some time investigating and marking the area with urine and anal gland secretions before mating. The cat shows a flehmen response, or gape, similar to that of ungulates (see Chap. 2). He calls to the female, circles her, and sniffs her genitalia. A nonreceptive female will actively, even violently, rebuff a male. When a female is receptive, the male approaches her from the side and behind and grips her neck in his mouth. He then mounts with the front legs, then the hind, and rubs her with his forepaws. Intromission follows a forward stepping with arched back and pelvic thrusts (Fox 1975). Ejaculation occurs seconds after intromission, and intromission usually lasts less than 10 seconds. The penis is covered with numerous small spines that apparently cause an intense stimulation as evidenced by the loud copulatory cry of the female with intromission. With retraction of the penis the female rolls and claws at the male. The male will often lick his penis after copulation. Copulation may occur every 10 to 15 minutes for several hours. A seasonal variation in sexual readiness with a decline in the fall is seen in the male cat under experimental conditions (Aronson and Cooper 1966).

In a feral situation, one estrous female will be surrounded by a group of males. Only one will be close to the female and will be the one who mates. Fifteen copulations will occur in 24 hours. The queen is not passive, however, and may run a short distance causing tumult among the males and sometimes a change in the tending tom, but usually the other males have no chance to mate unless the tending male leaves voluntarily. This pattern of one large older male and a number of smaller, attentive, but usually unsuccessful smaller or younger males, is seen in both farm and densely populated urban cat populations (Dards 1983; Liberg 1983).

In groups of cats living on farms, only one tom seems to be sexually active, whereas in house cats, sexual maturity and related activities, such as urine spraying and fighting, normally commence at 9 to 12 months. Young males in farm groups remain subordinate to the active male to an age of 18 months or more. They retain the squat urination position of the juvenile and show either no aggression or actual submission toward adult females and the dominant male. The mechanism of this maturation delay is not known, although it has been shown to be pheromonal in rodents (Vandenbergh 1979). These juvenile males tend to move away from their birth area before their 3rd year.

BEHAVIORAL ABERRATIONS. Reluctance of the male to breed a female is usually the result of the female's nonreceptivity and, thus, aggressive response toward the male's advances. Inexperienced males may be especially intimidated by the aggression of a proestrous female. In a laboratory, only one tom in three will consistently copulate with fully receptive queens. It is not surprising, therefore, that many visits to the tom are necessary before successful breeding takes place. Estrous females may indicate a mate preference by actively rebuffing one male or staying near another.

Males may mount other males as a sign of dominance, but this behavior is very unusual outside caged colonies. Dominance hierarchies are routinely adjusted via aggressive encounters. Kittens show little sexual behavior during development, unlike puppies in which mounting may be seen at 5 weeks. Thus, sexual maneuvers do not develop as part of the cat's communication repertoire outside a sexual context. Masturbation is also uncommon outside caged groups (Michael 1961) but can occur in pet cats, including free-roaming, castrated males and sexually active tomcats. Usually a furry toy or a bedroom slipper is the object chosen to mount.

CASTRATION. Castration is a widely accepted procedure for the pet cat. Prepubertal (from 6 to 8 months) castration generally eliminates sexual behavior. Fox (1975) points out that although androgens are secreted by testes by 4 months, mating behavior does not develop until 8 to 9 months. Castration is usually physically difficult until the testes are reasonably well developed at 6 to 8 months. Some owners object to a male that appears feminine and so delay castration until 12 to 14 months or the first serious fight abscess. Rosenblatt and Aronson (1958a,b) and Rosenblatt (1965b) point out that the effectiveness of castration in eliminating mating behavior depends on the previous level of sexual experience. Thus, owners should be advised to restrict the access of their cat to females until after surgery unless they do not mind the cat's continued sexual interests.

Hart and Barrett (1973) studied the effects of postpubertal castration on fighting, roaming, and spraying. Castration seems much more effective in reducing or eliminating these behaviors in the cat than it is in the dog. Eighty-eight percent of cat owners interviewed 23 months after having their cats castrated reported a rapid or gradual decline in fighting, 92% reported the same for roaming, and 87% responded favorably for spraying. The failure of surgery to eliminate these behaviors completely is probably due to the learned components of the behaviors. See Chapter 1 for treatment of spraying by neutered cats.

5

Maternal Behavior

BIOLOGICAL BASIS OF MATERNAL BEHAVIOR. Maternal behavior is characterized by sudden onset. One day a single cat spends her day eating, sleeping, grooming herself, and hunting or playing. The next day, she has become five cats, four of whom are kittens, and the original cat now spends almost all her time feeding and grooming the kittens. This behavior will gradually subside but, unlike other behaviors, is remarkably persistent. Aggression, for example, may be sudden in onset but does not persist very long. Other adult behaviors seem to have been rehearsed in play by the developing animal. Play includes elements of aggressive and sexual behavior, chasing and fleeing, but not of maternal behavior.

What then is the basis of maternal behavior? It is certainly an innate behavior pattern, although experience does play a role, as will be discussed later in this chapter. Even if a behavior is innate, that is, a genetically programmed response to a certain set of stimuli, the behaving animal must be physiologically prepared to respond to the appropriate stimuli. To determine the biological basis of maternal behavior, it will be necessary to investigate the hormonal basis of maternal behavior, as well as those features of the neonate that may release maternal behavior.

Maternal behavior can be induced in ewes with a combination of estrogen and progesterone followed by corticosteroids. Following this treatment, multiparous ewes are more likely to accept lambs than nulliparous ones (Le Neindre et al. 1979). This indicates that the combination of the proper hormonal milieu and the stimulus for maternal behavior, the neonatal lamb, plus prior experience as a mother can elicit maternal behavior. One of the sequelae of parturition is cervical stimulation. Artificial cervical stimulation applied for 5 minutes can elicit maternal behavior immediately in multiparous ewes pretreated with progesterone tampons and estradiol. Cervical stimulation will also cause a ewe that is already selectively maternal toward one lamb to be maternal toward another, alien, lamb (Keverne et al. 1982). Cervical stimulation will result in the reflex that stimulates oxytocin release. Oxytocin is released not only from the posterior pituitary into the bloodstream but also from the terminals of cells whose cell bodies lie in the periventricular area of the hypothalamus whose axons can stimulate in other parts of the brain the neural mechanism underlying maternal activities. Brain oxytocin levels increase at parturition, at suckling, and when the vagina is stimulated (Kendrick et al. 1986). Increasing oxytocin in the cerebrospinal fluid can stimulate maternal behavior (Kendrick et al. 1987).

Primiparous ewes routinely reject their lambs if they have been delivered by Caesarian section, which indicates the importance of the passage of the lamb through the vaginal canal. Multiparous ewes, however, will readily accept their lambs even if they have been delivered by Caesarian section, which indicates the importance of prior experience in ovine maternal behavior (Alexander et al. 1988).

After parturition, sensory factors will replace hormonal ones in the control of maternal behavior.

Neural structures in domestic animals have not been directly manipulated, so an understanding of the neurology of maternal behavior must be based on evidence from laboratory animals, primarily rats. A discrete area of the brain does not appear to be involved in maternal behavior. Lesions of the cortex (Beach 1937), especially the cingulate cortex (Slotnick 1967; Slotnick and Nigrosh 1975), septum (Fleischer and Slotnick 1978), hippocampus (Kimble et al. 1967), and lateral hypothalamus produce defects in maternal behavior. After lesions, nursing, but not retrieving, behavior may be abnormal; however, especially in adipsic and aphagic rats with lateral hypothalamic lesions, the significance of such findings is questionable. Deafferentation of the medial hypothalamus interferes with lactation so that oxytocin must be administered, but maternal behavior is normal (Herrenkohl and Rosenberg 1972). Lesions of the medial preoptic area or knife cuts that sever the lateral connections of the lateral preoptic area disrupt maternal behavior in rats (Numan and Smith 1984).

Studies of incorporation of gonadal hormones into cells of specific areas of the brain and studies of the effects of implantation of hormones into the brain have elucidated the neurohumoral basis of sexual behavior (see Chap. 4). Maternal behavior deserves a similar experimental treatment but has not yet received it. Two studies have shown that implantation of estrogen into the preoptic area (Braden 1965) or progesterone into the arcuate region of the hypothalamus (Roth and Lisk 1968) of parturient female rats depresses maternal behavior.

Despite the paucity of information available on the neuroanatomy of maternal behavior, two very interesting aspects of maternal behavior indicate that the central nervous system is very much involved. First, evidence exists that learning is involved in maternal behavior.

General Principles of Maternal Behavior

Learning. The evidence for the role of learning in maternal behavior is found mostly in higher primates. Monkeys that had been artificially reared made very poor mothers and very reluctant sex partners (Harlow et al. 1963). Apparently, a monkey must have been mothered to spontaneously be a good mother. Intriguingly, monkeys that neglected or even killed their first offspring exhibited normal maternal behavior after the second pregnancy. This aspect of maternal behavior has not been well investigated in domestic animals, but it is worth noting that most problems in maternal behavior are seen in primiparous animals. Ovine maternal behavior, in particular, seems to be more independent of physiological changes after the ewe has mothered one lamb. The quality of maternal behavior in artificially reared cats, dogs, or sheep has not been documented. When maternal behavior in beef and dairy cattle is compared, the beef cattle exhibit more maternal behavior. These animals, at least on the range, raise their calves, and adequate maternal behavior is necessary for their calves' survival. Artificial rearing of dairy calves has been practiced on many generations of cows, so few dairy cows have much experience at mothering or at being mothered.

Concaveation. The second aspect of maternal behavior that implicates the role of the higher portions of the central nervous system is the phenomenon by which the presence of neonates can induce maternal behavior in virgin females and even in males. This phenomenon is called *concaveation.* When exposed to rat pups daily for 7 days, virgin female rats, and even male rats, will begin to retrieve the young, lick them, and even huddle over them in the typical nursing position. Mice, as mentioned, will show similar behavior with no latency whatsoever, as long as the pups presented are only 1 or 2 days old. Maternal behavior can, therefore, be induced in these rodents in the absence of hormonal stimulation, although hormonal stimulation accelerates the appearance of maternal behavior. Experience also reduces the latency of the appearance of maternal behavior: parous rats begin to retrieve much more quickly than nulliparous rats (Beach and Jaynes 1956; Lamb 1975).

Stimuli That Elicit Maternal Behavior. What stimuli emanating from the pup are important in maternal behavior? In mice these stimuli are presumably olfactory because olfactory bulbectomized mice will cannibalize pups until the pups are old enough to have acquired hair and a more mouselike appearance (Gandelman et al. 1971).

Rats also show an increase in cannibalism following olfactory bulbectomy but not following peripheral anosmia. Some function of the olfactory bulb, in addition to odor perception, must be important. The vomeronasal organ may play a part in rodent pup recognition because ablation of the vomeronasal system reduces the latency with which virgin rats begin to show maternal behavior to rat pups (Fleming et al. 1979). It has been hypothesized that the vomeronasal organ serves to identify pups as "strange"; without the organ, all pups are familiar. Neonatal rats emit an ultrasonic call when cold or stressed, and the sound stimulus may elicit maternal behavior. The appearance of the newborn may serve as a visual stimulus, for Lorenz (1952) has hypothesized that the short forehead, cheeks swollen by sucking fat pads, and the erratic gait of the neonate elicit maternal behavior in a number of species, including humans. Little evidence for nonhormonal stimulation of maternal behavior in domestic animals exists, although "aunts" that care for other mares' foals have been reported in horse herds (Schoen et al. 1976), and prolonged exposure to an alien lamb may induce a ewe to accept it. Spayed cats will adopt and nurse kittens and may even lactate.

Adult rodents in the proper hormonal state are attracted to infants of their species. It also appears that infant rodents are attracted to lactating females. This is particularly true of rat pups in the 3d week of life. The attraction is olfactory, not based on the smell of milk, but rather on the smell of the products of cecal bacterial action. Lactating females produce large amounts of cecal feces, and it is these that attract the rat pups (Leon 1974). Suckling domestic animals are certainly attracted to lactating females. The attraction is, in part, based on olfaction, but the source of the odor stimulus has not been identified.

Weaning. Weaning is a distressing experience for most domestic animals, especially if performed before natural weaning would occur. Domestic species are often weaned very early, usually for economic reasons, as with piglets and calves, but sometimes only through custom, as with the horse. It has been noted that plasma growth hormone declines when rats are separated from their mothers (Kuhn et al. 1978), and this physiological change may explain the failure to thrive of early weaned animals of other species. Measurement of the hormonal and other physiological changes occurring in domestic animals of various ages as they are separated from their dams may provide a guide to the optimal time to offspring.

Summary of the Biological Basis of Maternal Behavior. To summarize the somewhat sketchy knowledge of the biological basis of maternal behavior, hormonal priming can lower the threshold for the initiation of maternal behavior that can be brought on and maintained as a response to the stimuli characteristic of the neonate even in the absence of the appropriate rise and fall of gonadal and pituitary hormones. Animals, especially those believed to be higher on the phylogenetic scale, can learn to be good mothers by having been mothered themselves as infants and by having been mothers previously. Vaginal stimulation and oxytocin release appear to be important in sheep; otherwise the hormonal and central nervous system control of maternal behavior in domestic animals is virtually unknown and is a field that demands more attention from biological scientists. Figure 5.1 summarizes the factors involved in maternal behavior.

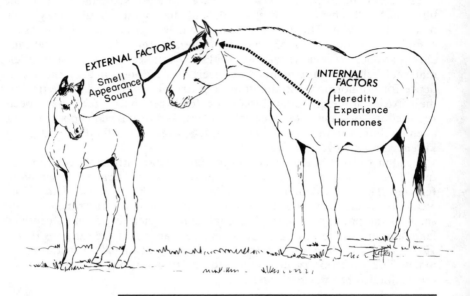

Fig. 5.1. Factors that influence the expression of maternal behavior. The horse is used as an example, but the influences on maternal behavior are similar in all domestic animals (Houpt and Wolski 1979, copyright © 1979, with permission of Veterinary Practice Publishing).

Imprinting. Imprinting is an ethological term that has entered the layman's vocabulary. Therefore, in discussing the presence or absence of the phenomenon in domestic mammals great care must be taken to define imprinting. As defined by Lorenz (1957), imprinting is a special process that (1) can only occur during a definite and short period of the animal's life; (2) is irreversible; (3) involves an attachment to an object that will later evoke adult behavior patterns, including sexual behavior; and (4) involves reactions to a particular object that can be generalized to all objects in that class, for instance, all humans or all duck decoys. Subsequent laboratory studies have revealed that in ducks imprinting does not appear to be irreversible or to influence any adult patterns of social behavior (Moltz 1960).

Ducklings and chicks will follow a moving object, even a red ball. This behavior is the following response. Eventually the following response ceases. Fear appears to strengthen the following response, which is much easier to produce in an environment strange to the fowl. Nevertheless, a duckling originally imprinted to an inanimate object or to a human quickly stops following the former object when placed with a maternal duck and her brood. Natural imprinting of a young precocial bird onto its mother may be irreversible, but many factors besides following her retreating form are involved in the interactions between the newly hatched fowl and its mother (Hess 1972).

OCCURRENCE IN DOMESTIC ANIMALS. Does imprinting occur in domestic animals? It would be expected to occur in precocial species so that lambs and foals are the most likely ones to imprint. Both of these species have a tendency to follow large moving objects at birth. Foals, in fact, may become attached to large stationary objects, such as a tree, and refuse to leave (Tyler 1972). The following tendency is not irreversible. For example, Cairns and Johnson (1965) showed that normally reared lambs that presumably followed their mothers rapidly began to follow dogs when the two species were housed together but lost the response after returning to the company of other sheep. Scott (1958b) hand raised a lamb that subsequently showed abnormal social and maternal behavior. The lamb's situation was quite comparable to that of the geese that were hand raised by Lorenz (1952).

Lorenz's geese also showed abnormal adult behavior that included directing their courtship behavior toward Lorenz rather than toward other geese. Apparently, a long-term association with another species, especially when the association includes feeding, can influence the animal's behavior and choice of sex partners. It is most probable that a critical period for socialization, such as occurs in the 7-to-12-week-old dog, exists for sheep, geese, and other animals; socialization, however, can occur long after the following response of imprinting has been established. Finally, the Sambrauses (1975) have shown that for goats, pigs, and other domestic animals to direct sexual behavior toward humans, the animals must have been isolated from conspecifics and have been in close association with humans.

MATERNAL BEHAVIOR OF DOMESTIC ANIMALS
Pigs

The Free-ranging Sow. Maternal behavior in sows can be divided into two prepartum behaviors, nest site seeking and nest building, and the postpartum behavior of nursing. One day before farrowing, free-ranging sows will leave their herds and their normal home ranges, traveling as far as 7 km. Each sow will build several rudimentary nests before selecting a final site at which she will dig a hole 10 cm deep and 1.5 m wide. She will bring grass and sometimes sticks to the nest (Jensen et al. 1987). The sow will spend more and more of her time building her nest during the day of parturition. Usually 3 to 7 hours elapse between the onset of nest building and farrowing. After farrowing, the sow remains with the pigs for 24 hours and will continue to spend most of her time with them for the next few days. She will defend the nest against her juvenile offspring and against other adults. Failure to defend the nest results in crushing of the piglets and a consequent three to fourfold increase in mortality (Newberry and Wood-Gush 1986). She will nurse the piglets every 45 minutes. Within a week the sow will rejoin her herd, and the piglets will follow. She abandons the original nest and uses the herd nest or constructs a separate one near the common nest. The sow weans the pigs at 14 to 17 weeks (Jensen 1986).

The Confined Sow. Modern husbandry practices have all but eliminated most porcine maternal behavior except nursing. Sows are placed in farrowing crates that prevent them from turning around or touching the sides of the pen, and, consequently, the piglets are protected from crushing. Crushing of piglets is reduced by farrowing crates but not eliminated. Losses can be further reduced by using a device that shocks the sow's belly when a piglet screams, but the sow may also be shocked when extraneous noises trigger the device (Friend et al. 1989).

The use of farrowing crates for nursing sows has resulted in much lower death rates for piglets because the sow can rarely crush or cannibalize her young, but farrowing crates prevent sows from building the elaborate nests used by wild and feral swine; all that remains of the nest-building behavior is a futile pawing at the floor of the pen. The restlessness, which increases linearly during the last 48 hours prepartum, probably represents attempts at nest seeking and nest building. Sows in a pen provided with straw or those in a semi-naturalistic environment build nests the day before and for several days after parturition (Hansen and Curtis 1980; Lammers and deLange 1986). Once labor begins, most sows lie down in lateral recumbency. The sow will swish her tail violently as she strains abdominally. Parturition usually takes 3 to 4 hours, but varies considerably with litter size and the condition of the gilt. When attending a farrowing sow, it is well to bear in mind the barnyard adage, "She has had her litter, or half of it." Most farrowings take place in the afternoon or night (Signoret et al. 1975).

When not confined, the sow will eat the placenta. The function of placentophagia remains unknown. It may be a recycling of nutrients or a form of defense against predators by removing odors. Kristal et al. (1986) have found that placentophagia enhances analgesia produced by other means in rats. Whether this occurs in domestic animals as well deserves investigation.

Pigs lick their newborn very little. Therefore, human attendance at parturition is recommended. Although most piglets begin to breathe and quickly struggle free from the fetal membranes, a few will not. The removal of membranes, clearing of the airway, and stimulation of respiration can save a piglet that would otherwise die.

Piglets make a most startling transition from fetal to independent existence. They may be apneic for 5 to 10 seconds after birth. Then they gasp a few times before beginning to breathe regularly. Their eyes and ears are open, and they are able to walk immediately, although their gait is staggering for the first few hours. The firstborn may be slow to find the udder, but later born pigs apparently respond to the voices of their littermates and quickly begin to seek the udder. Most piglets are nursing within 30 minutes of birth (Hemsworth et al. 1976). During farrowing and for some time afterwards piglets can suckle continuously, presumably because oxytocin levels are high. Thus they are rewarded for each suckle in the correct place, that is, on a teat.

The cues used by piglets to locate the udder are not definitely known but are probably related to temperature and resilience (Welch and Baxter 1986). The piglet is attracted to soft, warm surfaces. Suckling attempts are probably stimulated by tactile contact with a protuberance (the teat). Piglets rarely attempt to suckle on a haired portion of the sow; they will suck on the snout or the tip of the sow's vulva. Olfaction also plays a role because piglets are delayed in finding a teat if the udder is rinsed with an organic solvent (Morrow-Tesch and McGlone 1990). Piglets nose the udder and intersperse nosing with gapes, the behavior in which the piglet opens its mouth as if to grasp a teat. Larger pigs do more gaping, which may account for their success in reaching teats. The nosing behavior of lighter pigs declines more rapidly than that of heavier pigs (Rohde and Gonyou 1987). The piglet may find the udder, give a few inept sucks at a teat, and then make another circuit or two of the sow before it settles down to nursing. The firstborn pigs appear to use thermal, tactile, and olfactory cues to find the udder, whereas the later born probably respond to the nursing noises of their older littermates and walk straight to it. Social facilitation is strong in pigs at birth.

Experiments using artificial sows have revealed that the piglets are attracted to the voice of the sow and to either end of the udder but avoid the middle and quickly abandon teats that give no milk (Jeppesen 1982a). Competition during formation of the teat order is intense and only one-third of pigs end up on the teat they initially chose. Once a teat has been chosen and "won" by competition with other piglets, it is recognized by odor rather than visual cues (Jeppesen 1982b).

Nursing. Approximately 10 hours after the birth of the first pig, nursing becomes cyclic (Lewis and Hurnick 1985). Nursing bouts occur approximately every 40 minutes (Fig. 5.2). The interval between nursing is longer at night than during the day. Small litters nurse less frequently than large (Winfield et al. 1974). The sow ordinarily calls the piglets to nurse with a

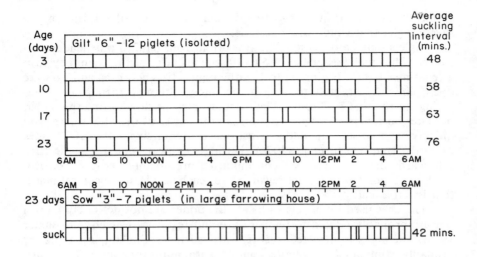

Fig. 5.2. Suckling frequency of piglets. Suckling frequency decreases with the age of the piglets in an isolated litter, but in a large farrowing house (*bottom*) social facilitation of nursing bouts occurs and older piglets nurse more frequently (courtesy Dr. Ron Kilgour, Ruakura Animal Research Station, Hamilton, New Zealand).

low-pitched rhythmic grunting. As the piglets begin to massage the udder with their snouts, the frequency of the sow's grunts increases from 1 per second to a peak of 10 per second. The peak of grunting frequency may correspond with oxytocin release and, therefore, with the letdown of milk (Whittemore and Fraser 1974). Rubbing of the udder of a lactating sow can induce her to lie down and begin to give the nursing call. Stimulation of the anterior half of the udder and, especially, rubbing of the nipples in that area can increase the grunting rate (Fraser 1973, 1975a,b).

Usually, udder stimulation is provided by the piglets. If they are removed from the udder before the rapid rate of grunting has occurred, milk will not be ejected. Rubbing of the belly calms even immature pigs and can be used to great advantage in handling swine.

As discussed in Chapter 2, piglets form a teat order during the first few days of life. The anterior teats are preferred, possibly because the sow responds to stimulation there with a vocal signal that the milk will soon be forthcoming (Whittemore and Fraser 1974). The more abundant milk supply of the anterior teats may also account for their popularity.

A nursing bout is divided into four phases: an initial massaging of the udder for one minute; a quiet phase during which the piglets' ears go back and they stop massaging, which may correspond to the peak of the sow's grunts; true suckling for approximately 14 seconds while the milk is ejected, during which the piglets' ears are back, their tails are tightly curled, and their front legs are rigidly extended; and a final massage phase that is quite variable in length, 2 to 15 minutes (Gill and Thomson 1956). Young piglets often fall asleep on the nipple or curled beside the udder, whereas older pigs will nose the udder and pull the teats for some time.

Not all nursing bouts are successful. In 22% of the nursing bouts, the sow may call the piglets, who approach and massage the udder, but no milk is ejected. Unsuccessful nursing usually occurs less than 40 minutes after a successful bout. The piglets leave the udder as they do after a successful nursing but return much sooner (Fraser 1977). Apparently, unsuccessful nursing bouts also occur in free-ranging sows.

The strong social facilitation and dependence on vocal communication exhibited by pigs can be used to practical advantage. If one sow in the farrowing house calls her litter to nurse, soon all the litters will be nursing. Nursing rates and weight gain can be increased by playing tape recordings of nursing noises to the sows at more frequent intervals than they normally nurse (Stone et al. 1974). A talented handler can imitate the noises and accomplish the same thing (Hartman and Pond 1960). Piglets can also initiate suckling by their calls and persistent nudging at the sow. If half of a litter has been fasted, the hungry piglets will induce the sow to lie down, and then all of the piglets will nurse, although the nonfasted piglets will eat less (Houpt et al. 1977).

Feral pigs live in small groups and keep the piglets in nests so that individual recognition of the young is probably not as critical as it is in herd-dwelling ungulates. Nevertheless, sows will reject strange piglets older than two days. The rejection is based on olfaction, for Meese and Baldwin (1975b) found that anosmic sows would accept strange piglets.

Other Behavior of the Suckling Pig. Like most newborns, piglets tend only to eat and sleep. They have some unique physiological problems that affect their behavior. Piglets are born almost hairless and with little subcutaneous fat for insulation. Their small size, lack of insulation, and low energy reserves leave piglets very vulnerable to hypothermia (Mount 1979). Piglets solve the problem of heat conservation behaviorally rather than metabolically. They huddle with their littermates, thus decreasing their surface area and, in effect, make one large animal from twelve small ones. If a heat source is provided, they will lie next to it. Unless the environmental temperature is over 80°F, a heat lamp should be provided (Titterington and Fraser 1975). Piglets that do not huddle or that persist in wandering away from the sow and the heat source have very likely been brain damaged at birth. A quick inspection of the farrowing house can enable one to determine whether the temperature conditions are correct: if piglets are sprawled on the floor in extension, they are warm enough; if they are crouched on their sternums with their legs drawn under them, they are too cold.

Clinical Problems

CANNIBALISM. Cannibalism occasionally occurs in sows; nervous primiparous gilts are the most likely offenders. The most common occurrence is immediately after parturition. In fact, many sows will bark at the first piglet that walks by their heads after parturition (Randall 1972). Farrowing crates successfully prevent cannibalism unless an unwary piglet walks right in front of the sow. The tranquilizer, azaperone 2.2 mg/kg, has been used to treat cannibalistic sows. A more common problem is seen in sows suffering from mastitis; a sow that is normally a good mother will attack her litter whenever they attempt to nurse. This early behavioral sign warns of disease before many physical signs can be detected. This type of behavior is seen in sows that become afflicted with mastitis late in lactation, after the farrowing crates have been removed. Sows that have the mastitis-metritis syndrome shortly after farrowing are usually too ill to protest when the piglets suckle. The failure of the sow to eat and of the piglets to gain weight is the best clinical evidence of the latter syndrome.

DEFENSIVE REACTION. Sows normally exhibit strong defensive reactions when their piglets are threatened. They give a crescendo of barks, open their mouths, and attack. Only when pigs are defending their young are they really dangerous. The use of the farrowing crate has had a definite effect on the maternal behavior of sows. The sows can do nothing if their piglets are handled or hurt despite the piglets' loud distress calls. In herds where the piglets are handled often and by many people, as in university or research institutions, the sows become accustomed to the distress calls of their pigs. A sow may even continue to sit on a piglet that is screaming loudly and eventually smother it.

CROSS FOSTERING. Cross fostering of piglets is not at all difficult. Pigs may be added from other litters at the time of parturition or later. The sow may be able to identify the strange pig by its smell, but if she is in a crate there is little that she can do. The strange pig can be smeared with the sow's feces to facilitate acceptance. It is often advisable to cross foster so that a runt will be competing with much younger pigs, or so that a large (more than 12) and a small (less than 3) litter will be recombined to form two medium-sized litters.

Sheep. Maternal behavior in sheep has an important clinical aspect because most lamb mortality occurs within the first week of life in range-reared sheep. Mortality rates are from 5% to 15%, rising to as high as 50% in huge flocks in bad weather, even in the absence of predators (Morgan and Arnold 1974). Abnormal or weak maternal behavior accounts for parts of these high losses and, perhaps, for some of the losses attributed to coyote predation. Most maternal rejection or simply poor mothering without absolute rejection occurs most frequently in young ewes and in those that had difficulty at parturition (Wallace 1949).

Parturition. Lambs may be born at any time of the day or night, with peak frequencies being noted at 9:00 to 12:00 AM and at 3:00 to 6:00 PM (Lindahl 1964). A few days before parturition the ewe withdraws from the flock, if on the range, and seeks some sort of shelter. In a pen she will seek a corner (Bray and Wodzicka-Tomaszewska 1974). Shelter seeking by the ewe improves the environment into which the lamb is born so that its chances of survival are greater, but the ewe is responding primarily to her own thermoregulatory needs so that shorn, but not unshorn, sheep seek shelter (Lynch and Alexander 1976). Allelomimetic behavior is so strong in sheep that some of the herd may follow her. Penned sheep become restless, circle, vocalize, and paw at their bedding 60 to 90 minutes before parturition. Grazing and ruminating ceases. The older the ewe, the shorter the time becomes between the onset of restlessness and the onset of labor. The interval between onset of labor and the appearance of the lamb can vary but is usually 30 to 60 minutes. Even before parturition, 20% of ewes show maternal behavior toward other lambs (Arnold and Morgan 1975). This prepartum maternal behavior results in lamb stealing (discussed later in this section). The amniotic fluid dripping from the vagina to the ground attracts the ewe. She will sniff and lick at bedding contaminated with amniotic fluid. This attraction to amniotic fluid can be used to predict parturition. Only ewes close to parturition will eat food contaminated with amniotic fluid (Levy et al. 1983). The attraction of the ewe to this fluid may keep her in the area where the birth will take place, ensuring that the lamb will not be abandoned before it can get to its feet (Smith 1965).

LICKING. When the lamb is born, the ewe begins to lick it while simultaneously emitting a special parturition call that is a very low-pitched gurgle or rumble. At times this call may be given before parturition. The call is heard only at parturition in domestic sheep but persists in the feral Soay sheep as a close contact call (Shillito and Hoyland 1971). Licking of the lamb can be very important in cold or windy weather because it dries and stimulates the neonate. While the lamb is recumbent, the ewe licks its head, even restraining the lamb with a front leg to prevent it from standing. Once the lamb is standing, usually within 30 minutes (Wallace 1949), the ewe continues to lick it but mostly on the hindquarters. If the ewe stops licking, the lamb gives distress calls (Bareham 1976). Finally, licking of the lamb by the ewe establishes the maternal-offspring bond, for the ewe will be able to identify her lamb by smell and taste. Usually the fetal membranes are licked off the lamb and ingested, but the placenta is not eaten (Fig. 5.3).

SUCKLING. Once a lamb has been born and has managed to stand and walk, it seeks the udder. The innate pattern to which lambs appear to respond is

Fig. 5.3. Ewe licking the head of her newborn lamb.

the curved underline of the ewe. Once the lamb has established contact with an underline, odor, texture, and temperature probably serve to guide it (Vince 1984). The warmest surface of the ewe is her woolless inguinal area; furthermore the lamb is attracted by the odor and resiliance of the inguinal wax (Vince et al. 1984; Vince and Ward 1984; Billing and Vince 1987a,b). Contact with the face of the lamb stimulates it to push its head up and forward. Contact with the lips causes it to open its mouth and protrude the tongue. Contact with its tongue causes the lamb to curl the tongue into the suckling position (Vince et al. 1984). This series of innate responses brings the lamb to the ewe, then to the udder, then to the teat, and finally, to suckle the teat.

Although lambs are able to stand within 1 hour of birth, it may take 2 to 3 hours before they find the udder (Wallace 1949; Alexander and Williams 1964). The ewe plays a part in the search. She may either facilitate or inhibit teat seeking. For instance, her licking activities orient the lamb, but she may thwart the lamb's attempts to suckle by turning away from it as it noses along her flank. Lambs whose dams are stanchioned take longer to locate the udder (Alexander and Williams 1964). The drive to suckle is inhibited, but not eliminated, by intragastric loads of milk, so hunger is not the lamb's only motivation (Alexander and Williams 1966). Once the udder has been located, the lamb uses visual cues to relocate it for subsequent sucklings (Bareham 1975) (Fig. 5.4).

Fig. 5.4. Initial orientation of the lamb to the ewe's underline but in the axilla rather than the udder (courtesy Dr. Martin Siegel Annandale, N.J.).

Once suckling has begun, it occurs with great frequency; twin lambs suckle 22 times during 16 hours of daylight and single lambs suckle 6 to 14 times (Doran 1943; Munro 1956). Newborn lambs may suckle for as long as 3 minutes in one bout; later the duration falls to 20 to 40 seconds. Frequency of suckling and suckling duration both decrease with age.

By the 5th week, both twins and single lambs are suckling 6 times per 12 hours (Ewbank 1964; Ewbank and Mason 1967). Triplets suckle less often and for shorter duration than singletons or twins. Probably because they are not receiving adequate nourishment from their dams, they are most likely to try to suckle an alien ewe. During the first 2 weeks the ewe will allow one twin to suckle without the other. Later the ewe will walk away when the lamb nudges her in the inguinal area and will refuse to let one twin nurse until the other is also present. If the ewe is lying down, the lambs will not only nudge her but will also jump on her back and paw at her in an attempt to make her stand. The ewes will call their 5-week-old lambs to them and then refuse to let them nurse. Such behavior encourages the lambs to stay in close contact (Ewbank 1967b). As suckling decreases, grazing by the lambs increases. The ewe, meanwhile, will graze farther and farther from the lamb. Although the ewe stays within 10 meters of the lamb the first few days, she will soon increase her distance from it. The lambs will begin to form groups of their own and by day 29 will spend 60% of their time with other lambs rather than with their mothers (Morgan and Arnold 1974).

Acceptance of the Lamb. The "critical period" during which a ewe will accept a lamb is the first several hours after parturition (Collias 1956; Smith et al. 1966). Normally a ewe will stay within 2 meters of its lamb for most of the first day (Lynch and Alexander 1976). If a ewe's lamb is removed immediately after birth and before she has licked it, the ewe will accept any lamb presented to her. Once the ewe has spent 20 to 30 minutes licking a lamb, her own or a substitute, she will not accept another. The importance of olfactory cues in the establishment of the bond is demonstrated by the fact that ewes will only temporarily accept strange lambs that have been rubbed with the ewe's placenta. Ewes can also be induced to follow their placentas (Collias 1956). Primiparous ewes will not accept a lamb whose wool has been washed free of amniotic fluid, whereas multiparous ewes will (Levy and Poindron 1987), indicating the importance of both amniotic fluid and of prior experience at mothering.

In hilly country, a newborn lamb may roll down the hill away from its birth site. The ewe may neglect the lamb because the odor at the birth site is more attractive than the lamb itself. In this situation, a weak lamb, or the weaker of a pair of twins, is most likely to roll away and not be licked. A more vigorous lamb will survive and seek the ewe. If too long a time elapses between parturition and the presentation of the lamb, the lamb may be rejected. A pair of twins were removed from their dam and returned in 4.5 hours; one member of the pair was accepted and the other was rejected (Collias 1956).

Mutual Recognition by the Ewe and Lamb

RECOGNITION OF THE LAMB BY THE EWE. Recognition of lambs depends on at least three senses: olfaction, audition, and vision. The wool of the lamb contains the odor used by the ewe to identify her own lamb (Alexander 1978; Alexander and Stevens 1981). It seems apparent that ewes base recognition of their lambs at a distance on vision and at close range, 0.25 m or less, on smell (Lindsay and Fletcher 1968; Alexander 1978); recognition cues are reinforced each time that the lamb nurses. During nursing the ewe sniffs at the tail and perianal area of the lamb. Olfactory bulbectomy eliminates the preparturient lip licking observed in Soay sheep as well as the licking of the newborn and normal lamb recognition. The bulbectomized ewe will accept other lambs indiscriminately (Baldwin and Shillito 1974).

Olfactory recognition presumably is based on the odor cues present during parturition, but the ewe also uses the voice and appearance of her lamb as cues to its identity. At least two senses must be impaired before ewes are unable to find their lambs (Morgan et al. 1975). Visual cues may be most important because ewes had more trouble finding a hidden lamb than a silent one (Alexander and Shillito 1977b). By changing the appearance of various portions of the lamb's body, it was found that maternal recognition was most impaired by altering the appearance of the head (Alexander and Shillito 1977a). Although it represents only 12% of the body surface, the head is apparently the area that the ewe uses to visually identify the lamb. There is also evidence that ewes use the color of their lambs to identify them. They reject their own lambs when they are dyed, and, if they do reaccept them, will choose lambs of the same color when their own is not available (Alexander and Shillito 1978). Laboratory experiments (Chap. 1) indicate that sheep can perceive color; the studies on lamb recognition indicate that the sheep use color vision.

The strong individual recognition of her lamb by a ewe, however, can break down. Multiple birth lambings of three, four, or more lambs are not uncommon, especially in Finnish landrace sheep. If several ewes and their "litters" are penned together, the ewes cannot distinguish their own lambs from the others; communal suckling of all lambs by all ewes results.

RECOGNITION OF THE EWE BY THE LAMB. During their first several days of life, lambs are not able to discriminate their mother from other ewes very well. A lamb separated from its mother will rush up to the nearest ewe and attempt to nurse, only to be butted aside. In experimental situations, given a choice between their own mothers, similar (same breed), or dissimilar ewes (different breed), lambs detect their own mothers faster and more accurately with age. They will continue to be more attracted to similar rather than dissimilar ewes (Shillito and Alexander 1975; Alexander and Shillito Walser 1978).

A number of experiments have been performed on sheep to determine which sense or senses are most important in individual recognition of the ewe by the lamb. Audition appears to be important. Arnold et al. (1975) demonstrated that lambs were more apt to approach ewes that were not their mothers when the voices of the ewes were muffled. New reproductive technology has allowed advances in understanding of recognition. Most Dalesbred and Jacob lambs born after embryo transfer to Dalesbred ewes could identify the ewe by her voice, whereas most of those born to Jacob ewes could not (Shillito Walser et al. 1982). Sonographic analysis indicated that there were more intersheep differences among bleats of Dalesbred ewes than among Jacob bleats (Shillito Walser and Hague 1980; Shillito Walser et al. 1981).

As lambs mature, visual cues become more important. A lamb less than 1 week old is not affected by a change in its dam's coat, such as shearing or blackening, but a 2-week-old lamb may hesitate to join a visually altered dam (Alexander 1977). Shillito (1975) also found that covering the pens in which the ewes were restrained slowed the approach of their lambs.

A critical period within the first few hours after parturition may exist for acceptance of lambs by ewes, but the lamb is not restricted by time for social attachments. The tendency to follow any large moving object is most marked during the first 3 days of life; for the next 3 days fear responses predominate, but from 6 days to 2 months lambs will continue to follow even an artificial sheep model (Winfield and Kilgour 1976). Lambs tend to follow large moving objects, but imprinting in the avian sense does not occur, for a lamb's attachment can be quite impermanent. Lambs can easily become attached to a goat doe or to a human who feeds them. Lambs that had been normally reared with ewes quickly formed attachments to dogs when one of each species was penned together. Within 8 weeks, the lamb would follow the dog, vocalize if the dog was removed, and even run a maze to be reunited (Cairns and Johnson 1965). After living in a normal situation for 4 months, the lambs no longer preferred dogs to sheep. Therefore, social attachments in lambs seem to be relatively easily formed and equally easily dissolved. This phenomenon of attachment can be used to bond sheep to cattle. The cattle deter coyotes from attacking the sheep (see Chap. 2, Predator Control).

Clinical Problems. Poor maternal behavior is often seen in ewes that have been in labor more than 30 minutes. The corticosteroid levels of such ewes are elevated, indicating that they are stressed (Bray and Wodzicka-Tomaszewska 1974). Poor maternal behavior may vary; the ewe may reject the lamb outright, but more frequently she will lick it in a desultory fashion only and nervously avoid the lamb's attempt to nurse.

CROSS FOSTERING. The problem of cross fostering of lambs is common. In general, older ewes will more readily accept lambs than younger ewes (Smith et al. 1966). Fortunately, if a ewe is exposed to young of its own species long enough, maternal behavior will occur. The existence of this phenomenon, concaveation, in other domestic species is questionable. The process may take weeks and the ewe should be stanchioned or somehow restrained so that the lamb will not be badly butted in the interval before maternal behavior appears. Tranquilization of the ewes will facilitate acceptance (Neathery 1971) but does not facilitate fostering of alien lambs onto ewes with their own lamb present (Tomlinson et al. 1982). Pherphenazine is the only tranquilizer that has been tested; another drug might be more effective. A variety of methods have been used to facilitate cross fostering. The time-honored method is to tie the skin of the ewe's own lamb to the lamb to be fostered. Quicker and easier methods of providing olfactory cues from the ewe's own lamb that do not depend on skinning a dead lamb include pouring amniotic fluid on the alien lamb, washing the lamb (Alexander et al. 1983a), and putting a garment worn by the ewe's own lamb on the alien (Price et al. 1984b; Alexander et al. 1985; Alexander and Stevens 1985). This method of transferring the familiar scent to the alien lamb is successful in fostering a second lamb onto a ewe.

Visual cues should also be altered. Either the lamb can be tied so that it cannot stand, thus mimicking the attempts of a neonate to stand, or visual cues can be eliminated. Stanchioning the ewe is effective because the ewe cannot move away from the lamb or butt it, and if her view of the lamb is also blocked (eliminating visual cues), fostering is facilitated (Price et al. 1984a; Alexander and Bradley 1985). Advantage should also be taken of cervical stimulation, to facilitate fostering. The technique of "slime grafting," in which the vaginal fluid of the ewe is rubbed on the lamb to be fostered, probably owes its success to the consequent vaginal stimulation in addition to, or instead of, the transfer of her odor to the lamb.

Breed differences in the frequency of abandonment of one of a pair of twin lambs are great. Merino sheep are much more likely to abandon one of their twin lambs than are Dorsets or Romneys (Alexander et al. 1983a).

MISMOTHERING. Mismothering, that is, maternal behavior directed toward a lamb that is not the ewe's own, is a common problem, especially in large flocks in which ewes are not penned separately for parturition. Up to 15% of lambs may be raised by ewes that are not their mothers. There are many combinations. A ewe's lamb may be born dead, and the ewe will steal another ewe's lamb. She may steal a lamb before parturition, have one of her own, and be credited with twins. Although "stolen" lambs may survive, it is impossible to accurately assess the productivity of a given ewe under these circumstances. A shepherd might cull a productive ewe and keep one that never produces twins but often "acquires" them (Welch and Kilgour 1970). The opportunities to mismother are increased when sheep are confined at parturition, but if cubicles are provided, the incidence of mismothering is considerably lower in those ewes that choose to lamb in them (Gonyou and Stookey 1985).

ORAL VICES OF ARTIFICIALLY REARED LAMBS. Artificially reared lambs suck one another's navel or scrotum and eat feces. Such behavior may cause injury or interfere with weight gain (Stephens and Baldwin 1971).

Goats. As parturition approaches, does, especially multiparous ones, leave the herd and seek a sheltered place, almost always near a vertical object, to kid. The does will defend this area, both before and for the first day after the kid is born. Parturition is most likely to occur during the day at a time when goats are generally inactive. Does will lick the kids for the first 2 to 4 hours and will vocalize frequently with a low-pitched bleat similar to the rumble of periparturient sheep. There seems to be a critical period for acceptance of the kid; kids removed at birth and presented to the doe 1 hour later may be rejected. It is particularly important that the kid has no contact with other goats before being accepted by its mother. Apparently, the doe recognizes her own scent or label on the kid and will reject a kid that smells of other goats (Gubernick 1980). The doe must have contact with the kid for more than 5 minutes to become not only maternally responsive, but also selective in that response, i.e., to accept only her own kids (Gubernick et al. 1979). Olfaction seems to be important; local anesthesia of the olfactory receptors of goats caused them to accept all kids, rather than only their own (Klopfer and Gamble 1966). The small ruminants seem unable to distinguish between species; lambs can easily be cross fostered onto goats and vice versa. The kids of does with udders transplanted to the neck region located the udder as quickly as the kids of normal does (Stephens and Linzell 1974). The kid seeks the udder and usually searches the axilla first because the doe is turned towards the kid licking it. Within the first hour most kids will have suckled (Hersher et al. 1963; Lickliter 1985).

Intensive maternal behavior is short-lived in goats because within a day the kid will have left the doe to hide and the doe will rejoin the herd or stay nearby if she can obtain adequate forage there (Lickliter 1984b; O'Brien 1984a). She will approach the hidden kid several times a day and call to it. The kid will answer and emerge to suckle as infrequently as twice a day. In the absence of a proper hide, a dim area with vertical sides and a roof, hiding behavior may not be recognized, but it has persisted in domestic as well as feral goats (Rudge 1970; Lickliter 1984a). Two-day-old kids can identify their mothers visually, apparently because they are more apt to be further from them than the nonhiding lamb that develops visual recognition only slowly (Lickliter and Heron 1984).

Clinical Problems

KID REJECTION. The importance of olfactory identification of the young is emphasized by the following case: a 1-week-old kid was castrated with an open castration technique in which incisions were made and the testes removed. When the kid was returned to the doe, she took one sniff and rejected it. The kid had to be hand raised. The smell of the fresh wound was probably responsible for the rejection response. Closed castration techniques should probably be used on suckling kids.

Cattle. In general, the practitioner should avoid trying to predict the time of parturition in any species, and the time of parturition in cattle is not an exception. Some of the signs of imminent parturition are relaxation of the sacrosciatic ligament, slackening of the tissue of the perineum and vulva, distention of the udder and teats, and mucous discharge from the vulva. Unless the afternoon body temperature is below 102°F, parturition is unlikely even in the presence of all the other signs (Ewbank 1963). The normally lower body temperature of cattle in the morning interferes with the predictive value of temperature (Dufty 1971, 1973).

Parturition. In a study of Hereford cattle, George and Barger (1974) found that 82% of all parturitions occur between noon and midnight. Parturition times are distributed throughout the 24 hours, but dystocias occur mostly at midday (Yarney et al. 1982; Owens et al. 1984–85). A greater proportion of births will take place during the day if cows are fed late at night (Lowman et al. 1981). Arching of the back and an elevated tail occur for 1 to 3 hours before the chorioallantoic membrane ruptures. When the membranes rupture, the cows often lick the fluid and tend to stay near the spot, now attractive to the cow, where the fluid fell. Ninety-five percent of all cows are recumbent for the actual delivery (Selman et al. 1970a). Approximately 100 minutes elapse from the rupture of the membranes to the birth of the calf. The placenta is eaten by 82% of the cattle. Parturition is longer in cows that give birth to large calves and in nervous heifers. In fact, labor may cease if nervous heifers are disturbed (Dufty 1972).

Bonding. Heritability of maternal behavior is low in cattle, but some breeds are more maternal than others. For example, Angus are more maternal toward and more defensive of their calves than Herefords and Charolais (Buddenberg et al. 1986). Contact between the cow and her calf for as brief a period as 5 minutes postpartum results in the formation of a strong specific maternal bond. Cows groom their calves during the early postpartum period, concentrating on the back and abdomen. If contact between the cow and her calf is delayed for 5 hours postpartum, 50% of the time the calf will be rejected. The critical period for formation of the cow-calf bond must be the first few hours postpartum. When the calf is removed after a brief initial contact, the cow vocalizes and is restless, but after 24 hours she can no longer distinguish her own calf (Hudson and Mullord 1977). Licking the calf occupies up to half the cow's time during the first hour postpartum; heifers lick less (Edwards and Broom 1982).

Suckling. The newborn calf shakes its head, snuffles, and sneezes. This behavior may begin during parturition as soon as the calf's shoulders are free of the mother's vulva. Some calves will remain motionless for up to 30 minutes after birth, but within an hour most calves can stand. It may take 20 to 30 minutes before the teats are located, and the cow's conformation may not provide the higher recess that the calf appears to seek. Passive transfer of immunity to calves is poor in cows that have had dystocias, presumably because the calves did not suckle as much or as often (Donovan et al. 1986). Up to one-third of the calves may not suckle within 6 hours of birth. This is particularly apt to be the case when the cow has a pendulous udder (Edwards 1982). Once the teat has been located for the first time, the calf will be able to locate it much more quickly at subsequent nursings. Calves usually can stand in 1 hour and nurse within 2 hours. As the calf nurses, the cow will lick the perineum, stimulating urination and defecation by her calf. All cows, not just immediately postparturient ones, will lick a wet newborn calf. This behavior illustrates the attractiveness of fetal fluids to cows that are not hormonally primed. Advantage can be taken of this behavior (licking of fetal fluids by cows) to stimulate respiration in the neonate if the dam is too weak to do so.

The majority of cattle do not leave the herd to calve but under certain circumstances do leave to take advantage of the environment. When trees or rocks are available, the cow leaves the herd and hides, but in an open pasture the risk of predation is less if she stays with the herd (Lidfors and Jensen 1988).

Lambs and foals are considered followers, but calves, like goat kids, are hiders although they may follow both strategies, depending on the environment. Were they living in the wild, they would remain hidden while their dams grazed. They would nurse infrequently, that is, when the cows returned. For these reasons, it is easier to move a cow by moving its calf than to move the calf by leading away its mother.

Nursing. When nursing, calves assume a particular stance with spread legs that lowers their shoulders, which allows them to butt upward at the udder. The butting appears to stimulate milk flow (Hafez 1975). Like other young ruminants, they nuzzle and lick along the cow, especially in high recesses like the axilla and groin, and will mouth any hairless protuberance as they seek the udder (Selman et al. 1970b). They appear to be confused when they encounter a hairless teatlike object that does not supply milk. They wag their tails while nursing, although not as fast as lambs. Newborn calves normally nurse 5 to 7 times a day. Each nursing session lasts 8 to 10 minutes (Hafez and Lineweaver 1968). Usually the number of nursing bouts decreases with age, but beef calves may actually suckle more frequently with age, possibly because the milk supply of the beef cow is small (Hafez 1975). Beef calves 0 to 7 months old suckle 3 to 5 times per day (Odde et al. 1985). The most regular nursing time is at daybreak, with other nursing bouts occurring between 9:00 A.M.–12:00 M., 5:00 P.M.–8:00 P.M., and 10:30 P.M.–1:00 A.M. (Walker 1962). Most suckling takes place during the day (Schake and Riggs 1969). Suckling bouts are long, approximately 6–12 minutes, and do not seem to vary with frequency of suckling (Herbel and Nelson 1966a; Hafez and Lineweaver 1968; Gary et al. 1970; Petit 1972; Someville and Lowman 1979).

Natural weaning has not been studied in cattle, except for Bos indicus, in which bull calves are weaned at 11 months but heifer calves much earlier, at 8 months. The cow apparently invests more of her resources in a son that can produce many offspring per year than in a daughter that can produce only one.

It is now possible to produce twins in beef cattle by embryo transfer.

These twins, like triplet lambs, suckle more often and are more apt to suckle from an alien cow. The smaller of the twins is most likely to suckle from an alien cow, and the cow suckled usually has a single calf (Price et al. 1981). The total time spent grooming twins was less than the time spent grooming a single calf (Price et al. 1984–85).

When calves are raised artificially on a nippled feeder, they show similar rates of suckling when the milk is similar in concentration to cows' milk, but suckling increases in frequency when the milk is diluted. The calves often stand touching the wall while sucking from the feeder, just as they would touch the side of the cow if they were suckling.

Clinical Problems. Suckling problems. Nonnutritional suckling is a very frequent problem when calves are raised in groups, especially if they are pail rather than nipple fed. Nonnutritional suckling can occur 78 to 300 times a day (Hafez and Lineweaver 1968). Skin irritation or even hernias can result from prolonged sucking by one calf on the umbilicus or sheath of another. Calves that engage in nonnutritive suckling often fail to thrive (Stephens 1974). The incidence of intersucking on British dairy farms is 13%. If the problem exists on a farm, as many as 30% of the calves and 11% of the adult cows may be affected. The vice apparently is socially facilitated (Wood et al. 1967). Most dairy farmers solve the problem by penning calves separately, but when this is not feasible, other procedures may alleviate the problem, such as muzzles, providing nipples for the calves to suckle, or applying unpalatable substances, such as creosote, to the part suckled, and changing to dry food rather than milk.

Calves weaned after 6 days are more apt to suck one another than calves weaned earlier. Self-sucking is a related problem. Various harnesses and even surgical procedures, such as tongue splitting, have been devised to deal with the problem. Calves that have not suckled for the first 6 days of life cannot learn to suckle (Finger and Brummer 1969). Younger calves or those that have suckled can learn to suckle from another cow.

CROSS FOSTERING. As noted previously, cross fostering can be accomplished by draping the calf with the skin of the cow's dead calf, if that is available. A more difficult problem is convincing a cow to accept a foster calf in addition to her own offspring. Fostering calves onto dairy cows is fairly easy because they have not been selected for maternal behavior that includes rejection of alien calves. Beef cows have been selected for these traits, and, hence, it is much more difficult to foster additional calves onto them. There are practical reasons for wishing to do so, because a well-fed beef cow has a milk supply large enough for two calves. Dairy calves can be fostered onto these beef cows, and when raised in this manner will suffer far less from maternal deprivation and from the respiratory and enteric diseases to which artificially reared calves are susceptible (Kiley 1976; Hudson 1977).

It is difficult to persuade the cow to accept the dairy calf. Although the cow-calf bond is presumably formed when the cow first encounters the fetal fluids and the newborn calf, a beef cow may reject a dairy calf even when its own calf has been removed immediately after birth and when the foster calf is rubbed with fresh amniotic fluid. The cow apparently can still discriminate between a newborn and the older, larger, more active foster calf, probably on the basis of visual cues; blindfolding the cow may help (Kiley 1976). It is possible to substitute one calf for another by removing the cow's own calf after 48 hours postpartum, leaving the cow with no calf for 3 days, and then returning her own calf plus the alien calf. The bond to the original calf has been broken, but maternal responsiveness persists (Kent 1984). Placing a jacket worn by her own calf on the calf to be fostered is helpful as is the same technique when applied to sheep (Herd 1988).

Much obviously remains to be learned about the basis of maternal bonding in cattle. Some cows will grudgingly foster the dairy calf and mother their own calf. Others will become promiscuous mothers that allow four or more calves to suckle (Kilgour 1972). The maternal bond has been broken in these cases and replaced by nondiscriminative tolerance. The promiscuous mother may suffer teat or udder damage when too many calves suckle. There is also the danger that mastitis may be passed from nurse cow to nurse cow by the calves. Cows that live in a group with their foster calves continuously present are more likely to be maternally selective than those cows that are exposed to the calves only for nursing periods twice a day (Perez et al. 1985).

Although most modern dairy calves are weaned at 1 or 2 days of age, alternatives exist. Cows that are allowed to suckle their own calves for 10 weeks produce more milk than cows whose calves are removed. Although the suckled cows produce a large supply of milk, some of the total goes to the calf, and they do not make up in marketable milk for all that the calf consumes. Calves allowed to nurse only twice a day gain more weight than those fed from buckets and those with continuous access to their dams (Everitt and Phillips 1971).

Horses

Parturition. The onset of parturition in mares is heralded by waxing of the udder, but, as in other species, the length of time between the appearance of udder waxing and the appearance of the foal may be quite variable, up to 21 days. Body temperature is lower the day prior to parturition (Shaw et al. 1988). The mare will walk more and stand less the evening of parturition. Berger (1986) has found that only primiparous mares leave the herd to foal; multiparous mares remain with the herd. In the first stage of labor, which lasts for about 4 hours, the mare is restless and will crouch, straddle, and urinate. The smell of fetal fluids is attractive to parturient mares. The mare will exhibit the flehmen response in response to amniotic fluid that is expelled. Sweat will appear on her elbows and flanks (A. Fraser 1974).

During the second stage of labor, the mare will lie in lateral recumbency. The second stage is very violent and very short in horses, lasting less than one- half hour. For that reason it is important to have a veterinarian in attendance when complications are expected. There will not be time for professional help to reach the mare if problems develop in the course of labor.

Mares are notorious for their ability to thwart observation of their parturition. Although most mares foal at night, some will wait until they are released from their stalls in the morning to foal in the solitude of the pasture. Thus, parturition appears to be under some type of voluntary control in such mares, but the evidence is all anecdotal.

Postparturient Behavior. The foal is ordinarily delivered in such a way that the mare need only to turn her head to meet her foal muzzle to muzzle. The establishment of the maternal offspring bond is still uninvestigated in horses. It may be based on olfaction, as the mare licks the fetal membranes and then the newborn; licking behavior is confined to a few hours after parturition. The foal is usually on its feet and able to nurse within 1 hour and is able to locomote quite well within 4 hours of birth.

Licking, as well as sniffing, is concentrated first on the head of the foal and later on the hindquarters, particularly the perianal area. The rate of licking decreases markedly during the first hour postpartum. Although the period for bond formation has not been identified in horses, the first hour is probably critical for the mare to learn to recognize her foal selectively. The foal appears to take much longer, perhaps as long as a week to recognize the mare. The foal will follow any large moving object. The mare is usually very aggressive toward other horses and sometimes toward people for the first day or two after foaling. This behavior serves to keep away other horses that the foal otherwise might follow. Standing and suckling occur within the first hour after birth for pony foals and within the first 2 hours for Thoroughbred and American saddlebred foals (Rossdale 1967; Waring 1982; Campitelli et al. 1982–83).

Nursing. Foals suckle four times per hour at 1 week of age and gradually decrease the frequency to once per hour by 5 months (Tyler 1972; Carson and Wood-Gush 1983a,b; Crowell-Davis 1985) (Fig. 5.5). The length of the nursing bout is about 80 seconds during the first week and decreases with age until 9 months, when the frequency of suckling decreases, but the length of each bout increases. The mare usually flexes her hindleg on the side opposite the foal, possibly conserving energy by shifting her weight to the stay apparatus (Winchester 1943) (Fig. 5.6). Nursing also occurs after any separation, even a very brief one, or after the foal has been frightened. When a foal approaches its dam to nurse, it often shakes its head and nickers. Foals turn their heads sideways to nurse, especially as they grow larger.

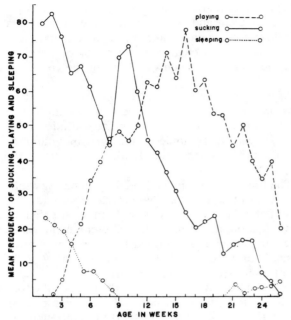

Fig. 5.5. Changes with age in frequency of sucking (solid line); playing (dashed line); and sleeping (dotted line) of free-ranging pony foals (Tyler 1972, copyright © 1972, with permission of Academic Press).

Fig. 5.6. Identification of the foal. The mare sniffs the anal area of the foal. The hindleg on the side opposite the foal is flexed to facilitate nursing (Houpt 1977a).

Colts may suckle more frequently than fillies when food is a limiting factor because sex differences in suckling have been found only in poor environments (Duncan et al. 1984; Berger 1986). Nursing is often terminated by the mare, not by aggression, but by simply walking away from the foal. This behavior occurs most frequently during the first month of the foal's life (Crowell-Davis 1985). The foal may be forced to practice following the mare as a result of her behavior. The more skilled the foal is at identifying and following its dam, the more likely it is to survive. Mares do aggress toward their foals during nursing, but the aggression appears to be a response to bunting of the udder and does not prevent or shorten suckling.

Weaning occurs at 40 weeks when the mare is about to foal but is prolonged beyond a year if the mother is not pregnant (Duncan et al. 1984). Under domestic conditions of abundant food, foals continue to suckle for 3 or 4 years, even when they are larger than their mothers. Although nonnutritive suckling is usually associated with the deprivation of the suckling experience, orphaned or newly weaned foals will often attempt to suckle from nonlactating mares, suckle the sheaths of geldings, and investigate the inguinal area of any horse.

Mares seldom venture far from their foals throughout the first few months. Foals spend several hours per day lying down. During these periods, the mare remains with the foal, either grazing in circles around it or standing next to it. This behavior, which probably protects the foal both from predators and from becoming lost, wanes as the foal matures. When the foal is awake, it is responsible for maintaining contact with its dam (Crowell-Davis 1986). The mare is within 5 yards of the foal 94% of the time the first week and 52% of the time the 5th month of the foal's life. When the foal is on its feet it is responsible for proximity. The typical equine family group will travel in the following order: mare, most recent foal, yearling foal, and then the other offspring in the order of increasing age (Schafer 1975).

Mutual Recognition. The roles of the vision, hearing, and smell senses, in the mare-foal bond is complex. The neighs (or whinnies) of the separated mare and foal are impressive, and horses use these calls to locate one another. The more frequently a mare neighs, the more quickly her foal will find her. Neighs are not specifically recognized by the foal, but Tyler (1972) felt that nickers, low-amplitude nasal sounds, were specifically recognized. Changing a mare's or a foal's appearance by hooding, blanketing, and bandaging it does not interfere with recognition, yet visual cues must be involved because both foals and mares have difficulty finding each other when one is in a closed stall. They orient toward whinnies but need visual confirmation of the dam's or foal's presence.

Olfaction is also important. Mare and foal sniff one another's heads, and the mare sniffs the anal region of the foal (see Fig. 5.6). Masking olfactory cues with a strong odor greatly retards location of mares by foals, especially in the absence of visual cues (Wolski et al. 1980). Thus, olfactory and possibly vocal cues (nickers) are used for close-range identification, whereas other vocal cues (whinnies) and visual cues are important for more distant communication of identity or presence (Fig. 5.7). Visual cues are probably not vitally significant, because many blind mares have successfully raised foals even in seminaturalistic conditions.

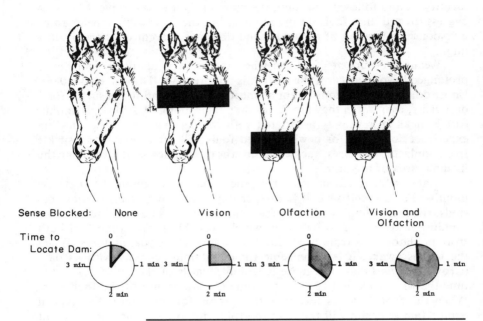

Fig. 5.7. The effect of masking vision or olfaction or both on the time that a foal takes to locate its dam (Houpt and Wolski 1979, copyright © 1979, Veterinary Practice Publishing).

Clinical Problems

MISMOTHERING. Mismothering can occur in equids, although it is much more rare than in sheep. A female mule adopted and successfully raised a foal; only examination of the foal's karyotype revealed that the foal was a Shetland pony, one of a pair of twins born to a pony mare in the same pasture as the mule (Eldridge and Suzuki 1976). Mules may be particularly prone to this behavior because a mule also stole a calf from its mother and raised it. Mules have successfully raised Thoroughbred foals born to them after embryo transplant, but a donkey foal was rejected (Shaw and Houpt 1985).

FOAL REJECTION. Three types of foal rejection occur immediately after foaling: (1) rejection of suckling; (2) fear of the foal; (3) attacking the foal. All three forms are more common in the primiparous mare, but the third form may occur again and again.

1. Many mares lick the foal and appear attracted to it but will not tolerate suckling and will kick the foal if it persists. These mares can usually be treated successfully by either tranquilization and/or by milking the mare while holding the foal next to her and feeding the milk to the foal from a bottle held in the mare's inguinal area. The newborn foal normally nurses every 15 minutes so this exercise should be repeated many times to nourish the foal, transfer colostral antibodies, and teach the mare that milking relieves tension on the udder. Gradually the foal should be encouraged to suckle the teats, and less and less restraint should be applied to the mare until the pair can be left alone safely.

2. The fearful mare tries to escape from her foal and may injure it by running over it. She will kick when the foal approaches her. Several behavioral methods can be used to stimulate maternal behavior. Turning the mare and foal out in a paddock so that both the mare and the foal can avoid one another may lessen the mare's fear. Adding another horse may stimulate the aggression periparturient mares normally show, and this may be followed by acceptance of the foal. A large dog has been used for the same purpose, that is, to stimulate maternal defensiveness.

3. The most dangerous mare is the one that actively attacks her foal, usually biting it in the withers and throwing it across the stall. She will not have licked the fetal membranes or the foal. She will also kick if the foal approaches her. These mares act toward a foal as foal-killing stallions do. Tranquilization is usually not successful, but passive restraint, immediate punishment of aggression, and administration of oxytocin and progestins can be. This problem can be seen in any breed of mares but is most common in Arabians, which indicates a genetic component. Passive restraint is best obtained by placing a pole across a box stall so that the mare cannot move sideways, forward, or backward. She can still bite, but the foal can escape and soon learns to avoid her head. She can kick forward, cow-kick, but a persistent foal can suckle. Usually the mare will accept the foal after a week or two of restraint. They tend to merely tolerate their foals and some may relapse after a few weeks.

Pain, such as that associated with passing of the placenta, may result in aggression toward the foal. Pain lowers the threshold for aggression. Removing the source of pain will eliminate the aggression.

Maternal rejection may occur some time after foaling. Too much disturbance of the mare and foal has been implicated, as has changing the odor or appearance of the foal. A common clinical problem is rejection of the foal that has been separated from its mother for medical treatment. The foal's odor has been changed and its appearance may have been altered by clipping and bandaging. Allowing the mare to have visual contact with the ill foal may help to prevent rejection even if the foal is too weak to suckle for many days.

Redirected aggression also occurs. A mare may be aggressing against another horse and then bite or kick her foal. More frequently, the foal is simply kicked accidentally during a fight. Occasionally a mare may aggress against her own foal when an alien foal approaches — a recognition failure.

Some mares do not reject their foals but do not respond vocally to them when they are separated. The foal may be injured because it will approach other mares and be attacked or will try to jump a fence or gate to return to the stall where it apparently believes the mare is. Although this situation would be presented clinically as accidental trauma to the foal, it is the result of poor maternal behavior.

Mares can be vicious when protecting their foals, but they can also be vicious when weaning them. A mare may bite not only the foal but also a nearby person when the foal attempts to nurse. This behavior may be caused by mastitis or injury to the udder or may be of unknown cause.

Cats

Parturition. Gestation is 63 to 66 days in cats. Distribution of births is seasonal with the greatest number of litters born in the summer and the least in autumn and early winter (Prescott 1973). Cats rarely deign to use the boxes carefully provided for parturition by their owners. Most cats will choose a cavelike place, such as a closet or a linen cabinet, inevitably on the best percale. A cat is fairly oblivious to her kittens during parturition so that moving the kittens to a more suitable location will not entice her from her chosen spot. Parturition in the cat is characterized by a great deal of licking by the queen: self-licking mostly directed at the belly and genital area, licking of the fetal fluids from her body or the floor, and licking the kittens. The queen is responding to the fluids rather than to the kittens at this time.

The queen is typically very restless, and a normal protocol will comprise lying down, sitting up, licking the vulva, squatting, bracing lordosis, circling, walking around the cage, lying down again, rolling, licking of a kitten, and so on. The pattern is not consistent, and the behavior of the queen varies with the endogenous stimuli (from the uterus) and exogenous stimuli (from the birth fluids and kittens). The uterine contractions of labor can be distinguished from fetal movements because the raised hind legs of the queen will usually flex with the former (Schneirla et al. 1963).

Most cats prefer solitude, although some highly socialized ones seem to be content only when the owner is present. Cats, with the exception of Siamese, are usually quiet during parturition. The restless behavior of the queen stretches the umbilical cord of the newly delivered kitten. When the placenta is delivered, the queen will eat it and part of the cord with the same tilt of the head and pronounced chewing motions that she shows when consuming prey. In the process of eating the placenta, the queen stretches the umbilical cord so that there is little bleeding when the cord itself is severed. It is rare for the eating to extend to cannibalism of the kitten. The interval between kitten births can be as long as 1 hour, even in normal births, but is usually much shorter. As mentioned, the queen seems unaware of the kittens between her bouts of licking them; she may inadvertently step on them while she paces and ignores their cries. The bursts of activity are interspersed with periods of fatigue.

When the last kitten has been delivered, the queen directs her attention to her litter. She lies down with an encircling motion, positioning her legs to form a U around the kittens. For the next 12 to 24 hours she rarely leaves the newborns and then only for brief intervals to eat, drink, and eliminate. Each time she returns, she arouses the kittens by licking them, after which she encircles and then nurses them.

Nursing. The cues used by the neonate to find the mammary glands are unknown in this species, as well as in the other domestic species. The kittens probably use temperature cues but also the mobility and responsiveness of the adult cat to locate her. Blind and deaf kittens apparently use smell and, probably to a greater extent, tactile sensations to locate the nipple. Kittens with anesthetized tongues can find the nipple but cannot suckle; kittens with anesthetized lips cannot locate the nipple but can suckle (Shuleikina 1976). Olfactory bulbectomy also eliminates the ability of kittens to find the nipple (Kovach and Kling 1967), but damage to the olfactory bulb eliminates more than the sense of smell alone.

Most kittens are suckling within an hour or two of birth. Ability to initiate suckling disappears after 22 days of age if the kittens have been fed intragastrically. Kittens can learn to nurse from artificial nipples, and prolonged experience (1 to 3 weeks) delays, but does not abolish, the kitten's ability to initiate natural suckling (Rosenblatt 1965b). Milk reinforcement is not necessary for suckling, as intragastrically fed kittens will initiate suckling on the teats of a nonlactating cat as rapidly as on a lactating cat, even on repeated trials (Koepke and Pribram 1971).

Kittens locomote by pulling themselves along with their front legs while paddling with their weaker hindlegs. As they crawl forward, they turn their heads from side to side. When they encounter the nipple, they pull their heads back and lunge forward with open mouths. Eventually, the nipple is secured in the mouth. The position of the mother facilitates locating the mammary region. The responses of the kittens to the areola and nipple appear to be innate, almost reflexive (Tobach et al. 1971).

Kittens have determined a teat order by the 2nd day, which is usually, but not always, followed. The largest kittens, however, do not appear to be those that acquire the best-producing glands, in contrast to the teat order determination in piglets (Ewer 1959). In some litters, no regular teat order is formed. Once the teat order is established, the kitten can use the presence of its sibling on either side to help guide it to the proper nipple, although sometimes the kittens obstruct more than facilitate each other's progress. Kittens massage the udder with treading motions of their paws. Treading or kneading persists in the adult cat, presumably as a pleasurable activity or in pleasurable situations.

The feline nursing period has been divided into three stages: stage 1, 1–14 days—mother initiates nursing; stage 2, 14–21 days—both mother and kittens initiate nursing; stage 3, 22–35 days—kittens initiate nursing. In the hormonally primed queen, the suckling of kittens is probably tactilely pleasant. However, as kittens grow older and their feeding demands become more persistent, the female develops an approach-avoidance type of behavior toward her kittens; she discourages their attempts to nurse from the 3rd week onward by moving away. Although in a home situation she could easily escape from the kittens, in a cage she can only jump to a shelf to which the kittens can soon gain access as their locomotor skills improve. Another ploy by the mother is to lick the kittens vigorously, thus preventing them from nursing. Figure 5.8 illustrates the change with time in the queen's relationship with her kittens. In a two-kitten litter, weaning occurs later (Martin 1986). In the one-kitten litter, these stages are not as discrete. The queen spends less time with a single kitten initially (30% compared with 70% of her time), but she continues to allow it to nurse long after a larger litter would have been weaned. Apparently, a single kitten is less attractive than several kittens but also less aggravating. From the 4th week on, "barn" cats will begin to bring live or dead prey to their kittens (see Chap. 2), but these females sometimes allow intermittent nursing until the next litter is born.

Even prior to the opening of their eyes, kittens can return to their nest from several feet away. Apparently, they use olfactory cues, because washing the floor between the kitten and the nest prevents it from finding its way back (Schneirla et al. 1963).

Experiments conducted on artificially reared kittens revealed that they followed a path produced by their own body odors to find the nipple of the brooder before their eyes were open. The kittens could also make tactile discriminations under these experimental conditions, learning to choose a nipple with bumps on it over one with concentric ridges when the former was associated with milk reinforcement (Schneirla et al. 1963).

Feline retrieval behavior is quite different from the canine form. Queens retrieve their kittens in response to auditory, not visual, signals. The more the kitten vocalizes, the more apt the queen is to retrieve it (Haskins 1977). The queen usually picks the kitten up by the scruff of its neck, though occasionally she grasps the skin of the back of the head or even the kitten's whole head. Queens are able to lift and even jump several feet with quite large kittens. In fact, the peak of kitten carrying occurs when the latter are three weeks old. Picking cats up by the scruff is an effective way to

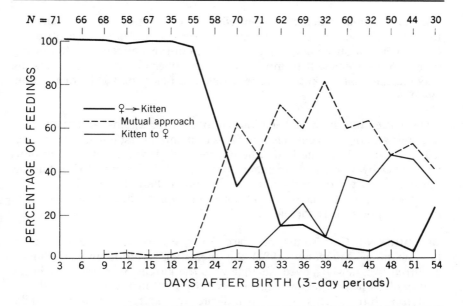

Fig. 5.8. Approach of the queen to her kittens and kittens to the queen during the first 8 weeks of the kittens' life (Schneirla et al. 1963, copyright © 1963, with permission of John Wiley and Sons).

establish dominance over even an adult cat, probably because the cat is being treated like a kitten.

Grooming. Grooming plays as important role in feline maternal behavior as in most other species. Queens lick their kittens frequently and in particular lick the perineum to stimulate urination and defecation for the first 2 to 3 weeks of life. In common with most carnivores, female cats ingest the kittens' urine and feces for several weeks postpartum, thereby keeping the nest clean.

Acceptance of Kittens. Cats will accept kittens that are not too much older than their own at the time of parturition. Maternal behavior persists much longer in cats than in ungulates, so a queen whose kittens were removed at birth will accept one kitten weeks later and encircle it. Three kittens will be avoided or actually attacked under the same circumstances (Schneirla et al. 1963). In general, species that produce litters are more willing to accept foster young than are those that produce singlets or twins, probably because the mothers of litters do not discriminate between individual offspring.

Clinical Problems. Cats present few clinical problems of maternal behavior. In fact, the efficiency of feline reproduction is much more of a problem. Occasionally, a queen may reject her litter, but this happens less frequently than in other species. It is important to remember that time spent suckling may indicate that the kittens are receiving insufficient milk (Mendl and Paul 1989).

INFANTICIDE. Tomcats have been known to kill their kittens, an abnormality of paternal, not maternal behavior. This is an interesting phenomenon studied quantitatively in the lion, where infanticide is routine after a pride is taken over by a new male or males (Schaller 1972; Bertram 1975), but has not been reported on in the domestic cat other than anecdotally.

Cannibalism by the queen, although infrequent, does occur, usually at parturition or shortly thereafter.

MISMOTHERING. Cats sometimes care for one another's kittens or communally nurse (Macdonald 1981). An interesting variation on this behavior occurred in a newly spayed cat that stole the kittens of another cat in the household. The problem was easily solved by shutting the natural mother in a room with the kittens. This case indicates that maternal behavior is independent of ovarian hormones and/or is stimulated by a dramatic decline in estrogen and progesterone levels that occurs at spaying and at parturition.

Because cats socialize with one another almost exclusively as kittens, orphan kittens should probably be placed with the litter of another lactating queen if possible. The nutritional, but not the social, development of orphan kittens will be adequate if they are artificially fed. Sometimes older kittens will show sucking abnormalities. They may suck on one another, on a human finger, on cloth, and on wool. The latter is a particular favorite of Siamese (see Chap. 8). Occasionally, this behavior can be related to early weaning, but more commonly no explanation is immediately obvious. The sucking is usually not injurious to the kittens' health and will gradually diminish in frequency.

Dogs

Parturition. The pregnant bitch gives little indication of her condition for the first 30 days. Late in pregnancy her activity will decrease, and her appetite will increase. In the last weeks she may wish to eat small, but frequent, meals as abdominal pressure increases. She may grunt each time she sits, especially if she is carrying a large litter. The slow development of the fetuses during the first half of pregnancy allows the bitch in the natural state to hunt and forage as well as ever. She is encumbered for only the last two weeks of the 60- to 63-day gestation period. At that time, she may urinate or defecate in the house unless she is taken out frequently.

In contrast to the more independent feline behavior, when a nest box is provided, bitches will usually use it. If nesting materials, such as strips of rags or paper, are provided, the bitch will make a nest. Restlessness, inappetence, and a drop in body temperature are the cardinal signs of impending parturition. Once labor begins, the bitch usually lies in lateral recumbency. As labor progresses, her hindlegs will twitch, and she will shiver; several strong abdominal contractions are visible just before the puppy is expelled. The fetal fluids appear to be attractive to the bitch. She will lick herself, any soiled bedding and, almost incidentally, the puppy. The pup's head, umbilicus, and perineum are the areas most frequently licked. The bitch cuts the umbilical cord with her molars, and then licks the stump. She may recut the cord if it is too long. If the placenta has not been delivered, she may extract it by pulling on the cord. The placenta is then nearly always eaten. She is usually silent as labor progresses, unless there is dystocia. Adult male dogs, if present, often whine (Bleicher 1962).

Pups are usually born at 30-minute intervals, but delays of a few hours are not pathological. Labor can be interrupted if the bitch is disturbed; the disturbance can be as mild as the entrance of another human observer. If she is resting between deliveries, the birth interval will be lengthened; if she is in the middle of labor, it may stop. Depending upon the temperament of the dog, 15 minutes to 1 hour are required for normal parturition to proceed (Bleicher 1962).

Although the bitch licks the neonatal puppy dry, she pays little more attention to it until all of the puppies are born. In fact she may step on the puppies and ignore their cries at this time. Once all the puppies are born, she will lie quietly and allow them to nurse. She helps to orient the puppies to her by licking. The puppies move forward by paddling and turn their heads from side to side, because they cannot lift their bodies from the floor yet. If they encounter a wall or any cold object, they will change their direction.

Nursing. Once puppies encounter the mother's body, they nuzzle through the fur, usually attempting to burrow beneath her. High-chested breeds like Shetland sheepdogs present much more of a challenge to the pup seeking a nipple than do flat-chested breeds like cocker spaniels. When the pup locates a nipple, it does not immediately grasp it but rather noses it from beneath. Unless it opens its mouth at the same time as the nipple moves past, it will not succeed in grasping it. Puppies improve markedly in the first 2 or 3 days of life in their ability to locate nipples. Once the nipple is located, the pup jerks up with its head, pushes at the mammary gland with its front feet, and arranges its hind feet to support itself against the mother (Rheingold 1963). All of this is not proceeding in a social vacuum; the other pups are also struggling for positions. The efforts of one will dislodge the other, and whining and scrambling will ensue. The supposedly serene maternal scene is, in fact, noisy and tumultuous. Puppies do not appear to have as definitive a teat order as cats or pigs. The inguinal mammary gland is preferred and is the one approached by a single pup (Fox 1972).

Puppies have two vocal signals, the whine and the grunt. The whine is emitted whenever the puppy is cold, hungry, or separated from its litter or mother. Whining will stop immediately if the puppy's head and neck are covered with a warm towel or if it is again placed with its litter. Puppies show a distinct preference for soft surfaces; they will spend more time on a cloth-covered rather than a wire-covered artificial mother (Igel and Calvin 1960). Textural as well as olfactory cues may help puppies locate their mother and her mammary glands. The grunt is apparently a pleasure communication that occurs when sought-after warmth or reunion is obtained. Despite the puppy's loud vocal response to separation, the bitch appears to notice that a puppy is missing when she sees it rather than when she hears it (Bleicher 1962).

Bitches lick their puppies a great deal. Licking serves three functions in puppies; two are common to other species, and one is unique to dogs. The licking arouses the pups to eat, and, when it is directed at the anogenital area, it stimulates urination and defecation that would otherwise not occur spontaneously. The bitch keeps the nest area clean by consuming the urine and feces of her puppies for at least the first 3 weeks of their lives. The third function of licking is retrieval. Bitches seldom carry their puppies. Instead they lead them back to the nest by licking the pup's head. The pup will orient toward the bitch and move toward her. The bitch will back toward the nest, while continuing to lick the pup that follows, until the nest is reached. Licking can be reinstated by substituting young (2- to 3-day old) puppies for older ones (Korda and Brewinska 1977).

Weaning. During the first few days, the bitch spends most of her time with the pups. Nursing reaches a peak at the end of the first week. The amount of time that she spends with them decreases with time (Fig. 5.9). The undisturbed litter will be weaned gradually by the bitch; weaning usually will be complete by 60 days. During early lactation, the bitch always approaches the puppies to initiate a nursing bout. By 3 weeks, the puppies have opened their eyes and can locomote well. They then approach the bitch and initiate most of the nursing bouts. Bitches rarely punish their puppies until the 3d week. Even during the weaning process the punishment is mild enough; it may momentarily deter the pup's attempts to nurse, but it will not inhibit it from trying again. The punishment may consist of a growl, a snarl, or an inhibited bite. The level of aggression of the bitch toward her puppies increases and the number of nursing bouts per hour decreases after the puppies' 3d week (Wilsson 1984–85).

Some bitches may regurgitate food for their pups during the weaning process at 4 to 6 weeks (Martins 1949). This behavior, commonly seen in wild canids, helps to maintain adequate nutrition in the young during the transition from a milk diet to a raw meat diet, when their powers of mastication and digestion may not have matured enough to enable them to survive on the raw meat (Fox 1972). Wolf pups beg for regurgitated food by

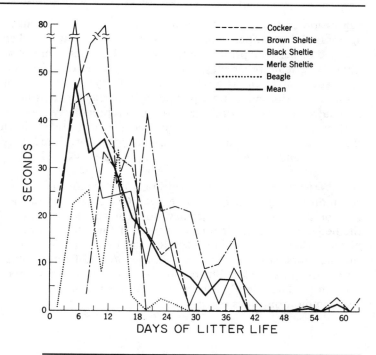

Fig. 5.9. The decrease in licking of the puppies by the bitch as the puppies mature (Rheingold 1963, copyright © 1963, with permission of John Wiley and Sons).

licking at the mouth of the adult. This behavior is occasionally seen in domestic dogs and may be the basis of face licking and licking intention movements directed toward people by adult dogs.

Clinical Problems

PSEUDOPREGNANCY. Pseudopregnancy in the intact, nonpregnant bitch during the luteal phase of the canine reproductive cycle is so common that it may be normal behavior. Bitches do vary, however, in the severity of the behavioral signs. Some dogs show only slight enlargement of the mammary glands, whereas others go through pseudoparturition at 49 days postestrus and actually lactate. The pseudopregnant bitch becomes less active, mimicking the slowing of the truly pregnant animal. She may make a nest, usually either in her own bed or in some cavelike environment under a table or in a dark corner. She may adopt a toy, a leash, a shoe, or some other object to mother. She will not only place it in her nest and assume a nursing posture next to it, but she will often defend it. Serious problems with aggression can arise in the pseudopregnant bitch. She may be generally aggressive, but, more commonly, she will only attack when her nest is invaded or her "offspring" are threatened. Estrogen therapy will relieve the signs of pseudopregnancy, but ovariohysterectomy should be recommended. The bitch who shows recurrent cycles of pseudopregnancy is prone to uterine infections, pyometra in particular. It is advisable to wait until after the behavioral signs of pseudopregnancy have passed before performing the ovariohysterectomy, because protective behavior persists if the animal is operated upon while pseudopregnant. Pseudopregnant bitches can adopt and successfully raise foster puppies (Fox 1968).

MATERNAL REJECTION. Maternal rejection also occurs in bitches. It is most common in dogs that have undergone Caesarean sections and have been anesthetized during the time they would normally be licking and smelling the neonatal puppies. It rarely happens if the bitch has delivered a puppy normally before surgical intervention was necessary. The bitch may tolerate the first nursing better if some milk is expressed from the engorged glands before the pups suckle. She can be restrained while they nurse for the first time and sedated lightly if necessary. Attacking puppies after parturition is less common but does occur.

6

Development
of Behavior

One of the most pleasant aspects of owning animals is watching the young develop. Even in a world overpopulated with cats, kittens have not lost their attractiveness. The gangly foal and the playful kid are also very appealing. The veterinarian or animal scientist will be asked more questions about normal developmental behavior — when to take a puppy home and when to start various types of training — than about adult behavior. To correctly answer these questions, the clinician should be familiar with the neurological development and behavioral maturation of various species because owners spend more time observing infant than adult animals.

DOGS
Critical or Sensitive Periods. During the past 40 years, the behavioral concept of critical periods has emerged, a concept that has had a strong impact on practical dog handling. Recently, the term sensitive period has replaced critical period. A sensitive period is a time in the life of an animal when a small amount of experience (or a total lack of experience) will have a large effect on later behavior. The critical periods for dogs are defined as the neonatal period (1–2 weeks), the transitional period (3 weeks), the socialization period (4–10 weeks), and the juvenile period (10 weeks–sexual maturity) (Scott and Marston 1950; Freedman et al. 1961; Scott 1962). Socialization can occur in older animals, albeit with difficulty. For that reason, Bateson (1979) has suggested that the adjective sensitive rather than critical be used with these periods. The sensitive periods are not sharply defined and may vary among breeds that are fast or slow to mature. Cocker spaniel

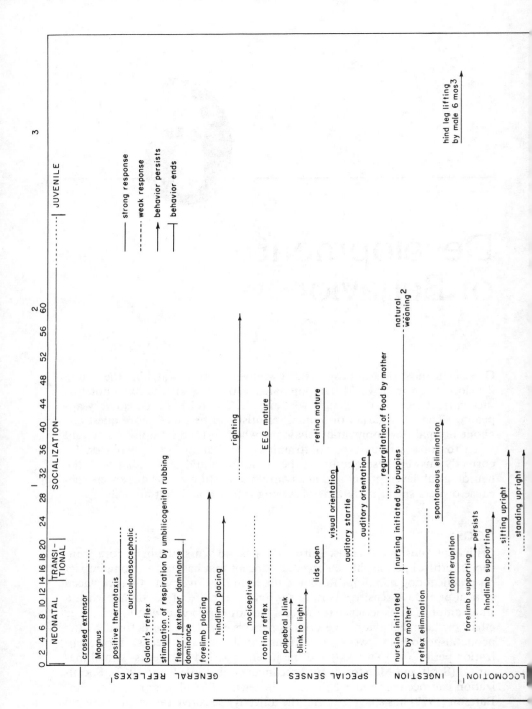

Fig. 6.1. Behavioral development of the dog. [1-6] Refer to references: Fox (1971);[1] Rheingold (1963);[2] Berg (1944);[3] Bleicher (1962);[4] Scott and Fuller (1974);[5] Fox and Stelzner (1966a,b).[6]

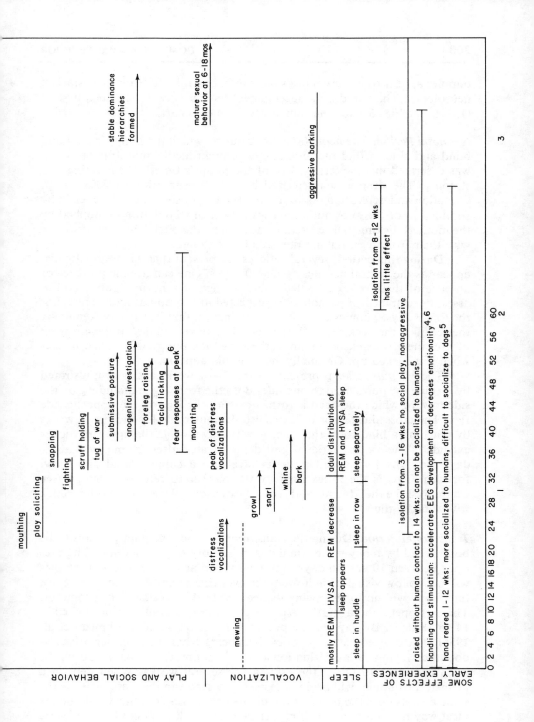

puppies appear to mature more slowly than basenjis, for example. Marked neurological changes during development have been well described by Fox (1971). See Fig. 6.1 for the canine development chart.

Neonatal Period. The neonatal period, during which the puppy is deaf and blind and able to find the nipple only through tactile and olfactory cues, was described in Chapter 5. Most of the puppy's time is spent eating and sleeping. The sleep is characterized by a high proportion of REM sleep. Urination and defecation do not occur spontaneously but can be elicited by stimulation of the anogenital area; usually such stimulation is supplied by the mother's licking. The excreta is ingested by the bitch. Puppies locomote with their front legs, pulling their hind legs along.

During this period, several reflexes are present that will gradually disappear as the central nervous system matures. One can use the presence or absence of these reflexes to determine the age of a normal puppy and to assess development if pathology is suspected in a puppy of known age. For the first 3 days, puppies show flexor dominance; that is, when picked up by the scruff of the neck they will flex their legs. From 3d day to the 4th week of life, there is extensor dominance; that is, the puppy will extend its legs when it is picked up. Gradually, normotonia appears.

The *Magnus reflex* is present for the first 2 weeks and is demonstrated by turning the pup's head to one side. Both the front leg and hind leg on the side toward which the head is turned extend; the legs on the opposite side flex. The crossed extensor reflex is also demonstrable for the first 2 weeks by pinching a hind foot; that foot is withdrawn while the opposite foot is extended. The rooting reflex is best demonstrated after the puppy is a few days old. The puppy will push its face into a cupped hand and crawl forward. This is the reflex used by the mother to retrieve a puppy (see Chap. 5). Like the Magnus and crossed extensor reflexes, the rooting reflex will wane by the 4th week.

Transitional Period. During the transitional period, the puppy begins to be bombarded by many more stimuli as its sensory organs develop. The eyes open between 10 and 16 days, although visual acuity is poor, and puppies will not follow visual stimuli when the eyes first open. As vision improves, the puppy will no longer swing its head from side to side as it locomotes. The ears open, and a startle response to auditory stimuli can be elicited at 14 to 18 days. By day 16, the puppy can localize sound (Ashmead et al. 1986). The crossed extensor reflex disappears from the front legs first, as does the Magnus reflex. Urination and defecation occur spontaneously; the bitch continues to ingest the excreta for several weeks. When kept in a kennel, the puppies will begin to leave the nest to eliminate and will use the same area as does the bitch. If the bitch is paper trained, this would be the ideal way to train a puppy, long before it can follow its mother outdoors. The puppy can support its weight on all four legs by 12 to 14 days, although normal adult sitting and standing will not be seen until 28 days. Tooth eruption begins during the transition period, and the pups will chew on one another and begin to clumsily play and growl.

Socialization Period. The 3d period, socialization, is the most important from a behavioral point of view. From the 4th to the 10th week, pups learn about their environment, about their littermates and mother, and about humans. Play begins and is most frequent in the socialization period. Canine play will be discussed in more detail later in this chapter. Dominance hierarchies are formed. Strong avoidance behavior develops and, by 8 weeks, fear reactions are seen.

This period is also most important from a clinical standpoint. Puppies weaned and removed from the company of other dogs before the period of socialization will, as adults, often be difficult to handle in the presence of other dogs. They will either be frightened of other dogs or, less commonly, too aggressive. They will not be able to play with other dogs and will be difficult to breed. A dog that has not had the opportunity to interact with other dogs, but has been with people, will be too human oriented. Male dogs may direct their sexual attentions toward humans.

Normally, pups are not weaned before 4 weeks of age, but they may be weaned even earlier if the bitch has died and the puppies must be hand fed. The nutritional requirements of the orphan puppy can be met, but many of the tactile and social requirements may not be. Hand-fed puppies usually will suck more on fingers and other objects than will normal puppies, indicating that the need to suckle has not been met despite scheduled bottle feedings (Ross 1951). It would be best to foster a litter of orphan pups onto another lactating bitch. If this is not possible, the litter should be kept together. A similar situation may arise if the bitch has lactation tetany, in which case the litter may have to be weaned quite early. Again, the litter should be kept together at least until the puppies are 6 weeks old.

The importance of social contact to the puppy during this period is demonstrated by their emotional reaction to separation from the litter. Six-week-old beagle puppies yelp 1400 times per 10 minutes when placed in a strange pen. Older and younger dogs are less disturbed (Elliot and Scott 1961) and, therefore, less vocal (Fig. 6.2). It is surprising that human contact is more effective than canine contact in alleviating separation distress in 4-to-8-week-old puppies (Ross et al. 1960; Pettijohn et al. 1977). By 7 to 8 weeks a fear posture, tail tucking, is first seen in beagle puppies (Davis et al. 1977). The puppy test used by many breeders at 7 weeks to predict adult behavior is based on the puppy's reaction to being startled, petted, picked up, and turned on its back, as well as on its willingness to approach and follow people; it is probably not reliable because dogs change relative to their littermates as they develop (Wright 1980).

Socialization to humans is equally important. A dog that has had little contact with humans until 14 weeks rarely becomes a good pet (Scott and

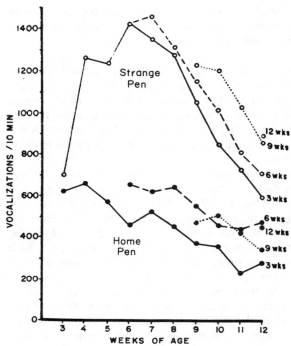

Fig. 6.2. The average number of vocalizations by puppies at different ages in two environments. Note that vocalization when alone in the home pen is uniformly lower than that in a strange pen. Puppies whose tests were begun when they were older than 3 weeks showed a slightly higher rate of vocalization, but the curves are parallel (Elliot and Scott 1961, copyright © 1961, with permission of the Helen Dwight Reid Educational Foundation, published by Heldref Publications, 4000 Albemarle St., N.W., Washington, D.C. 20016).

Fuller 1974). This reaction is, of course, typical of kennel-rather than family-raised dogs and is sometimes called kennelitis. Such a dog may be well socialized to other dogs but has had limited experience with humans. The dog, depending on its genetic background, may be overtimid or most difficult to dominate. Although normally socialized dogs find contact with a human rewarding (Stanley and Elliot 1962), the dog that has not been socialized with humans will not; it will, therefore, be a much more difficult animal to train.

A substantial amount of experimental evidence supports the sensitive period hypothesis. Most impressive is the effect of early isolation (Agrawal et al. 1967). Puppies completely isolated from the 3d to the 20th week of life are markedly disturbed. Their learning ability is impaired (Thompson and Heron 1954a,b; Melzack 1962). They are socialized to neither humans nor dogs, and their response to either species is fear. They are hyperactive and difficult to train. Beagles react more fearfully, whereas Scottish terriers are more hyperactive and show an impaired (higher) threshold to pain (Melzack and Scott 1957).

Partial isolation from 3 to 16 weeks has different effects on different breeds. Beagles become less active; terriers more active (Fuller 1967). Freedman (1958) found that the early environment of puppies had no effect on the reaction of some breeds to mild punishment (saying "No" and hitting with a newspaper). Basenjis ignored the punishment; Shetland sheepdogs were always inhibited by the punishment. Beagles and wirehaired fox terriers were inhibited only if they had not been punished in early life. Even 1 week in isolation will produce changes in the canine electroencephalogram (Fox 1971), although the changes are transient.

Complete isolation is rarely, if ever, imposed on puppies except for experimental purposes, but exclusive human, dog, or cat contacts do occur. Dogs raised with cats prefer the company of cats to that of dogs and fail to recognize a mirror image (Fox 1971). Hand-raised puppies or those weaned 3–5 weeks will approach a human much more quickly than dogs weaned at 8 weeks with little human contact before that time. Hand-raised puppies suck on fingers, but this behavior does not persist (Ross 1951).

Not only the presence or absence of human contact but also the quality of that contact will affect the puppy's later behavior. It has been shown in many species that early handling can influence emotionality in later life; the dog is no exception. Dogs were subjected to varied stimulation (exposure to cold, vestibular stimulation on a tilting board, exposure to flashing lights, and auditory stimulation) from birth to 5 weeks of age. The stimulated pups differed from controls in several physiological and behavioral parameters. They showed earlier maturation of the electroencephalogram, larger adrenal glands, and lowered emotionality, which enhanced problem-solving ability in novel situations, than the nonstimulated pups. Most interesting was the finding that stimulated pups were dominant over nonstimulated controls in a competitive situation (Fox 1971). Some effects of early stimulation are not permanent. Development is accelerated, but the nonstimulated animal eventually catches up. Not only social deprivation but also food deprivation during early life can affect later canine behavior. Elliot and King (1960) found that food-restricted puppies were more attached to their handlers during the period of deprivation and later showed increased eating rates and an increased intake of highly palatable food.

Neurological development continues during the socialization period. Placing reactions mature in the following order: chin placing, visual placing, contact forelimb placing, contact hindlimb placing, and tail placing (if a dog's tail contacts a surface, it will reach toward that surface with its hindlegs). By 6 to 9 weeks, all these reflexes are functional (Czarkowska 1983).

Juvenile Period. During the juvenile period a dog increases in size and competency in adult activities. Puppies begin to show adult sexual behavior at 4 to 6 months when they begin to show greater attraction to estrous bitches than to spayed ones. This attraction increases with age until 2 years when the dogs are fully mature (Beach et al. 1983). Although considered to be adult at puberty, most dogs do not mature socially until 18 months or later.

Conclusions. The conclusion to be drawn from the experimental studies and clinical observations of the critical periods of development of dogs is that dogs should be exposed to both dogs and humans during the socialization period. The exposure to both species should be pleasant because fear responses are also strong during this period at 5 to 10 weeks. Although the effects of isolation are most pronounced in dogs isolated during the socialization period, isolation, even semi-isolation, in a boarding kennel for several weeks any time during the first year can reduce the sociability of a dog and increase its fearfulness. The importance of human socialization to canine training is exemplified by the study of Pfaffenberger and Scott (1959) in which 90% of the dogs (mostly German shepherds) that were home raised from the 12th to the 52d week of life were trainable as guide dogs for the blind. Dogs that remained in the kennel for the same period failed the training program. Properly socialized dogs will be much more willing to work for a reward as simple as verbal praise or even reunion with the human handler.

Neurological Development. An old adage says that "One can't teach an old dog new tricks," but it is equally difficult to teach a very young one. As discussed in Chapter 7, 6-week-old puppies could not solve a barrier problem, and 4-week-old puppies (even after 13 days of training) could not remember the location of hidden food for more than 10 seconds, although 12-week-old puppies could remember for 50 seconds. Puppies less than 21 days old cannot learn to pull food into their cages with a ribbon (Adler and Adler 1977), and 5-week-old puppies took twice as long to learn a visual discrimination as 12-week-old ones (Fox 1971). Neonatal puppies, however, can learn to avoid an aversive stimulus (Stanley et al. 1974).

The poor performance of the young puppy is not surprising when the stage of development of its nervous system is considered. There is only 10% dry matter in the brain at birth. The adult percentage (19%) is not reached until the fourth week of life (Fox 1971). Myelin is almost completely absent from the newborn puppy's brain and appears gradually over the next 4 weeks. Conduction speed along nerves is related to the presence of myelin and the more rapid reactions of the month-old puppy attest to the myelinization of its central nervous system. At birth, the length and width of the canine brain are nearly equal. The increase in length of the brain with age is due to an increase of the frontal and occipital areas. A great increase in the complexity of the gyri and sulci also occurs (Fox 1971).

Sleep. Sleep shows many changes in duration, type, and posture with development. The newborn puppy spends most of its time (96%) sleeping except for brief nursing bouts. Most of the sleep in the neonate is rapid eye movement (REM) or stage 4 sleep. Only 1% is slow wave sleep (SWS). Rapid eye movement sleep is associated with dreaming in the adult human, but one wonders what the newborn puppy or human dreams of because their experience is limited to intrauterine life. Owners are often concerned by the twitching or hyperkinesia exhibited by normal puppies during the neonatal period. The newborn pup sleeps in a heap with its littermates. It has as much contact with them as possible. Huddling of puppies may prevent heat loss as the same behavior does in pigs.

A puppy removed from its littermates will waken and whine until it is either returned to the litter or placed on a soft, warm surface. Holding a puppy will often calm it, probably because of the warm body contact. As puppies mature, the percentage of time spent in REM sleep drops from 85% at 7 days to 7% at 35 days. Meanwhile, the percentage of time that the dog is awake has increased to 62% by day 35 and SWS occupies the other 30% of the 24 hours (Fox and Stanton 1967). By 3–5 weeks, puppies sleep in a row with side contact only. Later, they will sleep apart but may sleep against a wall for contact (Rheingold 1963). Even adult dogs often try to maintain contact while sleeping by curling against their owner or simply by lying on the owner's foot. Much of the nocturnal distress of the newly weaned or separated puppy can be alleviated by providing it with a warm "companion" like a hot water bottle or even an old, and preferably dirty, sweater with lots of olfactory stimuli.

Play. Puppies, kittens, lambs, and even foals are attractive not only because their foreshortened faces and awkward gaits inspire maternal, or at least protective, attitudes in humans (Lorenz 1952) but because they play. Play remains an enigma, not because we do not know how animals play but because we do not know why they play. Play appears to be important in the development of the social organization of animals, but that does not explain solitary play. Play may be important simply as a form of exercise or, perhaps, as a means of practicing and perfecting the skills necessary for the hunt, in carnivores, or the escape, in herbivores. None of the above reasons explain adult play and why it persists more in some species than in others. Finally, play is presumably pleasurable and may, therefore, be its own reward whatever the ultimate value to the organism may be.

Play in puppies begins when they are about 3 weeks old with mouthing of one another. The mouthing is concentrated on the head region of the opponent, which should not be surprising because the cranial nerves are most myelinated in the suckling animal. The biter and bitten will get maximal sensory input from play that involves the puppies' heads. As the puppies' strength improves and as their teeth erupt, the mouthings become genuine nips. Four-week-old pups may nip painfully, but the violent reaction of their littermates and, in particular their mother, to painful bites soon teaches them to inhibit the force of the bites. The early weaned or orphaned pup will not learn to inhibit its bites; it is up to the owner to punish, albeit mildly, the painful nip. The worst disfavor one can do a puppy is to wear heavy gloves when playing with it; the dog will not learn to play gently or to be submissive to humans.

Play Fighting. By 4 to 5 weeks, play fighting becomes more skilled as the puppies' motor and perceptual skills improve. Scruff holding and shaking or worrying appear. Pouncing, snapping, and growling occur in the course of play. The facial expressions of the adult dog replace the masklike expression of the younger pup. Tug-of-war is a favorite game with littermates or humans but may encourage oral vices. Wrestling bouts occur with the puppies alternating the standing-over and lying-on-the-back positions.

Sexual Play. Elements of sexual behavior appear at 5–6 weeks. The puppies will mount, clasp, and perform pelvic thrusts without regard to the sex of the partner. Male puppies, in particular, exhibit this behavior. Dogs deprived of all play experience and social contact as puppies can mate but are often misoriented when they mount and, consequently, achieve fewer intromissions (Beach 1968). The poor sexual performance of socially isolated dogs indicates the importance of play in puppyhood to normal adult behavior.

Characteristics of Play Behavior. It is important to the participants and to the observer to distinguish play from serious behavior, particularly fighting behavior. By 3.5 weeks, puppies can effectively signal that "what follows is play." The signal most often used is the play bow in which the dog lowers its forequarters and often paws at its own face while wagging its tail (see Fig. 1.3C). The play bow is an innate, not a learned display, for it occurs in hand-raised puppies (Bekoff 1977). The play face is distinguished by an open mouth and erect ears. Other signals are the exaggerated approach, repeated barking, approach and withdrawal, and pouncing and leaping. A submissive dog is more successful in soliciting play than is a dominant one. Perhaps the dominant dog is usually taken seriously by its subordinates (Bekoff 1974).

Play in the dog, as in all species, is characterized by actions from various contexts (aggressive, sexual, etc.) incorporated into unpredictable sequences in which the actions are repeated and performed in an exaggerated manner. A typical sequence would begin with a play bow, followed by an exaggerated approach, veering off, a chase, general biting, head shaking while biting, rolling and wrestling, reciprocal chasing, more wrestling, inhibited biting, rearing, and pushing with forepaws. Typical bouts last 5 to 15 minutes in puppies between age 3 and 7 weeks. The more play exhibited by young canids, the less true aggression is manifested; a comparison of the pups of three canid species showed that dogs were the most playful and coyotes least, whereas wolves were intermediate. An analysis of play and fighting can be used to identify canids of unknown genotype (Bekoff et al. 1975).

Play is not only valuable for the development of normal behavior, it is also diagnostically useful. Play occurs most often in the warm, well-fed, healthy puppy. The absence of play behavior in 3- to 9-week-old puppies is an indication of pathology. Social play is the most common form in dogs, but solitary play does occur. The dog pouncing upon and carrying a stick is an example. Tail chasing occurs in the absence of another puppy to chase. Games of fetch between owner and dog are the outgrowth of chasing play. If the game is not initiated during the socialization period, it is very difficult to teach (Scott and Fuller 1974), especially to a dog that is not genetically a retriever. Mouthing and chewing play is the beginning of the behavior that may persist as the "destructiveness in the owner's absence" syndrome (see Chap. 9). Destructiveness is an extremely difficult behavior to eliminate because it occurs when punishment cannot be applied immediately. Susceptible dogs (if it is possible to identify them) should be provided with a play partner on whom they can chew before the habit forms. Providing a companion once the behavior is a habit may often result in social facilitation so that both dogs will chew on the furniture (described in Chap. 9).

Exploratory behavior may be classified as a thrill-seeking type of solitary play. Exploratory behavior increases with age, not decreasing after 10 weeks of age as social play does. By 6 weeks of age, the puppy has mastered many of its social skills. It can signal play and aggression. It approaches another dog and investigates the inguinal area. It is beginning to form a dominance hierarchy. It eliminates in the same area as its mother and littermates. It can eat food and sleep alone. In the next week or two, it should be ready to become socialized with humans. See Figure 6.1 for the canine development chart.

Puppy Temperament Tests. Puppies that are exploratory at 6 weeks may not be at 12 weeks and vice versa. The same is true of social dominance (Wright 1980). These findings indicate that tests of puppies at 7 weeks, popular as a means of predicting adult behavior, are unlikely to be valid. The tests involve scoring the puppies' reactions to handling and their willingness to approach or follow people.

CATS. See Fig. 6.3 for the feline development chart.
Sensitive Periods. Critical periods have not been as chronologically well-defined in cats and occur earlier than in dogs, at 2 to 5 weeks (Turner and Bateson 1988), but they exist. A litter of kittens, that are born in a cranny inaccessible to humans will hiss when handled by humans at 2 or 3 weeks, whereas another litter of the same mother, that are handled daily do not react fearfully.

The most detailed study on the role of early experience in adult feline behavior was that of Seitz (1959), who separated kittens from their mothers at 2, 6, or 12 weeks. Kittens weaned at 12 weeks did not cry upon separation even though they had been living on their mother's milk alone. Kittens weaned at 6 weeks cried for a day or two. Those weaned at 2 weeks and fed by dropper cried for 1 week. As adults, the early weaned kittens showed the

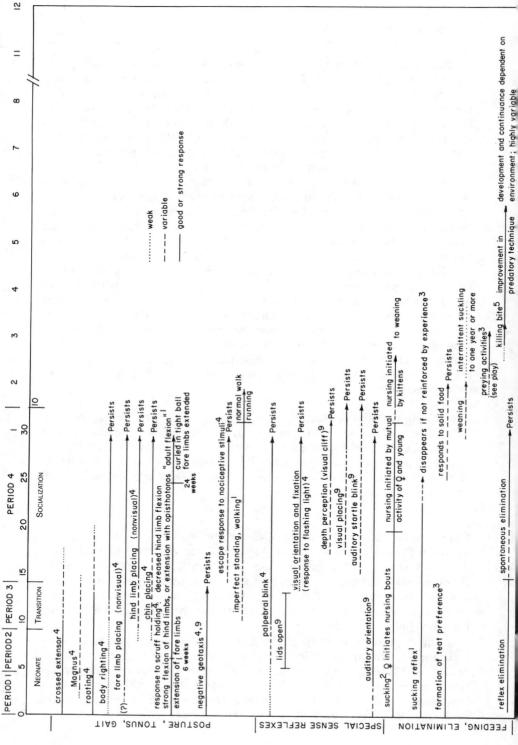

208

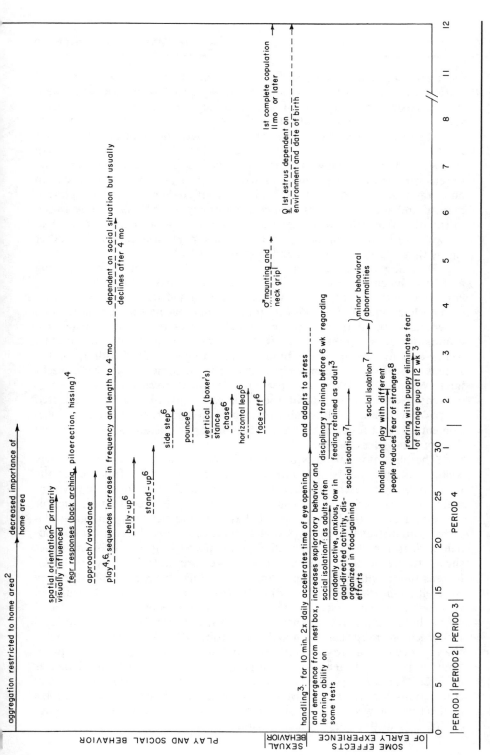

Fig. 6.3. Behavioral development of the cat. 1–10 Refer to references: Kling et al. (1969);[1] Rosenblatt and Schnierla (1962);[2] Fox (1975);[3] Fox (1970);[4] Leyhausen (1975);[5] West (1974);[6] Seitz (1959);[7] Collard (1967);[8] Villablanca and Olmstead (1979);[9] Turner and Bateson (1988).[10]

most random activity, such as trying to escape from a carrying cage and were most disturbed by novel stimuli. When tested with food, they were most persistent in trying to obtain food secured under a wire cover but least successful in competing with other cats for food. The early weaned cats were also the slowest group to learn to associate the sight of a light with food.

Konrad and Bagshaw (1970) also weaned kittens at 2 days of age. The kittens were fed by nipple and handled as little as possible. When tested in an unfamiliar room, the cats raised in the restricted environment explored, played, and approached less than conventionally reared cats. Kittens raised in isolation from 40 days of age spent more time close to another cat than did kittens raised communally, but all the cats spent more time with another cat as they grew older (Candland and Milne 1966).

Handling kittens each day for the first month accelerates eye opening and EEG synchronization. Such cats are more active and aggressive when confined, and quieter in a novel environment (Meier 1961). In another study, Meier and Stuart (1959) found that cats that had been handled and raised in a stimulating environment made fewer errors in a visual discrimination task. Kittens raised with their mothers and without handling were very slow to approach humans. In another study, handling of the kittens appeared to impair their ability to learn some tasks several weeks later (Wilson et al. 1965). It appears to be difficult to slow a kitten's development. Neither limitation of food nor severe hypoxic episodes affected kittens' development, although treatment with a goitrogen did delay development of solid food ingestion and locomotory skills and slowed physical development (Berkson 1968).

The physiological basis of the behavioral abnormalities seen in early weaned kittens may be inferred from the changes in the function of the visual pathways observed in cats reared either in the dark or in an environment in which they had no visual stimuli or very limited stimuli, such as horizontal or vertical lines. Both behavioral and neurophysiological evidence demonstrated that the visual system, especially the cortical components, does not develop normally; cats exposed only to horizontal lines show little response to vertical lines (Blakemore and Van Sluyters 1975). Kittens raised without an opportunity to see their front paws because they were either in darkness or wearing Elizabethan collars have difficulty in visual placing. They extended their paws appropriately but missed a small target that normally reared kittens reached 95% of the time (Hein and Held 1967). Kittens must learn to match paw position to target position. Similar changes occur in more complex behaviors: a cat that never had the opportunity to play as a kitten does not respond to the appropriate play signals as an adult. Kittens have adequate genetic capabilities to form the neuronal connections necessary for normal vision or social behavior, but the complex connections between cortical neurons form with visual or play experience during a "sensitive" or "critical" period.

Sleep. Sleep in kittens also shows a developmental pattern. For the first 3 weeks, the EEG cannot be correlated with the other behavior defining the different sleep stages, such as eye movements and muscle quiescence. Although the percentage of time kittens are awake remains constant, the percentage of active REM sleep decreases, while that of quiet sleep increases. Muscle twitching, which is characteristic of REM sleep, also decreases with age. The sleep cycles are much shorter than those of the adult cat. Kittens also pass directly from the awake state to REM sleep; adult cats almost always pass through SWS sleep before entering REM sleep (McGinty et al. 1977). It is not until 3 months of age that forebrain maturation and environmental influences mediate a mature sleep-wake cycle (Hoppenbrouwers and Sterman 1975).

Neurological Development. The neurological development of the cat has not been studied as systematically as that of the dog. The kitten shows a dominance of flexor tone for the first 2 weeks of life and then a dominance of extensor tone for the second 2 weeks. The motor cortex involved in forelimb movement develops during those first 2 weeks and cortical control of the hindlimbs develops in the second 2 weeks. This development is reflected in the locomotion of the kitten. It drags itself by its forelegs at first, but later the pushing movement of the hindlegs grows stronger (Rheingold and Eckerman 1971; Fox 1975). The eyes open at 7 days (6–10) and orienting responses to auditory stimuli develop a day or two before (Foss and Flottorp 1974). Between the 3d and 6th week, cats develop the ability to land on their feet (air righting).

Receptive fields are very similar in adult cats and newborn kittens, although the neonates showed a longer latency to respond (Rubel 1971). Visual acuity improves 16-fold between 2 and 10 weeks of age. The development of the cytoarchitecture of the sensory cortex is interesting; the cortical layers of the kitten brain are arranged in an orderly fashion with few dendrites linking the cells. The adult cat brain possesses disordered layers with many dendritic processes on the cells, which apparently pull the cells out of the original orderly alignment. It is hypothesized that the interconnecting dendrites form with increasing sensory experience.

Adult cats and dogs will respond to a silhouette of their own species as they would to a real animal. Five-week-old kittens do not even orient themselves to a cat silhouette, but 6-week-old kittens will, and the frequency approaches the adult level by 8 weeks. Adult cats are apparently threatened by silhouettes and will show piloerection toward a silhouette on its first presentation. Five-week-old kittens show no piloerection and 6-week-old kittens show very little, but 8-week-old kittens show the adult response to silhouettes (Kolb and Nonneman 1975). Hypothalamic stimulation does not elicit adultlike affective response with piloerection and enlarged pupils until 3 weeks, although motor responses, such as arching and jaw movements, can be elicited at 4 hours (Kling and Coustan 1964).

Adult cats show a unique behavior and facial expression, the gape, to conspecific urine. The components of the behavior are approach, sniffing and licking the urine, flicking the tip of the tongue repeatedly against the anterior palate behind the upper incisors, withdrawing the head from the urine, and opening the mouth and licking the nose (see Fig. 1.6). This response is not seen in kittens less than 5 weeks old and is essentially similar in frequency and performance to adult gaping at 7 weeks (Kolb and Nonneman 1975).

Kittens can make ultrasonic vocalizations. In general, the frequency limit and range fall with age. Deafened kittens produce vocalizations similar to those of normal kittens, which indicates that learning is not important; however, their calls are louder than those of normal kittens, which indicates that feedback through the auditory system normally occurs.

As kittens mature, they become more proficient at finding their way back to their home area. They also vocalize less (30 cries per minute at 1 day of age as compared to 17 cries per minute at 15 days) when placed on a cold surface. Once the eyes have opened, the kittens use visual cues to find their nest; before that time they use olfaction. Very young kittens will become less active and less vocal when placed on a warm rather than cool surface, but the calming effect of thermal stimuli is lost after the first week (Rosenblatt 1971; Freeman and Rosenblatt 1978a,b). Isolation produces most vocalizations (4 cries per minute) at 3 weeks of age; younger and older kittens vocalize less. Response to restraint remains high and unchanged (5 cries per minute) throughout development (Haskins 1979).

Play. Play in kittens is first seen at the beginning of the 3d week when the queen begins to wean the kittens by repulsing their attempts to suckle (see Chap. 5). Play in cats has been most thoroughly studied by West (1974, 1977), Caro (1980a,b,c) and Martin and Bateson (1985). Play in kittens begins with gentle pawing at one another. As kittens improve in coordination, biting, chasing, and rolling replace simple pawing. One kitten is usually in the belly up position with all four legs held semivertically. Social play increases from 4 to 11 weeks and then declines relatively rapidly (Fig. 6.4). At first, three or more kittens may play together, but by 8 weeks almost all play is between pairs of kittens. A reliable sign of play is the arched back and tail, but a definite play signal has not been defined in cats, although tail position and movement have been suggested (West 1974).

Play Periods. There are usually four play periods per day. Almost 1 hour a day is spent in play at 9 weeks of age. Most kitten play bouts begin with a pounce and end with a chase. In between, the kittens frequently face off, hunching forward with tails arched out and down. They may bat at one another. Kittens also assume a vertical stance in play, rear back on their hind legs, and sometimes stand up by extending the legs. Various leaps are seen also. Kittens are much more apt than puppies to paw rather than bite

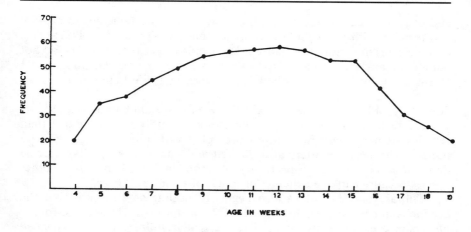

Fig. 6.4. The change in playing behavior of kittens with age. Social play reaches a peak at 11 weeks and then declines. Frequency refers to the percentage of the daily 90-minute observation period in which play was observed (West 1974, copyright © 1974, with permission of *Am. Zool*.

at one another. The prevalence of pouncing, stalking, and chasing in feline play may be evidence that it is practice for hunting. Play bouts may include 1 chase per minute. Play may occupy 9% of the kitten's total time and only 4%–9% of its energy expenditure, indicating that play may be important, but it is not calorically costly (Martin 1984).

Predatory Play. The mother plays an active role in the development of her kittens' predatory behavior. Not only do the mothers attack and eat prey in front of their kittens when the kittens are 4 to 8 weeks old but they also vocalize to attract the kittens' attention to it. After that, the mother defers to the kittens in that she rarely kills and almost never eats the prey. The kittens are more apt to interact with the prey if the mother has just been interacting with it than if a littermate has, which indicates that the mother has a greater influence (Caro 1980a,c). Kittens also learn other tasks better from watching their mother than from watching another cat (Chesler 1969).

　　The increase in predatory activity around 8 weeks is definite. At that time, most kittens will kill and eat mice, and more of their behavior is directed toward prey than toward playing with one another. Once the prey is dead and eaten, the kittens return to playing with one another, which indicates that the motivation to play is still present but is overridden by the motivation to hunt. Social play and predatory play are not correlated and probably are controlled by different systems (Caro 1980b). By 2 months of age those kittens that will be frightened rather than aggressive toward prey and other cats can be identified; the same kittens are reluctant to explore and to relax with people in a new environment (Adamec et al. 1983). This correlation is unfortunate because some people prefer that their cats do not hunt and many wish their cats to be less aggressive to other cats, but almost all owners want their cats to be friendly, even in a novel environment.

Sexual Play. Elements of sexual behavior are not seen in kitten play, but one sex difference exists in feline play. Males show more object contact than females; females with male littermates play with objects more than do females with no male littermates (Barrett and Bateson 1978). Play may be more important for intraspecies socialization in cats than it is in other species because cats are solitary for much of their adult lives.

Solitary play in kittens also begins to decline at 4 months, but the decline is much more gradual (West 1974). Kittens will chase small rolling objects or even a moving string. They particularly like to bat at suspended objects, such as window shade pulls or tassels. Many of the pounces and face-offs of social play may be performed by solitary kittens with "imaginary" playmates, a mirror, or their own shadows. Solitary play persists in many adult cats. Playfulness is a factor for which breeders should select because it enhances the pleasure a cat gives to its owner and to itself. Social play may also occur between species. Cats will often play with dogs with which they are familiar. Interspecies play consists mostly of chases by the dog and pounces by the cat.

Several factors may contribute to the decline of play in kittens. Subadult cats begin to sleep more during the day. Older cats tend to spend more time sitting quietly, but alertly. Male kittens show sexual activity by 4.5 months and attempt to mount and bite the scruff of females, who will reject these attempts until they reach sexual maturity a few months later. Young feral cats may also devote more time to finding their own prey. When canine and feline play are compared, dogs are found to chase, especially in a group, mouth, wrestle, shake, and indulge in solitary play more than cats. Cats stalk and ambush more frequently (Aldis 1975). See Figure 6.3 for the feline development chart.

Clinical Problems. Two common clinical behavior problems in cats are aggression between two or more cats in a household and rejection of the tom by the estrous queen. Both may be related to failure to socialize adequately to other cats as kittens. Kittens usually are removed from the mother at 6 weeks, long before the peak of playful interactions at 11 weeks. Cats that have remained with other kittens longer than 6 weeks may be more tolerant of other cats, including courting toms, as adults.

Playful behavior itself can be a behavior problem, particularly if it occurs in the middle of the night (Beaver 1980). It is most apt to occur when the kitten has been alone, and probably asleep, most of the day and has not had much opportunity to play. Punishment may inhibit the kitten's play, but it is more likely to simply move out of range and to continue racing about and knocking over objects. A scheduled play period in late evening is the best treatment. Cat toys also help. Two kittens play with one another but may become incompatible as adults.

The problem of playful aggression has been dealt with in Chapter 2.

HORSES

The Foal's First Day

First Hour. The perinatal behavior of foals has been described by Waring (1983), who studied American saddlebreds and by Rossdale (1967), who studied Thoroughbreds and ponies. The foal can move its head and legs immediately after birth. The suckling reflex appears within the first few minutes. The foal will suck on anything put into its mouth. Righting itself to sternal recumbency and the first attempts to stand occur within the first 15 minutes, but the foal will fail in a dozen attempts, so it may be an hour before the foal stands. Pony foals can stand at a younger age than those of the long-legged breeds. The foal begins to use all its senses within the first hour. It experiences tactile stimulation from its mother's licking and will begin to respond and orient to visual and auditory stimuli within the first hour. It will begin to communicate within the first hour by nickers to its mother and by snapping at any fearful object. The foal can walk soon after it can stand, although it will not be well coordinated for another few hours. As soon as the foal can walk it begins to search for the udder. It may attempt to suckle from the walls of the stall or from inappropriate parts of its mother as well as sucking when no oral contact has been made (vacuum suckling). The feature that the foal innately seeks is an underline, so it will attempt to nurse from the axilla as readily as the inguinal region. Defecation occurs within the first hour also.

The Rest of the First Day. Successful suckling is the major event of the foal's 2d hour of life. Although pony foals suckle within the 2d half-hour of life, a further 30 minutes is necessary before American saddlebred and Thoroughbred foals are able to suckle. By the 2d hour, the foal has also begun to follow its dam or any other large moving object. Lying down is another difficult task for the foal to master but is usually accomplished in the 2d or 3d hour. The foal will then sleep; a few foals will sleep standing up if they have not been able to lie down but will fall down if they go into REM sleep. By the 3d hour, the foal can also groom itself and gallop. Within the first day the foal can play, urinate, flehmen, and graze as well as communicate, suckle, and locomote; in other words, it is already a well-coordinated functional horse. See Fig. 6.5 for behavioral development of the foal.

The First Year. The ontogeny of the foal has been extensively studied in Welsh ponies (Crowell-Davis et al., 1985a; Crowell-Davis et al., 1985b;

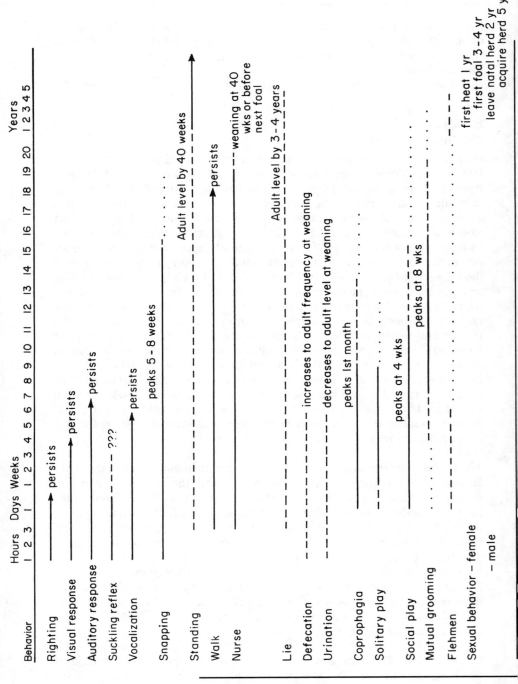

Fig. 6.5. Behavioral development of the horse. References: Duncan (1980); Waring (1982); Crowell-Davis and Houpt (1985a,b); Crowell-Davis et al. (1985a,b); Berger (1986); Crowell-Davis et al. (1987).

Crowell-Davis 1986), New Forest ponies (Tyler 1972), and Camargue ponies (Boy and Duncan 1979). There is little difference in the time budgets and rates of development of these three types of ponies even though the environments vary considerably in the amount of high-quality forage available. None of the ponies were stabled at night or managed to any great extent. Thoroughbreds and other horse breeds are more apt to be extensively managed; therefore, differences in time budgets are more apt to reflect artificial feeding and stall restraint rather than breed differences in ontogeny.

The Mare-Foal Bond. The distance between a mare and her foal is proportional to the age of the foal; that is, young foals are closer to their dams than older foals. The foal is responsible for this proximity in most circumstances; it follows the mother. This response changes when the foal lies down. Then the mare remains close to the foal, either stand-resting or grazing in circles around the foal (Crowell-Davis 1986). This behavior wanes as the foal matures, but one can almost guess the age of the foal by how close the mare remains while the foal sleeps. The mare follows the foal when the young foal ventures more than 10 meters away.

Grazing. Foals at first must spread and flex their legs to graze, especially if the grass is short; later their necks lengthen in relation to their legs, and they can graze more comfortably. Foals gradually increase the length of time that they spend grazing from 4 to 16 minutes per hour during the period from birth to 4 months. Thereafter the increase is more rapid, reaching adult levels (60%–70% of the time) at natural weaning (40 weeks).

   The development of feeding behavior is interesting because social facilitation plays such an important part. Foals graze only when their mothers are grazing (Crowell-Davis et al., 1985b) (See Fig. 6.6). This response illustrates the importance of providing creep feed for a foal in a location from which the foal can watch its mother eat. Drinking may not occur with foals on lush pasture. They obtain all their water needs from their dams' milk and moist grass and although they follow their dams to water, they do not drink. In arid areas foals do drink (Boyd 1980).

   Although adult horses will normally avoid feces (Odberg and Francis-Smith 1977), coprophagy is a normal ingestive behavior of foals, but its function is unknown. Consequences may be both negative and positive: the foal may ingest ova of parasites, but it may also be ingesting bacteria and protozoa that inoculate its gastrointestinal tract with the proper flora. The dam's feces are consumed in preference to those of another horse (Francis-Smith and Wood-Gush 1977; Crowell-Davis and Houpt 1985a; Crowell-Davis and Caudle, 1989).

Sleep. Foals rest either standing or lying. They spend a great deal more time lying than adult horses. The percentage of time spent in lateral recumbency

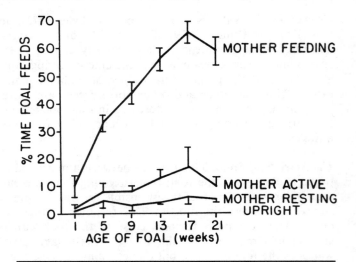

Fig. 6.6. The mean percentage of time the foals spent feeding when their mothers were feeding, active, or resting upright (Crowell-Davis et al. 1985a, copyright © 1985, with permission of *J. of Anim. Sci.*).

decreases with the age of the foal from 15% (first month) to 2% after weaning (Boy and Duncan 1979). Resting in sternal recumbency does not change very much throughout the foal's first 6 months (averaging about 15% of its day) and is still higher than adult levels in 2 and 3 year olds. There has been no electrophysiological study of foal sleep, but presumably they spend much of the time in lateral recumbency in REM sleep as do other young animals.

Standing and resting also occurs in foals. The foal usually stands beside the mother, often facing in the opposite direction to take advantage of her tail swishing to ward off flies.

Play. As the foal matures, it spends less time resting; it suckles less and grazes more. In between these activities foals play. For the first 2 weeks, play is solitary. Foals gallop away from and toward their mothers, which may be a form of exploration or even thrill-seeking behavior (Aldis 1975). Play in foals is one of the best examples of play as exercise. Seventy percent of locomotion in foals is in a play context (Fagen and George 1977). Foals at first play with their mothers by nibbling at their legs and mane. Later this will become true allogrooming. Social play with other foals gradually increases with age and solitary play declines, so by 8 weeks, solitary play is rarely seen (this behavior, of course, would not be true of foals that do not have companions) (see Fig. 6.7). In lone foals, solitary play persists, and social play may include dogs and humans (Schafer 1975). Foals may also

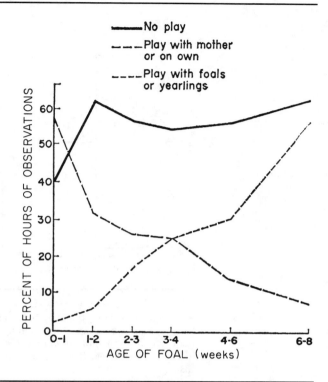

Fig. 6.7. Changes with age in foal's choice of play partners. As play with the mother decreases, play with other foals increases (Tyler 1972, copyright © 1972, with permission of Academic Press).

play with inanimate objects, such as twigs, tossing them into the air.

Although foals still spend more than half their time within 5 meters of their mothers, they are most apt to leave her to play (Crowell-Davis 1986). Sex differences in play are definite: colts mount and fight; fillies chase and mutually groom one another. Play in foals often centers about the head. Nipping of the head and mane, including gripping of the crest, accounts for the greatest number of play sequences. Rearing up and mounting is frequently seen, especially by colts. Properly oriented mounting is seen even in very young colts. Chases are a common play sequence (Schoen et al. 1976). Side-by-side nipping can progress to circle fighting in which each foal attempts to bite the tail and legs of another. Figure 6.8 illustrates the types of play seen in pony foals on pasture.

When colts do mutually groom, they tend to groom fillies rather than other colts (Crowell-Davis et al. 1986). Grooming the mare is part of the courtship behavior of the stallion. Certainly these sex differences in play may prepare the animals for their adult roles.

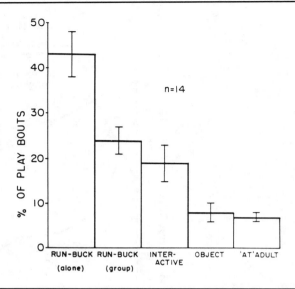

Fig. 6.8. Relative frequency of various types of play. The mean percentage of total play bouts by 14 foals in which the type of play was running and bucking alone, running and bucking as a group, interactive, manipulation of an object, or play at an adult. Standard error of the mean is shown as a vertical bar (Crowell-Davis 1983).

Flehmen and Snapping. Flehmen behavior is also much more frequent in colts than in fillies. It is probably investigatory behavior. Flehmen peaks during the colt's first month, possibly because his dam will be in estrous then and/or because he is still somewhat precocially masculinized due to his prenatal hormonal environment (Crowell-Davis and Houpt, 1985b).

The facial expression of snapping (also known as tooth clapping or champing) occurs almost exclusively in foals and subadult horses. This expression persists in zebra and donkeys as yawing, the mouth movements associated with estrus. In fact, the facial expression of snapping and that of yawing occur in the same circumstances, that is, an approach-avoidance situation. A colt is most apt to snap when approaching a stallion to whom he is apparently attracted but finds frightening (Fig. 6.9A). The estrous donkey and zebra mares, and rarely, a submissive, estrous mare (Woods and Houpt 1986) may be attracted to the stallion but may also be frightened. Snapping in foals is likely to occur when a stallion is courting the mare. The rate of snapping falls rapidly with age from once every 3 hours during the 1st month to once every 20 hours during the 6th month. The peak of snapping is during the foal's 2d month of life (Crowell-Davis et al. 1985a) (Fig. 6.9B).

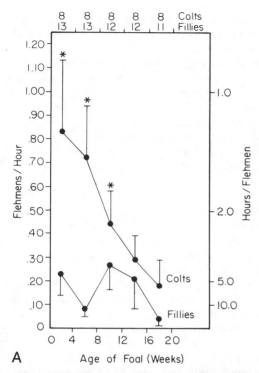

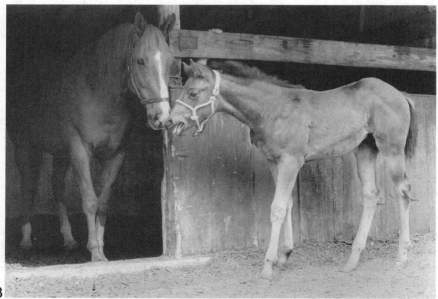

Fig. 6.9. (*A*) Rate of flehmen by foals of various ages. Colts exhibit flehmen more than fillies only during the first few weeks. Asterisks indicate significant difference (Crowell-Davis and Houpt 1985b, copyright © 1985, with permission of Academic Press). (*B*) Submissive snapping of the immature horse. Foal on the right approaches its dam while snapping (Wolski et al. 1980, copyright © 1980, with permission of Elsevier Science Publishers).

The Juvenile Period. Play and activity, in general, decrease with age; 2 and 3 year olds are more active than adults. Colts nearly always leave their natal herd and so do most fillies. The colts join other males, sometimes their older brothers, to form bachelor herds (Keiper 1985). A great deal of the young bachelor's time is spent play fighting. Those colts that remain in their natal herds do not have this experience and appear to be slower to mature. The peak of colt play occurs at 3 years. At 5 years, he will begin to show adult male behaviors, such as marking of urine and feces, true aggression toward other stallions, and driving of mares (Hoffman 1985). At 5 he is able to take over his own herd either by defeating an infirm harem stallion, replacing a dead one, or competing successfully against other bachelors for a filly that has left her natal herd.

Weaning. At least five different methods of artificial weaning are used: (1) removal of the foal from the mare and confinement of the foal by itself; (2) removal of the foal from the mare and confinement of the foal with another foal or foals; (3) interval weaning, in which the mare alone is removed from the pasture while the foal remains with the other, younger foals and their dams. The other mares will be removed gradually in order of their foals' ages. (4) separation of mares and foals into adjacent corrals for one week and subsequent removal of the mares; (5) Feeding mares and foals separately and gradually increasing the duration of separation. This method is particularly valuable for the owner of a single mare-foal pair. Foals appear to be slightly less stressed by weaning when they are confined with another foal, but problems can arise if one foal keeps another from feeding or is too aggressive (Houpt and Hintz 1982–83). In addition, separation of the two foals may also be stressful. Weaning by the fourth method, in which the foals can see and hear their mothers, but cannot make direct contact (or suckle) appears to be less stressful than methods in which the mothers cannot be seen, probably because weaning is more gradual (McCall et al. 1985). Weaning from the mother as a food source occurs before weaning from the mother as a social companion.

PIGS. Pigs are intermediate in their development at birth. They can walk, albeit unsteadily, within a few minutes of birth; they can see and hear. However, their brain development is not complete at birth, and some homeostatic mechanisms, such as temperature regulation, are not yet mature.

The neurological development of the pig has not been extensively studied despite the fact that such a study would have considerable clinical application. Piglet mortality is very high—approaching 20%. Some of this mortality is because of infectious disease, but a considerable number of deaths are "accidental." The piglets may wander out of their pen and drown in a gutter or simply become chilled and die from exposure, or they may wander in with older pigs that will maul them to death. Other piglets are crushed by the sow even though she is in a farrowing crate. Normal piglets do not wander; they stay close to their littermates, the heat source, and the sow's udder. Piglets, especially those among the last delivered, may suffer various degrees of brain damage as a result of hypoxia during birth. If these brain-damaged pigs could be identified and hand reared or otherwise given extra protection, piglet mortality would fall.

Sleep. Piglets sleep for 16 minutes of every hour. Like other young animals, they spend a lot of time in REM, when they assume a crouching position in sternal recumbency rather than lying in lateral recumbency as they do in non-REM sleep. The number of REM bouts decreases with age, but each bout remains of similar length (Kuipers and Whatson 1979).

Teat Order. Nursing behavior has been described in Chapter 5. The teat order is formed on the first day. The peak of aggression occurs 1 hour after birth. By day 6 only 10% of the piglets change teats (Hemsworth et al. 1976). To reduce injuries to the sow's udder or to the piglets that may occur during the formation of the teat order, it is common to clip the incisors and canines of day-old pigs (Fraser 1975c).

Initially, the piglets favor teats situated near the angles formed by the sow's body and legs. At first, four or more teats will be suckled, but progressively fewer will be suckled until the order is stable at 1 week and each piglet suckles only one or two teats (Hemsworth et al. 1976; Rosillon-Warner and Paquay 1984–85).

In a seminaturalistic setting, piglets form a teat order and may even force piglets from another litter off their own mother's teats (Newberry and Wood-Gush 1985). Larger litters have a less stable teat order.

The number and duration of suckling bouts gradually decrease as the piglets mature, but natural weaning is rarely seen under modern farm management.

Weaning. Weaning begins at 5 weeks when the sow begins to aggress against the piglets, but the piglets continue to suckle for 80 days in a natural setting. Under modern management techniques, piglets are weaned at 5 to 6 weeks or even younger. By 2 weeks of age piglets can usually be induced to eat solid food, but solid food intake will not become substantial for several weeks unless the sow's milk supply fails. Piglets are particularly apt to eat solid food if the food is sweet because suckling piglets possess a well-developed sweet taste preference (Houpt and Houpt 1976).

Few researchers have studied the effect of isolation on piglet behavior, although many piglets are raised under these conditions (Braude et al. 1971) in attempts to make artificial rearing of piglets commercially feasible or in specific pathogen-free (SPF) pig production (Dunne and Leman 1975). To prevent piglets from sucking on one another and to prevent spread of gastroenteric diseases that plague artificially raised piglets, the piglets are housed separately. Baldwin (1969) noted that pigs reared without the sow defecate in the nesting area, whereas normally raised pigs do not. Piglets raised in germ-free isolators give distress calls almost continuously during handling and feeding; conventionally raised pigs give distress calls only when hurt (Noyes 1976).

Despite the apparent low level of maternal activity in sows, piglets separated from their mother for even a short time (a few hours) exhibit considerable distress. They vocalize with either squeals or closed-mouth grunts (Fraser 1974b) up to 21 times per minute. The vocalizations increase with the length of isolation. The vocalization changes to a higher-frequency quacking when the piglets can hear their mother's voice, which they can discriminate from that of another sow (Shillito Walser 1986). If the piglets are in a strange pen, they will make persistent efforts to escape and often urinate. The vocalizations are reduced if the piglets are isolated as a litter rather than individually. The effect of the presence of littermates is additive with that of the sow so that a litter placed in a strange pen with their mother give only closed-mouth grunts and a few squeals. The olfactory cues are not sufficient to prevent vocalizations because the presence of the sow's bedding has no attenuating effect on vocalizations (Fraser 1974b). If the separated litter is closely confined and provided with a heat lamp, they are much quieter, and weight losses, especially due to urination, are reduced. Similarly, cuddling a piglet will reduce the number and volume of its squeals.

Early weaning (at 3 to 4 weeks) of piglets is often practiced to decrease the interlitter time (Pond and Maner 1974). Early weaned pigs massage and nibble on one another, yet spend less time rooting or nibbling on other objects. If placed in cages, early weaned pigs dog-sit (on their haunches) seven times more frequently than piglets in straw-bedded pens (Van Putten and Dammers 1976), which indicates that flooring type, as well as age at weaning, influences behavior. Aggression is high when 3-week-old piglets are first weaned but decreases with time. The piglets spend approximately 70% of the daylight hours lying down, 13% exploring, and 9% feeding (Wood-Gush and Csermely 1981).

An investigation into the effect of early handling revealed that piglets handled daily were not larger than nonhandled littermates but were more aggressive when penned with strange pigs (Schoen et al. 1974). The effects of human handling depend on the quality of the handling and can have economic importance. For example, Hemsworth and his colleagues (1987) found that pigs treated pleasantly (stroked) grow more rapidly and had better feed conversion than those treated unpleasantly (pigs were shocked if they approached the handler).

Play. When sows leave the maternal nest and rejoin the herd, the piglets are gradually integrated into the social life of the older animals (Petersen et al. 1989). Play in piglets develops in the 2d week of life and wanes by 5 weeks

(Fig. 6.10). It should be noted that failure to play is of diagnostic value in determining the seriousness of neonatal pig disease.

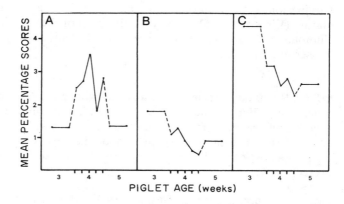

Fig. 6.10. The change with age in play fighting and solitary play in piglets. The graphs give the mean percentage of minutes in which early weaned piglets were scored as (A) biting littermates; (B) scratching their bodies; and (C) scampering during the first 3 weeks of life, expressed as a percentage of the number of minutes in which the animals were scored as active. Piglets were weaned at the end of the 3d week. The results are shown separately for the first 5 days after weaning.

Male piglets play more frequently than females, and play occurs more frequently between same-sex pairs than mixed-sex pairs. Play in piglets is characterized by play fights. These fights are usually head-to-head confrontations in which each piglet chews and roots at the other's shoulder and neck. In older pigs (3 to 4 weeks), chases and gamboling can be observed. At about the same age the typical porcine startle reaction (a woof and freezing behavior) can first be elicited.

Exploratory behavior is very pronounced in piglets and consists of rooting and mouthing anything that is new in the environment. In fact, oral manipulation of the environment is so much a part of swine behavior, persisting far longer than social play, and the sensory areas of the cortex that represent the snout and lips are so large in the pig (Adrian 1943), that one might profitably study the behavior and physiological effect of raising pigs in tactilely nonstimulating environments. Caged pigs can learn to play even in their very limited environments. They can learn to throw their water bowls from the racks with an apparently satisfying crash by using their prehensile lips to manipulate the springs that to hold the bowls in place.

This behavior can be used to advantage to reduce tail biting (see Chap. 2) by providing chains or rubber hoses for the pigs to chew.

Elimination. Pigs begin to eliminate only at the edges of their pens as they mature (Whatson 1985). This is the first sign of the pig's innate tendency to eliminate away from their sleeping area. Crowding or caging may result in indiscriminate elimination.

SHEEP. By 30 days of age, young lambs will spend 60% of their time with other lambs, but for the first few weeks they stay quite close to their dams. This behavior contrasts with the behavior of the kid, which is much more apt to stray. The lamb may depend on its proximity to its mother for protection, whereas the kid uses its own behavioral pattern, freezing, to protect itself.

Play. By 1 month of age, play is well developed in lambs (Morgan and Arnold 1974). Play in sheep begins with investigation of one another when the lambs are only a few days old. Play consists of intentional butting, vertical leaps, rearing up on the hindquarters, which may be a play signal, and twisting the forequarters and kicking. Lambs may center play around rocks or mounds so one lamb can butt others from above. The lambs push one another, lay their heads on one another, and mount each other. Male lambs mount much more than females and are the only ones to "nudge," raising a foreleg under the belly of another lamb while standing close behind it. Lambs form groups by a few weeks of age and rest, graze, and play together. Disportive or solitary play also occurs during which lambs gallop and leap. Adults may join the lambs in play. Lambs are especially playful in the evening, but by 4 months of age, play begins to wane (Grubb 1974b).

Weaning. Weaning does not exactly parallel decline in milk production. Milk production falls gradually, but suckling stops abruptly when milk production reaches a threshold (Arnold et al. 1979). In free-ranging Soay sheep, the ewe-lamb bond ceased just before the estrous period. The ewe lambs continued to follow their dams, but the ram lambs did not. None of the lambs slept touching their dams as they had done previously. The young sheep associated with their peer groups, the ewes remaining in the dams' home range groups and the rams wandering off to join a ram group (Grubb 1974b).

Social Relationships. The social relationships of sheep are formed in the first few weeks and appear to persist for life in the undisturbed flock. The ewe is followed by her lamb, and the lamb is submissive to her. Even as adults, daughters will follow the mother, and their lambs will be close at the heels of their respective dams (Scott 1945; Scott 1958b).

Sexual Behavior. Sexual behavior develops gradually in ram lambs. Despite the sexual elements of play, 6-month-old lambs rarely investigate a ewe in estrus. At 10 months they investigate ewes in estrus, and at 13 months will mount; but only at 17 months is the complete mating sequence, including copulation, seen in all rams (Winfield and Makin 1978).

Sleep and Activity Patterns. Sleep can occupy up to 40% of the lamb's day, but only 15% of the adult's (Ruckebusch et al. 1977). Grooming behavior, scratching with the hooves or teeth, occupies as much as 9% of the lamb's time, but far less of the adult's (Grubb 1974b). Figure 6.11 illustrates the changes in the activity patterns of lambs as they mature.

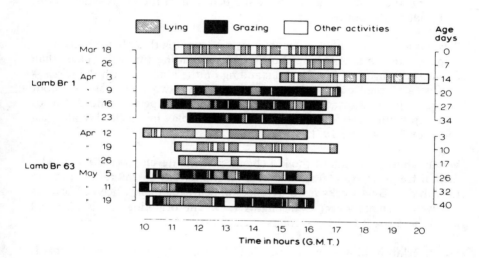

Fig. 6.11. Activity patterns of lambs. Note that grazing time increases and lying time decreases with age (Grubb 1974b, with permission of Athlone Press).

GOATS. Kids appear to be much livelier than lambs. Even in the confined environment of a pen they will play a game that can only be described as "king of the mountain," as each kid leaps onto the highest available horizontal surface, which may be an overturned bucket or even another animal.

Lickliter (1987) has described the changes in activity patterns of goat kids. Lying decreases markedly around 4 weeks from 60% to 70% of the time to 30% for the next few months. Ruminating begins at about 4 weeks as they begin to graze. The change to a herbivorous ruminant accounts for the biggest change in behavior during development. Presumably this is true of all ruminants and contrasts with the more gradual changes in the foal. Standing increases somewhat at the time that lying decreases. By 15 weeks grazing is the predominant activity, and behavior of the kid is synchronous with that of its mother.

Play. Other types of play are short bursts of running (less than 15 seconds), leaping into the air, kicking, butting and mounting (more in males than females), mouthing objects, and standing on the hindlegs with the forelegs against a vertical object. Play occurs most at dawn and dusk in 1-hour periods separated by a period of rest. Restraint in a pen or other forms of play deprivation are followed by a rebound of play in which play may last for 3 hours (Chepko 1971).

Vocalizations. The call of the kid changes during development. The fundamental frequency falls from 600 kHz on the 1st day to 250 to 350 kHz on the 5th day. By 4 weeks sex differences appear; the male voice is 3 octaves below the female voice. Call lengths vary. The orienting call is 1 second long, the distress call 1.5 seconds (Lenhardt 1977).

Social Relationships. The pattern of older animal dominant over younger is true of a stable herd of goats but not if strange adult animals are mixed (Stewart and Scott 1947). The dominance that an adult doe shows over an alien kid is apparently never challenged by the kid even when it becomes adult.

CATTLE. The development of calves has not been studied in great detail. In particular, beef calves have not been studied, so most of the descriptions of play and ontogeny are based on Brahma or Masai cattle (*Bos indicus*) or on the primitive Maremma cattle (*Bos primigenius taurus*).

Calves, like other ruminants, must change in a few weeks from simple-stomached milk-drinking animals to grazing ruminants. Rumination increases to occupy 7 hours per day by 7 weeks of age in calves kept on pasture. Sleep time declines to 4 hours per day. Calves tend to sleep in groups or "kindergartens." This attraction of calves for one another results in the calves spending more than half of their time 15 meters or more away from the mothers. The mothers will leave their usual sleeping areas to rest near their calves (Reinhardt et al. 1978a). When calves that had been suckling were separated from their mothers, they spent more time with other calves. This effect is due to social rather than nutritional needs, because if the dams remained, but nursing was prevented by cloth placed around the udder, the calves did not shift their social contact to their peers but instead maintained close contact with their mothers (Veissier and LeNeindre 1989).

Play. Play in calves has been studied by Brownlee (1954). Calves, as well as older cows, have a special play vocalization, the baa-ock. Calves play in a variety of ways, often trotting or galloping with the tail elevated. They often buck, kicking up and to the side with both hindlegs, but these kicks are not aimed at anything in particular. Directed kicking, which is playful rather than aggressive, may be aimed at a stationary or moving target. Calves also play by making noise with inanimate objects, such as buckets or latches. They may be increasing their auditory stimulation level just as children increase their vestibular stimulation level on slides, swings, and merry-go-rounds (Aldis 1975). Play in calves occurs more frequently in younger animals. At 7 weeks about 3 minutes per day is spent in play (Roy et al. 1955). Solitary play is more commonly seen in young calves. Social play, especially frontal butting and mounting, predominates in the older calf. Play tends to occur during grazing bouts (Vitale et al. 1986). Calves head butt with each other or with inanimate objects. This is the same action used by adult cows in dominance interactions. They may prance, paw the ground, or gore as adult bulls do, and even threaten human attendants. Soft snorting noises may also accompany play. Mounting behavior is also seen in calves. Mounting and pushing play decrease with age while butting increases.

Sex differences are apparent in bovine play. Male calves play more than females, the same pattern seen in most animal play. Exploration peaks later than play; yearlings are more apt to investigate a novel object than older or younger cattle (Murphey et al. 1981). Males mount, push, and exhibit the flehmen response more than females, but butting and social licking are seen equally in both sexes. Mounting and pushing by bull calves are usually directed toward other males, but the flehmen response is directed toward females (Reinhardt et al. 1978b).

As in other species, play can be used as a diagnostic criterion, for calves play more when well-fed and healthy than when malnourished or ill. They also play more often in fine weather than in foul. Play is stimulated in calves by any change in the environment. They play most when let loose from confinement, after gaining access to new terrain, or even when new bedding is placed in their stalls. A new pen mate, the arrival of a human attendant, or even the stimulation of scratching their backs may set off a play bout in calves.

Activity Patterns. Weaned calves raised in individual pens spend their time as follows: standing, 40%; ruminating, 28%; feeding, 22%; grooming, 5%; and drinking, 2%. Such calves quickly learn to anticipate feeding times and become restless at those times. Many calves are able to make contact despite being penned separately, by making tongue contact through the opening where they are fed (Kilgour et al. 1976). Calves raised in single pens are most likely to associate with the calves in adjacent pens when they are released on pasture. The singly raised calves are submissive to group-raised calves and rarely associate with them. There is no difference in weight gain between the groups (Broom and Leaver 1978). Dominance hierarchies are not stable even though artificially fed calves may compete fiercely for access to their feed (Canali et al. 1986). Calves raised in isolation for the first 10 weeks of their lives had higher cortisol values when stressed as yearlings than calves raised with the opportunity to interact with other calves (Creel and Albright 1988), which indicates the long-term effects of stress during development.

Conclusion. Researchers have undertaken very few behavioral studies of development in cattle and other ruminants. Despite the detailed knowledge about growth rates in these food-producing animals, next to nothing is known about the daily activity of young ruminants or how their relationships to their environment and their peers change with maturity. The change in suckling patterns with age is discussed in Chapter 5. Some general statements can be made. Young ruminants are born in an advanced state of development; they are true precocial animals. They can stand and walk within a few hours of birth and can apparently see and hear.

The drastic changes that are taking place in management of young ruminants have resulted in much higher incidences of infectious disease. The role of behavioral stress in the etiology of neonatal pneumonias and diarrheas remains unknown. The economic value of these animals dictates that more information is needed on their behavior in both naturalistic and highly artificial environments in order to best advise those who undertake lamb or veal operations.

7

Learning

TYPES OF LEARNING. Learning can be classified into two main types: classical conditioning and operant conditioning. Classical conditioning was first demonstrated in dogs by Pavlov. In classical conditioning, an unconditioned stimulus is paired with a conditioned stimulus until the latter causes the same response as the former. An unconditioned stimulus (UCS), such as the sight of meat, which produces a response such as salivation, is paired with a conditioned stimulus (CS), for example, the sound of a bell. The stimuli are paired repeatedly until the conditioned stimulus alone elicits the response; the dog salivates when the bell is rung.

Classical Conditioning. Classical conditioning in farm animal behavior can be illustrated by many examples. Perhaps the best illustration is the release of oxytocin in response to the jangling of milking equipment; oxytocin contracts the myoepithelial cells of the mammary gland causing milk letdown. Cows normally release oxytocin in response to suckling on the teats by the calf or squeezing of the teats by the negative pressure of the milking machine or the hands of the milker. When a cow has been milked in the same environment a number of times, the sounds of the approaching machinery and milk cans will have been paired with the milking process, and those noises alone will elicit oxytocin release (Ely and Petersen 1941). This type of classical or Pavlovian conditioning, therefore, shortens the time required to milk a herd of cows. A pet animal's fear reaction to the smell of a veterinary hospital or the sight of a person in a white coat is often a classically conditioned response. The dog or cat responds to a painful stimulus (UCS) with the fear or escape response. The hospital and the professional staff are the conditioned stimulus. It does not take many pairings of these stimuli to produce a fear response whenever the animal encounters the conditioned stimuli. For those who become veterinarians because they genuinely wish to help animals, it is somewhat disheartening to discover that many of their patients turn tail and hide whenever they approach, sometimes even outside the veterinary clinic. The behaviorally oriented clinician will make every effort to reduce the painful and frightening incidents, especially in a young animal's first visit, so that a conditioned fear response will not develop and hinder future patient-veterinarian (not to mention client-veterinarian) relationships. Figure 7.1 shows classical conditioning of the goat.

A practical application of shock avoidance learning is now commercially available. Both dogs and goats learn that they will be shocked if they cross a buried wire. An auditory signal on the transmitter collar they wear warns the animal that a shock is forthcoming if they approach the buried wire (Fay et al. 1989).

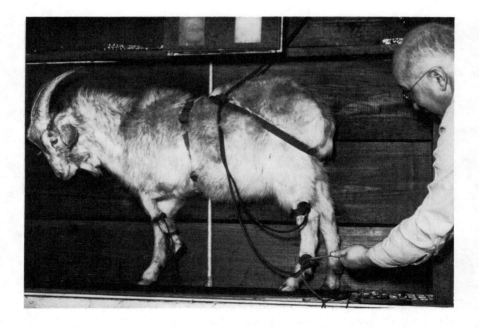

Fig. 7.1. Classical conditioning. A goat is conditioned to lift its leg when a metronome ticks by pairing the sound of the metronome with shock to the foreleg. The late Professor Liddell, who compared the rate at which various species of farm animals learned a classical conditioning, is shown.

Operant Conditioning. The second type of learning, operant conditioning, might be called trial-and-error learning. It became obvious to those interested in the theoretical basis of learning that classical conditioning, important as it is, cannot account for all learned behavior. It is highly unlikely that circus horses, for example, could be taught to walk on their hindlegs with their forelegs on another horse's rump by pairing the conditioned stimulus (the trainer's voice command) with some unconditioned stimulus. Instead, the animal operates on its environment to obtain a reward or reinforcement.

Operant conditioning was first demonstrated by Thorndike (1911) using cats as experimental animals. Hungry cats were placed in slatted boxes. Food was available outside the box within sight and smell of the cats. At first, the cats struggled vigorously to reach the food. Eventually, some of them, by chance, pulled a latch string that opened the box door. The cats were then free to consume their rewards. Each time a cat was replaced in the box it took a shorter time to escape, making fewer and fewer extraneous motions, until eventually it pulled the latch string immediately upon being placed in the box.

Operant conditioning is also called instrumental learning because the behavior is the instrument by which the reinforcement is obtained. A laboratory example of instrumental learning is a rat in a "Skinner box," named for the psychologist who popularized the technique (Skinner 1938), in which an animal presses a bar to obtain food, water, or electrical stimulation of its brain.

Shaping. In teaching an animal an operant task, one can wait until the animal performs the desired activity and then reward it as Thorndike (1911) did, or one can speed up the process by "shaping" the behavior. If, when teaching a dog to heel, the trainer first rewards the dog for staying within a yard of his side, then a foot, and finally only when the animal walks quietly exactly beside the trainer, he is shaping the dog's behavior.

Circus horses, for example, are usually shaped by being rewarded for a simple task like trotting around the ring, then for trotting close to another horse. Later the horse is rewarded for rearing on command, perhaps with special urging with a whip to get the first rearing motions. At last, the horse can be induced to rear and put its legs on the horse in front of it.

Animals that are trained to perform complicated and relatively unnatural tricks are usually reinforced with food rewards and reinforced for each correct response at first. The same techniques are used to teach chickens to play baseball or to teach pigs to put giant coins in a bank (Breland and Breland 1966). Figure 7.2 illustrates one of the many uses of operant conditioning in the laboratory. The dog is trained to press a pedal for a food reward.

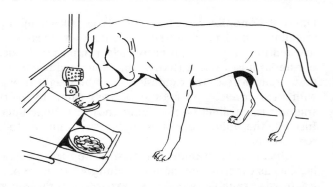

Fig. 7.2. Operant conditioning. The dog pushes a pedal (operates on the environment) to obtain a dish of food (Houpt and Hintz 1978, copyright © 1978, with permission of Veterinary Practice Publishing).

Reinforcement Schedules. When operant conditioning is employed, a variety of schedules of reinforcement can be used. The animal can be rewarded, for example, after every response, after every ten, or after every twenty responses. These schedules are called *fixed ratios,* or FR1, FR10, and FR20, respectively. This technical detail is important because the higher the fixed ratio, the faster the animal will respond; and, even more important, the longer it will take for the response to be extinguished. The animal will go on responding for some time after it is no longer rewarded. Owners inadvertently put their animals on high FR schedules in many situations. For instance, a dog barks while its owners are eating. They may have given it food once or twice when it barked until they became annoyed at the behavior. The dog continues to bark while the owners try to ignore it. Finally, they relent and give it some food; they have just increased the ratio. Dog and owner may adjust to this new level, but often the adjustment is temporary and the level of response (barking) needed for reward increases again. Dogs have been trained in the laboratory to bark 33 times for each small food reward and cats to meow 15 times (Salzinger and Waller 1962; Molliver 1963). The problem dog at the dinner table may continue to bark hundreds of times even though its owners do not give it any more food. Another type of reinforcement is called *fixed interval* (FI). In this case, the animal is rewarded for a response that occurs after a certain period of time has elapsed since the last reward. Animals do have a good time sense, and their rate of responding will slow down after a reward and then increase sharply just before the end of the time interval. If animals can learn to respond in this manner, it is not surprising that they learn to expect their owners home at a given hour. Another variant of reinforcement ratios is the progressive ratio in which, for example, the animal must respond once for the first reward, twice for the second, four times for the third, and so on. The number of responses per reward increases progressively. This technique is used to measure the strength of preferences for food.

A more long-lasting response, that is, more difficult to extinguish, follows a variable-interval reward (VI) schedule. Here the reward follows the first response after 1 minute has elapsed, then after 5 minutes, then after 3 minutes, and so on. The highest response rate follows a variable-ratio reward (VR). The owners of the barking dog may find themselves rewarding their dog on this type of schedule if they inconsistently reward the barking, depending, perhaps, on their own mood on particular evening or on which family member gives in to the pet. A high rate or volume of barking may contribute to the owners' giving in sooner, but the owners are probably not counting barks; they are merely "holding out" for as long as possible, a war of nerves the dog invariably wins.

Obviously, however, a variable-ratio reward schedule is to be recommended in routine animal training; it is suggested that owners supply verbal or food rewards sporadically during a training session rather than after every trick, or every few tricks (FR), or, say, every 5 minutes of a training session (FI).

One of the most difficult tasks for the animal trainer is to get the animal to understand the experimenter's instructions (Gleitman 1974). Dogs, for example, have remarkable olfactory acuity and can be taught to detect gas leaks, hidden narcotics, and a fatty acid at very low concentrations, but Becker et al. (1962) found it extremely difficult to get dogs to learn to turn right in response to an olfactory cue. It is also important to know what is rewarding for an animal. Dogs learn faster when the reward is simple contact with a passive person than when the reward is stroking or picking up (Stanley and Elliot 1962).

An animal will learn for both positive and negative reinforcement. Positive reinforcement is a reward, usually food but sometimes social interaction, for performing a response. Negative reinforcement is something aversive applied *until* the animal makes the response. One pulls on the horse's mouth until it stops; the rat is shocked until it moves to the other side of the cage. Many field dog trainers use negative reinforcement varying from a pinch to a shock collar in training. One of the most difficult concepts for owners to understand is the difference between negative reinforcement and punishment (Borchelt and Voith, 1985b). Punishment is something that occurs after an action as a consequence. The dog chews the slipper and is hit for it. Timing is very important. Punishment will not decrease the frequency of the behavior unless it occurs when the animal is misbehaving or within a second or two of the termination of the behavior. Most owners feel that they can punish a dog hours after it has chewed a slipper or eliminated in the house and are surprised when the dog does not learn.

Biofeedback. For many years it was assumed that operant conditioning could involve only voluntary actions, those mediated through skeletal rather than smooth muscle. They felt the somatic, but not the autonomic, system could be trained, but Miller (1969), using a curarized animal, demonstrated that such autonomic functions as heart rate, blood pressure, and pupil size could be conditioned. From this research has sprung the whole concept of biofeedback. Humans were encouraged to control their hypertension or gastric acidity by learning to turn off a light or buzzer that signaled too high a level. Although these techniques have not fulfilled their early promise, they may still have application to veterinary practice. Cats that had been taught to decrease the amplitude of the electrical activity of their brains were found to have seizure thresholds that were much higher than those of untrained cats. Dogs could learn to increase hippocampal theta waves to avoid shock (Black et al. 1970). Although time and financial considerations make such techniques impractical, dogs whose seizures cannot be well controlled with drugs might show a decrease in frequency of epileptic convulsions if brain wave amplitude were conditioned.

Testing Perception. Both operant and classical conditioning can be used to test perception in animals. In classical conditioning, a sound, for example, is paired repeatedly with a shock to the leg; eventually sound alone is presented. If the animal can hear the sound, it reacts as to the shock with an increase in heart rate or leg flexion. If the animal cannot hear the sound, it will not react; therefore, classical conditioning can be used to determine sensory thresholds of animals.

Operant conditioning also can be used to test perception. The procedure is longer, but positive rather than negative reinforcement can be used. For instance, the presence or absence of color vision in animals can be deduced anatomically by the ratio of rods to cones in the retina and biochemically by the presence of pigment, but only behavioral techniques will determine whether or not an animal can distinguish one hue from another. The animal is trained to press a panel for a food reward and then to press a panel when provided with one visual stimulus and not to press a panel when provided with another. The animal's ability to make shape or color discriminations can be tested in this manner (Sutherland 1961).

Habituation, Imitation, and Taste Aversion. Other types of learning appear that are difficult to classify, such as taste aversion, imprinting (see Chap. 5), habituation, and imitation.

Habituation. Habituation, considered the simplest type of learning, is the long-term, stimulus-specific waning of a response, or learning not to respond to stimuli that tend to be without significance in the life of the animal (Thorpe 1963). An example of habituation is a horse's response to traffic on a road beside its pasture. When first put in the field, it will react to traffic on the nearby road; later it will not. Other examples of habituation are found in pigs that soon ignore a sparkler over their feed trough (Dawson and Revens 1946) or dogs that are not repelled after a few exposures to a dog repellent.

Imitation. Animals can learn by imitation, by observing others. This form of learning has been most thoroughly studied in cats and horses and will be discussed in the section on learning in those species.

Taste Aversion. Taste aversion, or bait shyness, is the process by which an animal learns to avoid a food that it associates with illness, particularly gastrointestinal malaise. This form of learning has long been recognized by those attempting to rid farms of rats. When first used, a poison usually kills many rats; but after the first application, very few rats are killed. The animals that survive will no longer eat the bait. The same phenomenon occurs when rats are exposed to radiation and at the same time offered a novel food. They soon avoid the food that they associate with radiation sickness (Garcia et al. 1974).

Three unique characteristics of taste aversion differentiate it from classical and operant conditioning. First, it appears to be specific for taste and olfaction; other stimuli like visual or auditory cues will not be avoided. Rats will learn to avoid the taste of saccharin but not a blue solution. On the other hand, birds readily learn to avoid novel colored foods; avian species, which possess few taste buds, apparently depend more on sight than on taste for food identification. Second, the illness must originate internally, it must be a general or gastrointestinal malaise. External injury, like that from an electric shock to the feet, is not a sufficient stimulus. Third, the novel taste and the illness can be widely separated in time, and learning will still take place. This is in contrast to both operant and classical conditioning, in which the stimulus and the response must be close together in time for learning to take place.

Taste aversion has three uses, one experimental and two practical. Experimentally, taste aversion can be used to determine what substances an animal can taste or perceive. An animal will show an aversion to a substance at concentrations far below those at which it would show preference or aversion if the taste had not been paired with illness. Taste aversion has been used to teach coyotes to avoid lamb—a practical application. Repeated pairing of lamb with injections of lithium chloride (LiCl), which produces nausea and vomiting, resulted in a definite aversion to live or dead lambs by the coyote (Gustavson et al. 1974). Whether taste aversion techniques can be used on a large-scale or individual basis to reduce livestock predation by wild coyotes and stray or feral dogs remains to be demonstrated.

Although many species can form taste aversions in the laboratory, the rat is the only species in which there is good evidence that the ability contributes to "nutritional wisdom," for animals can learn that diet A, which is deficient, causes them to feel ill whereas diet B, a nondeficient one, makes them feel well. The animals have learned which diet is safe, and they must learn because they have no innate preference for a nutrient in which they are deficient (Rozin and Kalat 1971). The exception is salt hunger; nearly all species, herbivores in particular, show what is apparently an innate and very accurate preference for sodium salts when that mineral is deficient in their diets (Denton 1967) (see Chap. 8).

Learned Helplessness. A phenomenon called learned helplessness that has considerable application to practical animal training has been discovered in dogs and cats (Maier and Seligman 1976). Normal, naive dogs, when first placed in an active avoidance situation in which impending shock is signaled, at first escape the shock, once it has begun, and later avoid the shock by performing the necessary task, such as jumping over a barrier, during the signal before the shock begins. Dogs that previously have been exposed to unavoidable shock act quite differently. They not only fail to learn to avoid but also fail to escape; they simply sit and take the shock. These experimental findings indicate that the same form of aversive stimulus should not be used first as inescapable punishment and then later as negative reinforcement that the dog should learn to avoid. Improper use of the popular shock collars or lockets activated by buried fences may produce learned helplessness in dogs, and any form of inescapable punishment may inhibit later learning.

PHYSIOLOGICAL BASIS OF LEARNING

Anatomy of Learning. The search for the engram or the "learning center" has gone on for many years, and, although our knowledge of the process of learning and memory has increased greatly, the central nervous system (CNS) site most essential for learning remains uncertain; however, the hippocampus does appear to be very important in humans for consolidation of memory (formulation of long-term memory from short-term memory) (see Squire 1986). Loss of hippocampal tissue impairs feline learning (see Table 7.1). Those areas of the cerebral cortex that are directly involved in control of motor movements and those that receive sensory input from the body, the visual system, or the auditory system have been well mapped for various species. The remainder and the bulk of the cerebral cortex comprises associative areas and is involved in memory. Electrical stimulation of the temporal lobe of conscious humans elicits vivid recall of a specific event, yet extirpation of that area does not abolish that memory (Penfield 1958).

Memory is stored in the same portion of the brain that receives and analyses the information. Removal of areas of the cortex involved in processing visual information will result in deficits in visual memory. If the memory depends on many senses such as olfaction, vision, and touch, then the more cortex that is removed, the more severe the deficits in learning and retention will be (Lashley 1926). The older the memory trace, the more widely it may be distributed across the cortex. Although split-brain cats can learn, as will be discussed later, they learn slowly and forget quickly, which is a further indication of the importance of the cortex as a whole in the learning process (Sechzer 1970).

For some forms of learning the cortex is not necessary. The cerebellum appears to be involved in classical conditioning of blinking (Norman et al. 1977). Decorticate dogs can be classically conditioned (Culler and Mettler 1934). The major difference between the responses of intact and decorticate dogs is that the latter give very generalized escape reactions to a conditioned stimulus, whereas the former give discrete motor movements, such as leg flexion. It may even be possible to condition a response of the hindlegs of a dog whose spinal cord has been severed in the cervical region (Shurrager and Culler 1938).

Two aspects of learning may be affected by brain lesions: acquisition of learning and retention of a previously learned task. For instance, frontal lobectomy in a dog abolished retention of a previously learned task (housebreaking) but did not prevent the animal from reacquiring the learning (Allen et al. 1974). Prefrontal lesions have very subtle effects on learning in dogs. Dogs were trained to differentiate between two tones. In one task, the dogs were trained to press a bar for a food reward when one tone sounded, but no reward was given when the other tone was sounded. Medial prefrontal lesions profoundly affected the dogs' performance. They often barpressed when the nonreward tone was given, apparently because of lack of cortical inhibition. In another task, the dogs obtained a food reward if they pressed a bar in response to one tone and refrained from pressing the bar in response to another tone. Lesions of the dorsolateral prefrontal area severely affected performance, although medial lesions did not (Dabrowska 1971). Stimulation of the tegmentum or caudate nucleus of cats impairs learning (Thompson 1958), but this is due to effects on the cortex.

Not all lesions impair learning; lesions of the ventromedial hypothalamus actually improve learning of a conditioned avoidance response in cats (Colpaert and Callens 1974), which may be because of increased responsiveness rather than increased ability to learn. Electrical stimulation of the hippocampus, amygdala, or center median in cats and rats disrupts longterm, but not short-term, memory (Wyers and Deadwyler 1971), possibly because stimulation of these structures causes changes in electrical activity of the brain. Obviously, the anatomy of learning is not well known.

Tables 7.1 and 7.2 summarize present knowledge of the effects of brain lesions on different learning tasks in dogs and cats. The only conclusion that can be reached is that some areas of the brain are more important in a particular kind of learning than others, which probably indicates that different sensory modalities and different types of motor output are involved in different learning tasks. Care must always be taken to separate effects on

Table 7.1. Effect of brain lesions on learning of various tasks by cats

Site of Ablation	Task	Effect[a]	Reference
Cortex	Tactile habituation at spinal cord	↑	Uretsky and McClearly (1969)
	Conditioned salivation to light and odor	↑	
	Conditioned flexion to shock	↑	
Neocortex	Habituation of postrotatory nystagmus	↑	Hernandez-Peon and Brust-Carmona (1961)
Unilateral decortication	Conditioned motor response to auditory stimuli	↑	Bromiley (1948b)
All neocortex except sensorimotor (associative cortex)	Instrumental lever pressing for food reward	↑	Buser and Rougeul (1961)
Prefrontal cortex	Passive avoidance delayed response	↓	Wikmark and Warren (1972)
Prefrontal cortex	Delayed alteration	↑	Divac (1972)
Prefrontal cortex	Time discrimination	↑↓	Rosenkilde and Divac (1976)
Caudate nucleus		↓	
Prefrontal cortex	Instrumental lever pressing for food reward	↑↓	Rosenkilde and Divac (1976)
Neostriatal–prefrontal	Delayed alteration	↑	Oberg and Divac (1975)
Frontal cortex	Operant conditioning	↓↑	Olmstead et al. (1976)
	Maze learning		
	Spatial alteration		
	Black-white discrimination		
	Passive avoidance		
Orbitofrontal	Delayed response	↑	Thompson (1968)
Orbitofrontal	Visual discrimination	↑	Warren et al. (1962)
	Position		
	Size discrimination		
	Pattern discrimination		
Unilateral striate cortex and section of optic chiasm and corpus callosum	Visual discrimination (brightness and pattern)	↑	Sechzer (1970)
Ventral part of temporal region	Conditioned limb flexion	↑	Goldberg and Neff (1961)
	Tone pattern discrimination	↑↓	
	Simple tone discrimination		
Sensorimotor cortex	Instrumental lever pressing for food reward	↓↑	Buser and Rougeul (1961)
Visual area		↑	
Temporal occipital cortex		↑	
Somatic sensory areas I and II	Roughness discrimination	↓↑	Benjamin and Thompson (1959)
Media visual cortex	Spatial alteration	↓↑	Winer and Lubar (1976)
	Reversal of spatial lateration	↑	
	Active avoidance	↑	

Table 7.1. (*continued*)

Site of Ablation	Task	Effect[a]	Reference
Prestriate cortex	Visual discrimination	↑	Warren et al. (1962)
	Position	↑	
	Size discrimination	→↑	
	Pattern discrimination	↑	
Ectosylvan gyrus	Discrimination of cutaneous stimuli	↑	Glassman (1970)
Middle suprasylvian gyrus and anterior lateral gyri of precruciate cortex	Active avoidance	↑	Johnson and Thompson (1969)
Lateral and posterolateral gyrus (areas 17 and 18)	Visual discrimination	↑	Winans (1967)
Posterior temporal	Pattern discrimination	↑	Campbell (1978)
	Visual tracking	→↑	
	Orienting to cat silhouette	↑	
Basolateral amygdala	Pattern discrimination	→↑	Campbell (1978)
	Visual tracking	→↑	
	Orienting to cat silhouette	→↑	
Septum	Active avoidance	→↑	Hamilton (1969)
	Passive avoidance	↑	
	Position habit reversal	↑	
Caudate nuclei	Operant conditioning	↑↑	Olmstead et al. (1976)
	Maze learning	↑↑	
	Spatial alteration	↑	
	Black-white discrimination	↑	
	Passive avoidance	↑	
Posterior association cortex	Two-cue visual and auditory discrimination	↑↓	Moore and McCleary (1976)
Fornix	Object discrimination	↑	Cornwell et al. (1976)
Marginal and splenial gyri	Visual discrimination	↑	
Auditory cortex	Visual discrimination	↑	Zucker (1965)
Septal-limbic area	Visual discrimination	↑	
	Passive avoidance	↑	
	Active avoidance	↑	
	Punished extinction of active avoidance	↑	
Entorhinal cortex	Passive avoidance	↑	Entingh (1971)
	Active avoidance	↑	
	Punished extinction of active avoidance	↑	
	Position reversal	↑	

Table 7.1. (*continued*)

Site of Ablation	Task	Effect[a]	Reference
Total hippocampus	Active avoidance	→	Entingh (1971)
	Pattern discrimination	→	
	Passive avoidance	→	
Ventral hippocampus	Active avoidance	→	Andy et al. (1967)
	Pattern discrimination	→	
	Passive avoidance	→	
Dorsal hippocampus	Active avoidance	→	Andy et al. (1967)
	Pattern discrimination	→	
Hippocampal isolation	One-way avoidance	→	Uretsky and McCleary (1969)
Medial cortex	Passive avoidance	↑	Lubar (1964)
Septalo-limbic area	Active avoidance	↑	Lubar (1964)
	Passive avoidance	↑	
Cingulate gyrus	Active avoidance	↑	Lubar (1964)
Limbic and cingulate area	Passive avoidance	↓↑	McCleary (1961)
Cingulate gyrus	Active avoidance	↑→	McCleary (1961)
Subcallosal area	Passive avoidance	↑→	
	Active avoidance	↓↑	
Ventromedial hypothalamus	Passive avoidance	↓↑	Colpaert and Callens (1974)
Dorsolateral geniculate nucleus	Visual discrimination	↑	Winans (1967)
Mammilothalamic tractotomy	Avoidance visual discrimination	↑	Thomas et al. (1963)
Mesencephalic reticular formation	Conditioned salvation to light and sound	↑	Hernandez-Peon and Brust-Carmona (1961)
Brain stem transection formation	Conditioned response to auditory stimuli	↑	Doty et al. (1959)
Spinal cord (2d cervical)	Tactile habituation at spinal cord	↑	Hernandez-Peon and Brust-Carmona (1961)

[a]Key: ↓ = impaired ability; → = no effect; ↑ = improved ability.

Table 7.2. Effect of brain lesions on learning of various tasks by dogs

Site of Ablation	Task	Effect[a]	Reference
Cortex	Conditioned motor response to sound	↑	Culler and Mettler (1934)
	Conditioned response to light	↓	Girden et al. (1936)
Neocortex	Conditioned motor response to sound and light	↑	Bromiley (1948a)
Prefrontal	Auditory discrimination	↑	Konorski (1961)
	Delayed response	↑	Lichtenstein (1950a,b)
Medial prefrontal	Conditioned suppression of eating	↑	Dabrowska (1971)
	Differential symmetrical reinforcement	↑	Soltysik and Jaworska (1967)
Lateral prefrontal	Conditioned suppression	↑	Dabrowska (1971)
	Differential symmetrical reinforcement	↑	
	Differential asymmetrical reinforcement	↑	
	Conditioned suppression	↑ ↓	Soltysik and Jaworska (1967)
Premotor and caudate area	Delayed response	↓	Konorski (1961)
Sylvan gyrus	Auditory discrimination	↑	Konorski (1961)
Anterior part of 1st and 2d temporal gyri	Visual discrimination	↑ ↓	Konorski (1961)
Posterior part of 2d and 3d temporal gyri and gyrus fusiformes	Auditory discrimination	↑	
	Auditory discrimination		
Amygdala and hippocampus	Visual discrimination	↓	Fuller et al. (1957)
	Delayed response	↓	

[a]Key: ↓ = imparied ability; → = no effect. ↑ = improved ability.

243

learning from effects on performance, which may be due to motor malfunction and not to memory loss.

One clue to the anatomical changes that may occur at the microscopic level with learning can be obtained by experiments on animals that have been deprived of sensory stimuli. The occluded eye of a kitten will not have normal connections with the visual cortex even later when vision is permitted (Blakemore and Van Sluyters 1975). Mice reared in the dark have fewer spines on their pyramidal cells than normally reared mice, and conversely, rats reared in an enriched environment have a visual cortex that is thicker than that of conventionally reared laboratory rats (Horn et al. 1973). Even acquisition of a simple learning task can increase the number of presynaptic vesicles of the sensory neurons (Fillenz 1972).

Biochemistry of Learning. Learning appears to take place in two stages: the formation of a short-term memory, and consolidation or formation of a long-term memory trace. Although the time required for formation of long-term memory appears to vary with the species tested and the task learned, the stages are readily separable. Such procedures as electroconvulsive shock (Andry and Luttges 1972) or anesthesia (Chute and Wright 1973) can inhibit short-term memory formation, indicating that a short-term memory is an electrical event.

The biochemical basis of memory has been studied most profitably in a monosynaptic reflex of the invertebrate sea slug (*Aplysia*). Remote as that may seem to the problems of teaching a turn on the forehand to a horse or a drop on recall to a dog, the same physiological processes are probably involved. Short-term memory involves the acquisition of environmental information through chemical transmitter signals (such as serotonin) that bind to surface receptors. Short-term memory is the result of the following steps: (1) The receptors act through transducing proteins that (2) activate amplifier enzymes (such as adenylate cyclase) that (3) elevate the levels of intracellular messengers (such as cyclic adenomonophosphatase) that (4) activate protein kinases that, in turn, (5) modify target proteins (such as those in the potassium channels) modulating neuronal excitability and transmitter release. Cytoplasmic signals such as cyclic adenosine monophosphate (cAMP) generated by transmitter-mediated processes are inducers of gene expression, if the signal persists long enough. The induced protein of these genes is the basis of long-term memory (Goelet and Kandel 1986). The qualitative or quantitative changes required for changes in the genome in addition to changes in membrane channels remain unknown. Drugs, such as puromycin, that inhibit protein synthesis, interfere with long-term memory, which also indicates that consolidation may involve protein synthesis (Barondes and Cohen 1966; Flexner and Flexner 1971; Wallace 1975).

Neurotransmitters. Another nonanatomical approach to learning is the hypothesis that neuronal discharges are averaged statistically, and an invariant temporal pattern is associated with a specific memory (John 1972). Evidence supporting this hypothesis has been obtained in cats (Kleinman and John 1975).

Other hypotheses are that conduction rate increases with learning or that sensitivity to neurotransmitters increases with learning and decreases with forgetting (Deutsch 1971; Matthies 1974). Several neurotransmitters have been implicated in learning, as in other behaviors. It seems unlikely that one neurotransmitter subserves one behavior. The most actively studied putative neurotransmitter is acetylcholine (Stark and Boyd 1963; Bartus et al. 1982). Anticholinergic drugs have a profound effect on both long- and short-term memory (Deutsch 1971); for example, intracerebral administration of carbachol, a cholinergic substance, inhibits an operant response in cats (Nagy and Decsi 1976). In addition, both norepinephrine (Randt et al. 1971; Anlezark et al. 1973) and dopamine (Zis et al. 1974) have been implicated in learning, but this evidence is not as strong. Some of the pituitary peptides, vasopressin, adrenocorticotropin (ACTH), and melanocyte-stimulating hormone (MSH) have been implicated in the learning process (Marx 1975; Wallace 1975; Wimersma Greidanus et al. 1975). Serotonin increases in the thalamus, hypothalamus and midbrain, but decreases in the frontal cortex during training of a visual discrimination. Norepinephrine increases in the piriform lobe and midbrain, whereas dopamine increases in the neostriatum (Kitsikis and Roberge 1981; Vachon et al. 1984). These changes in different parts of the brain indicate the difficulty and complexity of interpretation of the relation of neurotransmitters and learning.

Sleep and Learning. The importance of sleep to learning has been investigated, but there have not been many studies on domestic animals. REM sleep increases when an animal is learning a complex task and deprivation of sleep impairs retention (Smith 1985). Being asleep does not preclude retrieval of learned information. Dogs were trained to press a panel with their paws for a food reward. Even while the dogs were sleeping, as indicated by the EEG records, there were both central (evoked cortical potentials) and peripheral (changes in the electromyogram [EMG]) of the limb with which the dog would respond if awake (Sasaki and Yoshii 1984).

COMPARATIVE INTELLIGENCE. Which animal species is the smartest (Bitterman 1965)? This question is commonly asked of veterinarians by lay people. Although knowledge concerning the IQ of the pig may not contribute much to one's medical or husbandry skills, a well-informed discussion of the facts and the pitfalls involved in assaying relative intelligence will be appreciated by the questioner. Furthermore, in species that are commonly trained, such as dogs and horses, many of the behavior problems revolve around learned tasks or, more frequently, tasks not learned. A dog that refuses to be housebroken or a horse that runs out of a jump are good examples. Volumes have been written on training horses and dogs; only the underlying principles will be discussed here.

Methods of Measurement

Brain Weight/Body Weight Ratio. An anatomical approach to intelligence can be used. There may be a correlation between brain size and intelligence (Rensch 1956). Elephants have larger brains than humans; possibly in self-defense, humans commonly use brain weight/body weight ratios for comparative purposes. The brain weight/body weight ratios in decreasing order are human 2%, cat 1%, mongrel dog 0.5%, rat 0.3%, goat 0.3%, horse 0.1%, and pig .05% (Davis et al. 1975). Figure 7.3 illustrates the brains of domestic animals.

Fig. 7.3. The relative brain size of various domestic animals. From top to bottom, *right row:* horse, cow, and pig; *left row:* dog, sheep, and cat.

The brain weight/body weight ratios of various breeds indicate that increases in brain weight are not linearly related to increases in body size — the smaller the dog, the higher the brain weight/body weight ratio. The toy poodle and Pekingese have ratios of over 1%, whereas the Saint Bernard and great dane have ratios of 0.2%; the medium-sized breeds such as the cocker have a ratio of 0.6% (Bronson 1979). A power function of .27 defines the brain weight/body weight scaling of most animals. In dogs the formula is brain weight = 0.39 body weight$^{.27}$.

Pigs and most other animals have suffered by domestication in this regard. Pigs have been bred for larger bodies, so wild pigs tend to have larger brain weight/body weight ratios. The fallacy of using as labile a parameter as body weight to judge intelligence is manifest when one realizes that the malnourished pig, which shows definite intellectual impairment (Barnes et al. 1970), has a greater brain weight/body weight ratio than the well-nourished pig. This oddity occurs because the brain appears to be spared when the rest of the body is stunted by malnutrition (Widdowson and McCance 1963). The inferiority of the intelligence of women to that of men purported by many male scientists and thinkers of the nineteenth century was supported by brain weight/body weight ratios similar to those just mentioned. Gould (1978) reviews the work of Broca and his disciples and shows how the same figures they used to prove male superiority could prove just the opposite, even if only such obvious factors as age at death, cause of death of the donors whose brains were examined, and the "sexual mass" (the differences in muscle mass and body fat) of the two sexes were evaluated. The value of these ratios in assessing comparative intelligence between species should thus be looked at skeptically.

Learning Rates. Intelligence might be measured by comparing learning rates of various species on the same task. Many confounding factors may invalidate this approach also. The task must be physically possible for all species tested. A task requiring manipulation with the forelimbs, which a rat could perform with ease, would be nearly impossible for any ungulate. Scaling also presents a problem. Is a 60-foot maze appropriate for a cow if a 6-foot one is appropriate for a cat? Similarly, one must be careful not to confuse measuring athletic ability with measuring intelligence. A dog that can run fast may complete a task before a slower animal that actually made fewer errors. The task to be learned should also be within the normal behavioral repertoire of all species to be tested. A cat can easily be taught to pounce on an object; a cow rarely performs such actions.

Classical Conditioning. Liddell and Anderson (1931) and Liddell et al. (1934) used classical conditioning to measure comparative intelligence. They compared the number of trials necessary to produce leg flexion in response to the conditioned stimulus, the sound of a metronome. The unconditioned stimulus was a shock to the foreleg. Dogs were most easily conditioned. Pigs were the most easily conditioned of the farm animals, followed by goats, sheep, and rabbits.

DELAYED RESPONSE METHOD. Several experiments comparing intelligence in a variety of species were done early in this century (McDougall and McDougall 1931). For example, Hunter (1917) used the length of time an animal could remember which of three boxes identified by a brief illumination held the food reward. This technique, called the delayed response method, revealed that a rat could delay its response for 10 seconds, a raccoon for 25 seconds. Children 2.5 years old could delay their responses for 25 minutes, and dogs could delay for 5 minutes. In other experiments researchers found that cats can remember for 6 minutes, adult dogs for 18 minutes, and goats for 30 minutes, although the goats had a more intense signal to remember than the other species (Soltysik and Baldwin 1972). Of two horses tested by Grzimek (1949), one could remember for only 15 seconds, the other for 60 seconds. The delayed response time is quite variable, however; the delayed reaction time for cats varied from 18 seconds to 16 hours, depending on the test and the investigator (Maier and Schneirla 1964).

MULTIPLE CHOICE METHOD. Hamilton (1911) used a multiple choice method to test comparative intelligence. The animals could escape from the apparatus through one of four doors; the correct or unlocked door was never the door that had been unlocked on the previous trial. Humans were superior in this test, followed by monkeys, dogs, cats, and the one horse tested. The horse engaged in stereotyped behavior, making repeated attempts to escape through the door that had been unlocked on the previous trial. Monkeys did as well as pigs in choosing doors when the correct response was the second door from the end, but an orangutan did poorly (Yerkes 1916).

AVOIDANCE RESPONSE METHOD. Willham et al. (1964) have suggested another method to measure learning ability both among and within species. An avoidance response is taught using a shock for the unconditioned stimulus and a buzzer for the conditioned stimulus. The animals must cross a barrier either to avoid or to escape shock.

Cats took 12 trials to learn to avoid shock by jumping on a shelf (Davis and Jensen 1976). Dogs needed only 4 trials to learn to avoid shock, pigs 10, and horses 8.

These results do not necessarily mean that dogs are more intelligent than horses and that horses are more intelligent than pigs. A much higher level of shock was used on the dogs and pigs than on the horses, which may have increased or decreased the learning rate (Solomon and Wynne 1953; Kratzer 1969; Haag et al. 1980).

Even very young animals have been tested. Newborn kittens cannot learn to escape an aversive stimulus (an air blast) (Bacon 1973), whereas puppies can (Stanley et al. 1974).

MAZE LEARNING. Gardner (1945) compared the learning ability of cattle, sheep, and horses and found that the horses and cows learned visual discrimination better than sheep. Maze learning has been used to assess species differences in learning ability. In a series of maze tests using the Hebb Williams maze shown in Figure 7.4, children made fewest errors; dogs, cows, goats, and sheep made approximately the same number of errors;

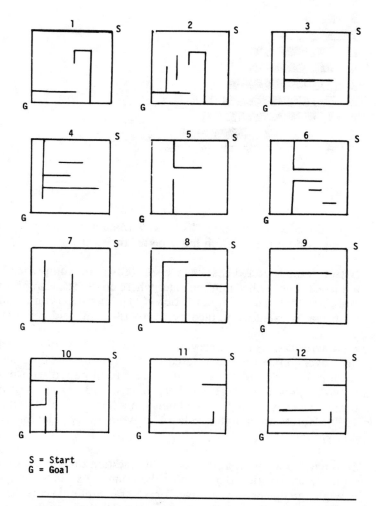

S = Start
G = Goal

Fig. 7.4A. The Hebb-Williams maze. Partitions can be rearranged to form 12 or more barrier problems. Each day the animal has a new problem to solve with 10 trials on each problem. The number of errors are used to compare with that of other species (McCall et al. 1981), copyright © 1981, with permission of *J. of Anim. Sci.*).

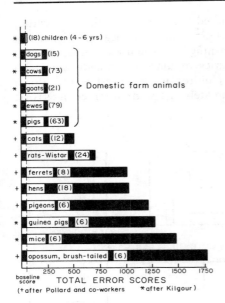

Fig. 7.4B. Comparative maze learning in various species of animals. Children made fewest errors and opossums the largest number of errors in a Hebb-Williams maze (courtesy Dr. Ron Kilgour, Ruakura Animal Research Station, Hamilton, New Zealand) (after Pollard et al. 1971).

pigs and cats made more. Karn and Malamud (1939) found that dogs learned a double alteration maze better than cats.

OBJECT PERMANENCE. Cats have been shown to comprehend object permanence. They will watch the place where an object disappeared from view and go to that place to find the object (Thinus-Blanc et al. 1982). This level of insight is usually obtained by 12- to 18-month old children.

TASTE AVERSION. As noted previously, the ability of various species to form a taste aversion to food associated with poisoning has been measured. Pigs readily form a taste aversion to a sweet food paired with poisoning after only one poisoning, whereas sheep, goats, and horses learn only after repeated pairings of the food and poisoning (Houpt et al. 1990; Zahorik et al. 1990). Horses are not able to learn to avoid a very palatable food (Zahorik and Houpt 1981).

Summary. One might also answer the question on comparative intelligence with the observation that each of the domestic species has apparently had enough intelligence to survive for several million years on its own and several thousand in man's service. Nevertheless, studies of comparative intelligence can, if performed properly, answer some questions about the role of particular learning abilities in the survival of a species.

LEARNING IN DOMESTIC ANIMALS

Pigs. Pavlov apparently felt that pigs could not be even classically conditioned, but this much-maligned species is perfectly capable of learning to salivate in response to a bell (Sutherland 1939) or to increase heart rate in response to a metronome tick (Marcuse and Moore 1944, 1946; Moore and Marcuse 1945). In fact, Liddell et al. (1934) reported that pigs were easier to classically condition than small ruminants or rabbits. Early work also indicated that they could run a maze for a food reward (Myers 1916) or choose one of a series of doors to gain access to food (Yerkes and Coburn 1915). Difficulty was encountered in teaching a concept such as "center" to the pigs; they could choose the middle of three, but not of five, doors (Yerkes and Coburn 1915).

Sex, Breed, and Age Differences. In the years since the pioneering studies described above, pigs have been used in many types of learning situations to validate or cast doubt upon concepts developed using laboratory species. Pigs learn to make more correct responses when trials are spaced in time rather than grouped. As had been shown in other species, 4 trials a day for 10 days produces learning superior to that shown by pigs given 40 trials in 1 day (Karas et al. 1962). Classical conditioning occurs more quickly if the interval between the unconditioned stimulus and the conditioned stimulus is 2 seconds (Noble and Adams 1963). Although pigs usually require a food reward, they will swim a maze when the reward is reunion with their littermates. The speed and accuracy of solving a water maze is not correlated with the pig's ability to learn to avoid shock by jumping over a small barrier (Hammell et al. 1975).

On a given task, sex and breed differences exist (Willham et al. 1963). Durocs learn avoidance more quickly than Hampshires (Willham et al. 1963, 1964). Palouse make fewer errors in visual discrimination than Pitman Moore miniature swine (Klopfer 1966). Artificially reared pigs make fewer errors in visual discrimination tasks than sow-raised pigs (Lien and Klopfer 1978). Yorkshires perform better in a T-maze than Poland Chinas; crossbreds are intermediate. Females perform better than males (Wieckert and Barr 1966).

Although season of testing, body weight, and age of dam do not account for variation in avoidance learning (Willham et al. 1963, 1964), pigs from large litters ran a maze more quickly than pigs from small litters (Wieckert and Barr 1966). Apparently, when the reward for completion of the maze was return to the company of littermates, social isolation was more severe and the reward of reunion greater when the litter size was larger.

Operant Conditioning. Pigs will not perform an operant response for some types of sensory reinforcement, such as pig noises (Baldwin et al. 1974), but will for others, such as light (Baldwin and Meese 1977b) and brain stimulation (Baldwin and Parrott 1979). Older pigs (40–150 days) learn to avoid shock less well than younger pigs (3 weeks) (Kratzer 1969) (Fig. 7.5).

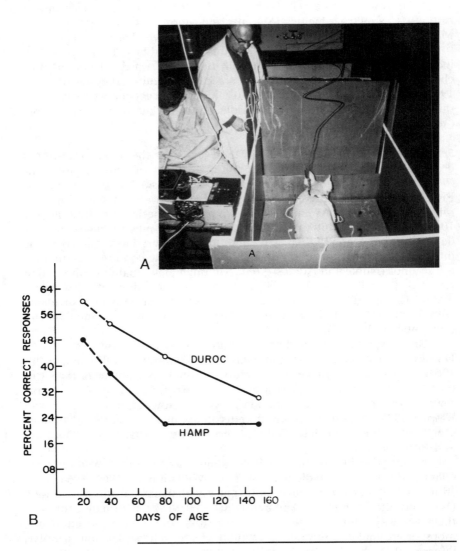

Fig. 7.5. Avoidance learning in pigs. (*A*) The pig must jump the barrier when a buzzer sounds to avoid a shock. The late Ulric Moore who performed some of the early experiments on learning in pigs is shown. (*B*) The ability to learn to avoid shock decreases with age in Duroc and Hampshire pigs (Kratzer 1969).

Operant conditioning of pigs is used for a variety of purposes: on the farm, in animal acts, and in the laboratory. Feeders with hinged covers that the pig must open with its snout are a good example of a very simple operant response that pigs learn rapidly, just as most farm animals learn to operate automatic watering devices. Pigs have been trained to perform such amusing tasks as putting giant pennies in a large "piggy bank" (Breland and Breland 1966), and pigs are often found in European circus acts.

Consideration of the anatomy and normal behavior patterns of pigs indicates that it would be much easier to teach them to manipulate their environment with their snouts than with their feet. Consequently, pigs have been taught to push a panel with their snouts for either a food reward (Baldwin and Stephens 1970) or for heat in a cold environment (Baldwin and Ingram 1968).

To measure a pig's motivation for the reward, a *progressive ratio technique* can be used. (A progressive ratio is a schedule in which an animal must make one response for the first reinforcement, two for the second, four for the third, eight for the fourth, and so on, until the *breakpoint,* at which the animal stops responding). Kennedy and Baldwin (1972) have used a progressive ratio technique to measure the strength of preferences for various sweet solutions.

A conditioned anxiety reaction can be produced in pigs by training them to press a panel for a food reward and then adding a tone that signals a shock if the animal presses the panel (Baldwin and Stephens 1970). The pigs learn to inhibit panel pressing, although experimental neurosis is sometimes produced (Curtis 1937); tranquilizers do not reduce the inhibition of responding (Dantzer and Baldwin 1974b). Conditioned anxiety does lower heart rate (Dantzer and Baldwin 1974b) and increases endogenous levels of corticosteroids, but the rise in corticosteroids is not as great as when the pigs are exposed to cold or chased with a goad (Baldwin and Stephens 1973).

Visual Discrimination. Although pigs are better at spatial (right compared to left) than at visual discrimination tasks and must be taught to discriminate visually before 20 weeks of age (Klopfer 1961), pigs do have color vision similar or slightly superior to humans and can discriminate between wavelengths of light differing by only 25 mμ (Klopfer 1966). Pigs have difficulty in reversal learning of either a visual or spatial discrimination; they abandon the original correct response, but their performance remains at chance levels for many trials. Overtraining, training carried beyond criterion on the first trial, does improve reversal learning (Wesley and Klopfer 1962; Klopfer 1966). Most visual discrimination tests involve pressing one of two panels for the reward. If the animal has learned the discrimination, it will get a reward after 6 responses on FR6, but, if it has not learned, it will take many more responses, an average of 18, to get a reward by chance. Pigs may fail to learn a visual discrimination not because they are unable to either learn or to see but because they are willing to work very hard (respond many times) for one food reward. They will tolerate being rewarded at a chance level, whereas another species will expend the minimum amount of energy.

Effect of Drugs. Dantzer and his colleagues have studied the effects of various drugs on learning in pigs (Dantzer and Baldwin 1974a,b; Dantzer 1976; Dantzer et al. 1976; Dantzer 1977; Mormede and Dantzer 1977a,b). Pigs apparently do not suffer from learned helplessness as dogs do (Mormede and Dantzer 1977b). Dantzer (1976) has found that the breakpoint is higher when the pig has been treated with the tranquilizer diazepam.

Effect of Malnutrition. Avoidance learning has been used to study the effects of malnutrition on brain function. Barnes et al. (1970) demonstrated that pigs previously malnourished but subsequently rehabilitated for several months show poorer avoidance learning than well-nourished pigs. Brain biochemistry, specifically cholinesterase, is altered in malnourished but rehabilitated pigs (Im et al. 1973).

Dogs. Table 7.2 summarizes the effect of brain lesions on learning in dogs.

Housebreaking. The first task that all house pets must learn is voluntary control of the anal and urinary bladder sphincters. There are probably as many methods for housebreaking dogs as there are books on dog training. Interestingly, both classical and operant conditioning methods have been recommended.

Immediately after a meal the gastrocolic reflex increases motility of the large colon and rectum. As a result, filling of the rectum will stimulate relaxation of the smooth muscle of the internal anal sphincter and the striated muscle of the external anal sphincter. If a dog is taken outside after every meal, the conditioned stimulus of being put outside will soon replace the unconditioned stimulus of the gastrocolic reflex. Some clinicians recommend the use of glycerin suppositories after a meal when the dog is taken outside. The principle is the same; the suppository is the unconditioned stimulus and is more reliable in action than postprandial defecation.

Operant conditioning is more useful in teaching voluntary control of urination for which there is no reliable unconditioned stimulus. Newspapers are spread all over the room where the puppy is confined. Gradually the area of newspaper is decreased; the puppy is placed on the newspaper when it squats to urinate and praised when it urinates in the proper place. The newspaper can then be laid outside the door to encourage the dog to go outside to urinate and to even whine to go outside. In other words, the puppy behavior is shaped by first being rewarded for a general action; later, the reward is contingent on more and more specific actions of the animal. It is often valuable to continue to reward use of newspaper in case the dog must be left indoors for longer periods than one can expect it to comfortably retain a bladder full of urine. (The owner must be consistent and not punish the animal if it uses the daily paper that has been inadvertently left on the living room carpet.)

Still another method takes advantage of the innate reluctance of animals to soil their sleeping quarters. The puppy is put in a small cage or crate and kept there except for trips outside to eliminate every hour or two. The premise is that the dog will not urinate or defecate in the cage, and once the animal can control its bowels and bladder for long periods it will generalize the control to the whole house and will no longer have to be closely confined. It would be asking too much of a young puppy, whose control is not good enough to wait an entire night without eliminating; in addition to being somewhat cruel, this may only condition the pup to accept urine and feces on its bed and on itself. This technique has merit for those who can stay home with their puppies or for an older dog that is still poorly trained.

Whichever method is used, the trainer must use appropriate and consistent rewards and punishment. Verbal praise and a perfunctory pat are ample reward; in fact, as already mentioned, Stanley and his colleagues (Stanley and Elliot 1962; Bacon and Stanley 1963) found that dogs will learn to perform better when simple contact with a passive person is the reward than when stroking and picking up is the reward.

Punishment of the animal for misbehavior must come as soon as possible after the offense. The unconditioned stimulus is the punishment and the response is avoiding the pain. Unless the conditioned stimulus, inappropriate urination or defecation, is closely paired in time with the unconditioned stimulus, the animal will not learn. If it is punished 10 minutes after it has performed its misdeed, it is more likely to be the trainer who becomes the conditioned stimulus than the act of elimination. Dogs who act "guilty" are those in whom this kind of conditioning has taken place. They have learned to associate the presence of a pile of feces in the house with a painful experience; they have not learned to associate defecation with the painful experience.

Although taking the puppy outside to eliminate is a good method of training, a common mistake that owners make is to put the dog outside for longer and longer periods. The dog does not know what it is to do outside, and, if it should eliminate, the owner is not there to praise it. Praising the dog when it returns to the door is a good way to teach the dog to return home but not to housebreak it. Still another complication occurs with the dog who is walked on a leash, but whose walks always end as soon as it eliminates. It learns to avoid eliminating in order to prolong the walks. Walk lengths should not be contingent on elimination. Dogs may also be afraid to eliminate when the owners are present because they have been punished when "caught in the act." The dog learns to avoid eliminating when the owner is present rather than to avoid eliminating in the house.

The problems of housebreaking have been covered in great detail because they represent 20% of the behavioral complaints of dog owners (Voith and Borchelt 1985a). Owners will tolerate many defects in their animals, but few will tolerate house-soiling, as a case of a Lhasa apso donated to a veterinary college illustrates. The dog suffered from lissencephaly (absence of normal gyri and sulci) (Fig. 7.6) and had many learning disabilities and visual deficits, but the owner's reason for donating the dog was its inability to learn bowel or bladder control.

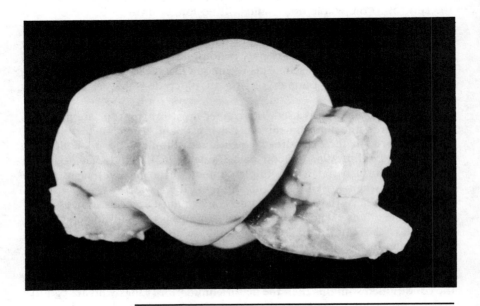

Fig. 7.6. An example of a learning deficit associated with organic disease of the central nervous system. The brain of a Lhasa apso terrier that suffered from lissencephaly and could learn neither bowel nor bladder control.

From the clinician's point of view, house-soiling in a previously well-trained animal indicates a medical problem. A dog with diarrhea will not be able to retain voluntary control of its irritated intestines. This is most apt to occur with problems of the large intestines. Dogs with chronic nephritis do not concentrate their urine, so they must excrete large quantities of dilute urine and will not be able to retain or limit their micturations as well as they could when healthy. Ulcerative colitis must be treated as both a behavioral and a medical problem; stress should be reduced and the appropriate drugs prescribed.

Breed Differences. There are breed differences in learning ability or, at least, in acquisition and performance of various tasks.

The most carefully controlled studies on breed differences in learning ability were carried out at Jackson Laboratory in Maine. Scott and Fuller (1974) reviewed the experiments, which compared learning ability in five breeds of dogs: cocker spaniels, beagles, wirehaired fox terriers, Shetland sheepdogs, and basenjis. These particular breeds were chosen because they did not differ much in body size, nor did any breed possess a breed-specific anatomical peculiarity, such as the achondroplasia of basset hounds. The five breeds were tested for their ability to learn tasks in three ways: forced training, reward training, and problem solving.

The forced method of training was used to teach the dogs to sit still on a scale, to heel on a leash, and to stay and jump on command. In all three types of tasks the cocker spaniels ranked highest in correct performance. The trainability of cockers in these situations is probably the result of selection within the breed for dogs that would crouch in response to a hand signal. Although cockers are not often used for hunting now, the behavioral predisposition remains. All the dogs learned to heel within the 10-day training period, but there were marked differences in the types of errors that dogs of the various breeds made in the early sessions. Basenjis fought the leash and often pulled ahead or lagged behind; Shetland sheepdogs interfered with the trainer, that is, tangled the leash around his legs. Beagles vocalized in protest (Fig. 7.7).

Reward training consisted of showing the dog a bit of fish in a box and then restraining the puppy behind a wire gate before allowing him to run to the box and eat the fish. The position of the box was changed to measure goal orientation compared to habit formation. Basenjis performed best on this test, probably because they could run the fastest. With the additional trials, all of the dogs reduced the time they took to reach the reward. When motor skills of the five breeds were compared, the basenji was also best.

Problem solving was also studied. The first type of problem the dogs were supposed to solve was a barrier or detour. The dogs were separated from a food reward by a wire barrier. When they learned to run around a short barrier, the barrier was extended and later formed into a U-shape to increase the difficulty of the problem. The dogs were only 6 weeks of age and had considerable difficulty solving the barrier problem that an adult dog could easily master. The puppies often yelped, and it was noted that they never solved the problem while yelping but engaged in stereotyped activity instead. Few puppies did well on the barrier test, but basenjis, which are already active at 6 weeks, did best. Puppies of other breeds still tend to be fat and clumsy at 6 weeks, and reacted to failure by going to sleep.

Another type of problem was a manipulation test in which the dogs were tested for their ability to pull a dish of food from under a box by pawing or pulling it out with their noses. Later, the dish was positioned so

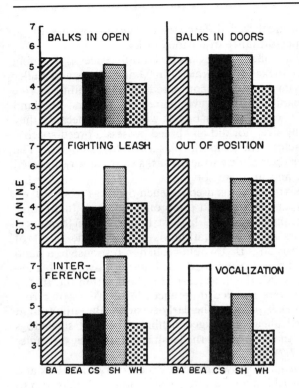

Fig. 7.7. Breed differences in response to leash training. BA = basenji; BEA = beagle; CS = cocker spaniel; WH = wire-haired fox terrier; SH = Shetland sheepdog. The higher the stanine score the more often the dog exhibited the behavior (Scott and Fuller 1974, copyright © 1974, with permission of University of Chicago Press).

that it could be maneuvered out from under the box only by pulling on a dowel and string attached to the dish. Again, the basenjis were the most successful. Most interesting was the effect of repeated failures on the dogs' performances. Although all puppies at first scratched and nosed at the box containing the food dish, those that had failed often took one look at the box and simply sat down to await the end of the trial, thus precluding any success.

When maze learning was tested in RLLRRR (right, left, left, etc.) or LRLL mazes, beagles did best. They completed the maze most quickly and made the fewest errors.

A final test of problem-solving behavior, when the dogs were 22 weeks old, was a delayed response test in which the dogs were shown a visual cue that indicated which side of a T-maze led to freedom. Before the dog was allowed to run the maze, a delay from 1 to 240 seconds was imposed. As in all problem solving, individual differences were great, but cocker spaniels could remember after the longest delay, and Shetland sheepdogs had the poorest memory.

Using similar tests Coon, (1977) tested a few dogs of many breeds. The results are shown in Table 7.3. The dogs were tested in their homes by the owners, and these facts alone indicate that there were many opportunities for factors other than the dogs' intelligence to affect the results ,but they are presented for interest. The low scores of very large and very small dogs indicate scaling problems as well. Nevertheless, the relative positions of the breeds is of interest.

The extensive studies at Jackson Laboratory indicate that care must be taken in comparing intelligence, even within a species, because breeds of dogs differ markedly in their relative performances depending on the task to be learned. Fox and Spencer (1967) have shown that dogs improve in their ability to make a delayed response. If puppies were trained for 13 days beginning at 4, 8, 12, or 16 weeks, all increased the interval over which they could remember during testing, but the 4-week-old puppies could never remember longer than 10 seconds whereas the 12-week-old dogs could re- member for 50 seconds. Sixteen-week-old dogs did not perform as well; they made many errors. Fox and Spencer (1967) explain the poor perform-

Table 7.3. Ranking of breeds of dogs according to scores received on problem solving following tests administered by the owner

standard poodle
bulldog
beagle
springer spaniel
golden retriever
basset
dachshund
great Dane
collie
Belgian tervuren   Australian shepherd
miniature poodle
pointer   German shepherd
toy poodle   airedale   silky terrier
cocker spaniel   schipperke   St. Bernard
schnauzer   Labrador retriever
boxer
Pekingese
whippet
Shetland sheepdog
Doberman pinscher
Lhasa apso
Irish setter
pug
West Highland terrier
Afghan
Siberian husky
Pomeranian
Chihuahua

Source: Coon (1977).
Note: The tests were modified versions of those developed by Scott and Fuller (1974). Some tests used a food reward (delayed response and manipulation); others involved ball following (detour) or ability of the dog to extricate itself from a towel wrapped around it. The scores were probably influenced by the dogs' motivation for food and for the ball as well as dog-owner interactions. Breeds shown on the same line had identical scores.

ance at 16 weeks by postulating a lack of inhibition at that age. Stanley et al. (1970) also studied the ontogeny of learning in dogs. They found that puppies less than 1 week old could be conditioned to suck more from a nipple when milk was the reward and to inhibit sucking when a quinine solution was the aversive stimulus although earlier studies had found that puppies less than 18 days old could not be conditioned (Fuller et al. 1950). Stanley et al. (1974) also found that puppies less than 1 week old can learn to escape from a cold stream of air and will even move from a comfortable carpeted surface to a hard cold one to do so (Fig. 7.8). Puppies less than 2 weeks old could also learn to choose a wire or cloth model that contained a nipple for a milk reward if given five tests a day, 2 hours apart.

It is much more difficult to train nervous dogs than calm ones (Woodbury 1943; Dykman et al. 1969). Angel et al. (1974) found that administration of a tranquilizer facilitated operant conditioning of a genetically nervous pointer. Petting does have a physiologically demonstrable calming effect. For example, dogs classically conditioned to expect a shock follow-

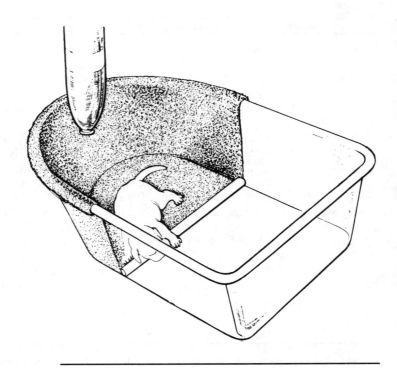

Fig. 7.8. Learning in neonatal dogs. The puppy learns to avoid the textured side, for which puppies have an innate preference, if that side is associated with a blast of air (Stanley et al. 1974).

ing a tone have a higher heart rate during the tone (Gaebelein et al. 1977), but heart rate declines if the dogs are petted during tone presentation (Lynch and McCarthy 1967).

Training is facilitated if the innate responses of dogs to auditory signals are used. Dogs will increase their activity to high tones and inhibit it to low tones (McConnell 1990), and this response is reflected in the signals shepherds use to signal their dogs (McConnell and Baylis 1985).

Because social facilitation is strong in dogs, slow dogs tend to run faster with another dog than they do alone (Vogel et al. 1950). This behavior is in contrast to that of cats, who run more slowly for a food reward in the presence of another cat.

Cattle. Few studies have tested learning ability in cattle, and most of those published described attempts to increase production or reduce labor on the farm.

Operant Conditioning. Kiley-Worthington and Savage (1978) have trained cows to come in to be milked when an automobile horn connected to a timer and the electric fence was sounded. Albright et al. (1966) trained cows to come into the barn in a given order, but the cattle reverted to their original order, probably based on dominance, as soon as the trainer was absent. Wieckert et al. (1966) also trained cattle to come to a feeding trough when an auditory stimulus was delivered to the cattle from a small timer-activated tape recorder attached to the cow's halter.

Offord et al. (1969) conditioned cattle to eat in response to an auditory stimulus and then attempted, unsuccessfully, to increase food intake in the cattle when they were free-feeding by playing the auditory stimulus. Dairy cows have also been taught to press a handle with their muzzles to obtain food and to make right and left handle discriminations. The most difficult task was to accustom the cows to lifting rather than lowering their heads for food, because they were accustomed to eating from the floor (Whittlestone and Cate 1973). Cattle have also been taught to use individual feeders that open only when the animal inserts the electronic collar around its neck into the feeder (Karn and Clanton 1974), and these systems are now widely used because cattle can be group housed yet feed themselves individually. Other cattle have been taught to enter a feeding stall when one signal, a bell, is presented and to leave the stall when another signal, a buzzer, was presented (Shaw 1978). Intake can be limited in group-housed cattle by thus teaching them to eat on command.

Moore et al. (1975) taught seven Jersey cows to press a panel for a food reward. The cattle learned to respond to various schedules of reinforcement: a fixed ratio, in which a certain number of presses resulted in the food reward; a fixed interval, in which food was delivered at intervals and only one panel press at the appropriate time was necessary for the food reward; a variable ratio; and a variable interval. The variable ratio schedule produces the highest response rate in cattle as in other species. Cattle can learn both radial-arm mazes and parallel-arm mazes, that is they learn that food will not be in a location they have previously visited even if visits were separated by as much as 4 hours, but not by longer intervals. Kilgour (1981) found that Jersey cows learned 12 detour problems in the Hebb Williams maze (Fig. 7.4A) and made few errors after four runs of a given detour test.

CONDITIONED AVOIDANCE. Many farmers teach their cattle to defecate in the gutter behind their stanchions rather than on the stall floor by running an electrified wire that shocks the cow whenever it arches its back to eliminate in the wrong place. Because cattle defecate 17 times a day, the cows learn quickly and the average fecal output of 60 to 80 pounds is deposited in the gutter. Cattle can also form a taste aversion to alfalfa pellets and corn (Fig. 7.9), but not a molasses grain mixture (Zahorik et al. 1990).

Cattle sometimes receive shocks through milking machines. Cows were taught to press a bar for food to determine how large a shock had to be before it disrupted a cow's feeding and other behaviors. They stopped bar pressing when a shock greater than 7 mA was applied to one teat or 6 mA to all four teats (Whittlestone et al. 1975). This type of conditioning can be used to determine what is painful to cattle.

A practical application of bovine learning is to train cattle to respect electric fences by confining them in a pen with a sturdy fence beyond the electrified wire. The cattle will then respect an electric fence, even in a new environment and will not be in danger of breaking through it (McDonald et al. 1981).

Visual Discrimination. One of the earliest studies of learning in cattle is one of the most extensive in that large numbers of cattle were tested. Gardner (1937a,b) studied the ability of cows to discriminate a box covered with black cloth, which contained feed, from two empty, uncovered, boxes. The errors per trial fell as the trials proceeded. Guernseys, brown Swiss, and Holsteins performed better than Jerseys, shorthorns, and Ayrshires. The cows had retained the learning when tested a year later. More errors were made when the cloth was moved below or, in particular, above the feed box. Calves have been taught to discriminate shapes and orientation symbols (Baldwin 1981). See sheep and goat sections for further discussion of visual discrimination in ruminants.

Effects of Age. When the ability to remember the location of a feeder was tested over a period of 5 days, heifers learned more quickly than older cattle, but cows after the second calving remembered the location best when tested 6 weeks later. In both tests, primiparous cows were intermediate in performance (Kovalcik and Kovalcik 1986).

Sheep and Goats. Liddell et al. (1934) have shown that sheep and goats can

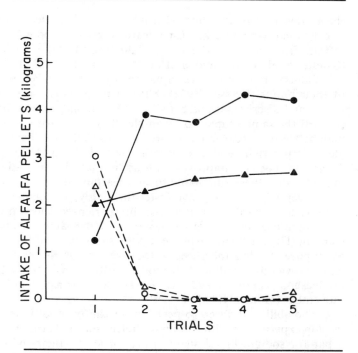

Fig. 7.9. Taste aversion learning in cattle. On day 1 (trial 1) the experimental cows (open symbols) were offered alfalfa pellets. After they had eaten the food, LiCl, which produces mild gastrointestinal malaise, was infused into the rumen through a fistula. The control cows (closed symbols) were infused with 0.9% NaCl. Four days later (trial 2), when the cows were offered alfalfa pellets, the cows that had been treated with LiCl ate almost nothing, but the controls ate more than they had initially. The experimental cows learned to avoid a feed associated with illness (Zahorik and Houpt 1977, copyright © 1977, with permission of Baylor University Press).

be classically conditioned to flex a front leg in response to a ticking metronome that has been paired with an electric shock, although the small ruminants did not learn as quickly as pigs or dogs. Liddell (1954) has also demonstrated that sheep can learn simple mazes. Cairns and Johnson (1965) have demonstrated that lambs will run a maze to be reunited with a cohabitant, whether the animal was another sheep or a dog.

Operant conditioning techniques have been used by Baldwin and Yates (1977) to study thermoregulatory behavior. The sheep learned to activate a heat lamp by sticking their muzzles through a slit to break a photoelectric beam. Unshorn sheep did not learn to turn on the heat at 5°C, but shorn sheep, deprived of the thermal insulation of their fleece, did. Sheep have also learned to press a bar for reward in response to a tone signal and to refrain from pressing the bar for 30 seconds after each reinforcement (Sandler et al. 1968; Sandler et al. 1969; Sandler et al. 1971).

Goats can learn to press a bar for the reward of electrical stimulation of their brains (Persson 1962). Sheep and goats can learn to make fairly fine visual discriminations between different shapes and different orientations of the same shape. For example, they can learn not only to discriminate between a circle and a square but also to discriminate between a triangle pointing right and one pointing left (Baldwin 1979; Baldwin 1981). Their learning rate improves with additional discrimination tasks, which indicates that these ruminants form a learning set or can learn to learn.

Sheep will learn to move faster through yards (chutes and alleys) with experience and will do better than inexperienced sheep months later, but training in moving a different way through the same yard is worse than no training (Hutson 1980). Apparently the sheep have to forget their previous experience. Giving the sheep a food reward at the end of a yard is an efficient way of speeding their movement through it but will not help if an unpleasant experience such as restraint is also associated with the same movement (Hutson 1985).

The ability of sheep to perform operantly conditioned tasks has been used in experiments with sodium-deficient sheep. These sheep learn to press a bar for sodium bicarbonate in proportion to their sodium deficiencies. Sheep can also learn to discriminate which of five feeders contain food (Zenchak and Anderson 1973). Sheep have also been studied by Gardner (1945) and Liddell (1926a,b), as discussed under comparative intelligence.

Horses. The horse, like the dog, is a species that is only useful to humans if trained. In fact, despite their aesthetic value, few horses are kept as pets unless they can be ridden or driven. Myriad volumes have been written on the training of horses, and it will be appropriate to mention only a few of the basic principles.

It is easiest to teach a horse a natural response; consequently, horses can be trained to race at a very early age. Two-year-old horses on the race track are very common; a horse under age five in a dressage class is a rarity. Most horse training is based on negative reinforcement — applying an aversive stimulus until the horse performs the response. The best approach to horse training is to try to substitute conditioned stimuli (a voice command or subtle pressure from the rider's legs) for unconditioned stimuli (the painful flick of a whip). In this manner, neck reining can replace direct reining. When punishment must be given, it should be applied as soon as possible after the misdeed. A slap on the pony's muzzle 30 seconds after it has nipped will only serve to make it head shy; a blow 1 second after the nip may inhibit further aggression.

Perhaps the best example of intelligence in horses is Clever Hans. This nineteenth-century Arabian stallion could perform mathematical problems by tapping out the answer with his hooves. He was able to give the correct answer whether or not his trainer was present. It was finally discovered that someone had to be present who knew the answer to the problem. The horse was able to perceive some subtle change in the person when he reached the correct number. Clever Hans was more clever at interspecies communication than he was at mathematics.

Operant Conditioning. A single horse was taught by Myers and Mesker (1960) to make a typical operant response, pushing a lever with its muzzle for .5 cup of grain. Once the horse had been shaped using continuous reinforcement, the schedule was changed to 3 and then to 11 responses for every reinforcement. The horse increased its rate of responding as the FR ratio (fixed ratio of responses per reward) increased. When a fixed interval schedule was imposed, the horse at first sulked by refusing to turn toward the lever, but later responded with the scalloped pattern typical of laboratory rodents on fixed interval schedules. The practical application of the study is that a horse will perform better when it is not rewarded for each performance.

Operant conditioning is used to evaluate drugs in equine pharmacology (Wood et al. 1989). The horses are usually conditioned to break a beam with their heads for food rewards. The rate of responding will be lower if a depressant, such as acepromazine, is given or will be higher if a stimulant, such as methylphenidate (Ritalin), is administered. The action of an unknown drug can be tested by comparing its effect on responses to that of other, known, drugs. Using this technique one can determine if the horse is being stimulated or depressed, that is, whether the horse's performance is being improved or worsened. Operant conditioning has also been used to measure environment preferences of horses (Houpt and Houpt 1988).

Visual Discrimination. Surprisingly few formal studies on learning in horses have been performed, despite the amount of training that is necessary to teach a Lippizaner to perform the courbette or to teach a riderless cutting horse to select a steer and separate it from the herd. Gardner's studies (1937c,d) revealed that a horse could learn to choose a feedbox covered with black cloth and containing food instead of two empty feedboxes. With increasing trials the number of errors decreased. The horses retained the discrimination when tested over 1 year later, but like cattle they found it difficult to choose the correct box when the cloth was moved to a position above or below the box. In general, the horses made slightly more errors than cows in a similar situation (Gardner 1937a,b). Using figures on the covers of feedboxes, Giebel (1958) was able to teach ponies to discriminate between many pairs of figures to obtain feed. If the symbols used were too similar to be perceived as different by the ponies, they would exhibit neurotic behavior, swaying from side to side over the boxes without making a choice. As in Gardner's study, the discriminations were well retained for long periods. Fat horses make more errors on a simple visual discrimination test than thin ones, which is probably a result of poorer motivation for the food reward rather than intellectual deficit (McCall 1988).

Maze Learning. Yearling quarter horses can easily learn a simple right- or left-turn maze in trials and could learn to turn in the opposite direction (reversal learning). Punishment did not improve performance (Kratzer et al. 1977). Using a similar maze, Haag et al. (1980) found that ponies who learned a maze reversal (to turn right rather than left) most quickly also learned to avoid a shock in the least number of trials. There was, however, no correlation between learning ability in either task and position in the dominance hierarchy. The maze is shown in Figure 7.10. Horses can also learn to form taste aversions (Zahorik and Houpt 1981). Learning in farm animals has been reviewed by Kratzer (1971).

Fig. 7.10. Maze learning in horses. The horse enters the maze and must make a right turn to leave the maze and receive a food reward.

Observational Learning in Horses. There have been two attempts to demonstrate observational learning in horses. In both cases the, young quarter horses watched another choose one of two feed buckets. In neither case did the horses that observed perform any better than those that had not had the opportunity to observe (Baer et al. 1983; Baker and Crawford 1986). This is particularly interesting because horse vices, in particular cribbing, are believed to be learned by observation. Tests of other types of learning might reveal that horses do learn by observation.

Age and Handling. Weanling foals learn to make the correct choice in the maze shown in Figure 7.10 with fewer errors than adult mares, but the latter move faster, so despite entering the wrong side, reach the food just as quickly as the foals. Orphan foals do not appear handicapped in their learning ability but move even more cautiously than normally reared foals (Houpt, et al. 1982).

A moderate, but not an extensive, amount of handling improves a young horse's performance in a maze learning test (Heird et al. 1981). Perhaps the most useful information is that handlers can predict trainability after working with a horse for 10 days, and, although another interpretation might be that handlers determine performance, these scores are correlated with trainability under saddle as judged by different people. The more emotional a horse, the poorer its learning ability is (Fiske and Potter 1979; Mader and Price 1980; Heird et al. 1986).

Cats. Whereas dogs were the animals used in the earlier experiments in classical conditioning (Pavlov 1927), cats were studied in the first experiment in operant conditioning (Thorndike 1911). Cats learned to operate on their environment to escape from puzzle boxes. They also learned to pull strings to which a piece of food was attached, selecting the one attached to the meat from among several others (Adams 1929). Manipulating strings is a task that dogs perform poorly (see the previous section); cats may be more anatomically than intellectually suited to the task. Cats can be classically conditioned (Harlow and Settlage 1939), and experimental neurosis can also be conditioned in cats, as in most species (Anderson and Parmenter 1941), by requiring them to discriminate between two very similar stimuli (auditory, in this case).

Discrimination. The cat's ability to learn discrimination has been used to great advantage by psychophysicists in studying vision. For example, color vision can be studied by teaching cats to discriminate between two symbols and then to discriminate between the symbols when they differ in no characteristic except hue. Cats can, in fact, make this discrimination but only after 1400 trials; they do have color vision, as both behavioral tests (Sechzer and Brown 1964) and electrophysiological studies (Cohn 1956; Ringo et al. 1977) attest. The color stimulus must be large (i.e., a big object) before the cat is able to make use of the hue. Nonneman and Warren (1977) used a two-cue discrimination to measure the salience or importance of various sensory modalities to cats. The animals were taught to feed from one of two feeders, the one with a buzzing noise and flashing light associated with it. Later one feeder flashed and the other buzzed; the cats went to the flashing feeder, indicating that auditory stimuli are less important to cats than visual ones when both are carefully equated for intensity.

Rewards. Unlike dogs, cats will not usually perform to be reunited with the experimenter. They will perform for food rewards. Feline finickiness can even interfere with the reward value of food, but, in general, cats will work harder for food rewards if the experimenter is the one who feeds the cats in the home cages (as well as in the learning situation). It is even more difficult to teach a cat to press a bar for water; water must be withheld for a week (Bailey and Porter 1955). Kittens will learn more quickly when the reward is freedom to explore a room than when the reward is food (Miles 1958). Kittens learn to make a light-dark discrimination more slowly than 35-day-old cats (Rose and Collins 1975).

Split-brain Lesions. Cats have often been the species chosen for learning experiments involving brain lesions, transections, or electrode implantation. One of the most fascinating series of experiments used cats with split brains, that is, brains in which the optic chiasm, the corpus callosum, the anterior commissure, and all other connections between the two cerebral cortices were severed. Cats thus prepared can learn one task with one eye and cortex and another task with the other eye and cortex. The cat can learn without confusion to choose the white circle instead of the black square with its right eye and to choose the black square instead of the white circle with the left eye. There is, of course, no transfer of learning (Sperry 1964; Pickens and Kelley 1967). Cats with split brains can learn, but learning is much slower than in intact cats, which indicates that transfer of learning from one side of the brain to the other is not essential for learning but greatly facilitates it (Sechzer 1970; Sechzer et al. 1976). Some of the experiments on the effects of brain lesions on various learning tasks are listed in Table 7.1.

Brain Stimulation. Cats can be classically conditioned when electrical stimulation of the cortex rather than the sound of a bell or some other sensory cue serves as the conditioned stimulus (Doty et al. 1956). Stimulation of the posterior hypothalamus is both rewarding and aversive to cats. A cat will learn to work to turn on and to turn off hypothalamic stimulation (Roberts 1958a,b).

Learning Sets. Cats are able to form learning sets, a skill once thought to be confined to primates. A learning set, or learning to learn, is the underlying principle by which a variety of related problems can be solved. Warren and Baron (1956) showed that cats could learn to solve a problem, such as choosing the object on the left when identical black squares were the stimulus, and would learn much more quickly on the next problem to choose the object on the left when white triangles were presented. After four problems the cats' errors fell to 36% of the original errors, and only 58% of the number of trials originally necessary were needed to reach criterion. Cats seldom show insightful behavior; they do not learn to move a light box under a suspended piece of fish to reach the fish (Adams 1929). Captured feral cats learn a discrimination more quickly than cage-reared ones (Wikmark and Warren 1972). These findings indicate that a varied environment or experience may lead to an increased learning ability in cats.

Imitation. Learning by observation or imitation takes place in cats. Cats watching a cat press a bar or jump a barrier to obtain food learned to press the bar or jump the barrier much faster than cats who did not observe a trained animal. Cats can also be misled. If the cats watched a cat that obtained food by simply approaching but not pressing the bar, they learned to bar press for food more slowly than nonobserving cats (John et al. 1968). Kittens can also learn by observation, and they learn more readily by watching their mothers than by watching another adult cat (Chesler 1969).

Both cats and dogs can be taught to make auditory discriminations and to lift the lid on a food pan when one pitch but not another is sounded. When the two sounds are too close to be discriminated (less than .33–.25 tone apart), the animals exhibit experimental neurosis. The cats respond to all tones as positive, and most dogs refuse to respond to any (Dworkin 1939). Early visual experience can influence a cat's performance on visual discrimination tasks (Zablocka 1975; Zablocka et al. 1975). Old cats (10 to 23 years) do not learn as well as younger cats. They are most apt to fail to learn, even after 1000 trials, if the conditioned stimulus begins too far in advance of the unconditioned stimulus (Harrison and Buchwald 1983).

A final note on performance of cats: whereas dogs tend to run faster when competing with other dogs, cats do not; in fact, they may refuse to compete for food in a runway situation (Winslow 1944a,b).

8

Ingestive Behavior: Food and Water Intake

Animals typically show a growth curve that includes a short dynamic phase in which weight gain per unit time is large and a much longer static phase in which there are no major gains or losses in weight but weight oscillates around a mean. Most domestic animals grow rapidly for several months or the first year following birth and then plateau at a mature body weight.

Animals treated by veterinarians fall into two general categories: food- and fiber-producing animals, and companion animals. The problems of ingestive behavior of animals differ from category to category. Animals used for human food rarely survive much beyond the dynamic phase of weight gain. The objective of livestock producers and the veterinarians who advise them is to maximize the animal's weight gain per unit time. The dairy cow presents a special case. She must be a good producer of milk, and, therefore, must increase her food intake, yet she should route that increased energy intake not to body fat but to milk production.

A quite different problem is presented in some companion species, especially household pets and horses. In a seminaturalistic setting, like a pasture, a horse can be fed ad libitum, but when presented with an energy-rich concentrate diet that it would not have encountered in the wild, the horse may overeat. Acute problems like colic or founder may result. Chronically, a simple shift upward in body weight, obesity, may result. Dogs and cats face a similar problem. They can maintain their body weights on relatively unpalatable, dry chow diets but may succumb to obesity or digestive upsets when offered a highly palatable meat diet ad libitum.

The physiology of the controls of food and fluid intake has been extensively studied by such diverse scientific groups as nutritionists, physiologists, animal scientists, and psychologists. Knowledge of the physiological mechanisms involved in hunger and satiety will enable caretakers to stimulate intake for maximum yield or to control body weight without inducing hunger in an animal that tends to obesity. The physiology of hunger will be reviewed here using the pig as an example; next to the laboratory rat, the domestic pig has been more thoroughly studied in this regard than any other animal. What is known about ingestive behavior in other species will then be discussed and the unique control of food intake in ruminants described.

The controls of food intake can be most conveniently divided into factors that stimulate intake and those that depress intake. Consideration must also be given to the mechanism by which these influences are integrated in order that animals not only survive but maintain an optimal and nearly constant body weight.

CONTROL OF FOOD INTAKE IN PIGS
Food Intake Stimulants

Defense of Body Weight. The constancy of body weight in the adult animal is believed to be the result of an innate set point for body weight or, more likely, a set point for total body fat stores. The set point is most easily detected when body weight is artificially manipulated. For example, if an animal is starved for a few days it will, when food is again freely available, eat more than it did before the fast and rapidly regain the weight lost during the fast. In some cases, pigs show compensatory increases in intake after food restriction; in other cases, they do not increase their intake over nonrestricted controls, but their weight gain is greater, presumably owing to increased efficiency or decreased energy output (Mersmann et al. 1987). When diets are diluted with noncaloric bulk, intake rises so that caloric intake remains constant (Owen and Ridgman 1967; Miller et al. 1974). Conversely, if the animal is force-fed, it will gain weight but will decrease its voluntary intake while being force-fed (Pekas 1983). Set point is also maintained by adjustments in energy output that can be varied by both changes in motor activity and by metabolic changes in heat production.

The set point of body weight is more difficult to determine in the young, rapidly growing animal. The set point will have been changed. The starved piglet may remain stunted for the rest of its life, as conclusively shown by Widdowson (1971). Similarly, overfeeding of piglets, as may occur in a small litter, may result in a permanently higher set point of body weight. If a young growing pig is fed 120% of its normal intake intragastrically, it will grow faster, which indicates that appetite, rather than genetic growth potential or gastrointestinal fill, limits the rate of growth (Pekas 1985).

In response to changes in consumer preferences, pigs are now selected for leanness, so the set point for body fat is low, and an obese pig is rarely seen on the farm. There are, however, genetically obese strains of pigs maintained for experimental purposes (Martin et al. 1973b; Hetzer and Harvey 1967). Obesity can also be produced by dietary manipulation in young meat-type pigs (Gurr et al. 1977). The number of fat cells did not increase in these pigs exposed to a palatable, high-carbohydrate diet, but the fat cells were larger, that is, they contained more fat.

In general, intake can be stimulated in the adult animal by any manipulation that lowers body weight or body fat stores beneath the set point. The neonatal piglet can respond to fasts of short duration with an increase in food intake, but longer fasts may produce irreversible hypoglycemia, coma, and death.

Low Environmental Temperature. Food intake is also stimulated by cold environmental temperature. This thermostatic control of food intake is part of body temperature regulation. When more energy must be applied to maintain body temperature, more energy is taken in (Ingram and Legge 1974). Those areas of the brain that are involved in temperature regulation, in particular the anterior hypothalamus, are also involved in the increase in food intake observed in the cold. Changes in temperatures of the brain are not correlated with the initiation and termination of meals; and thermostatic eating is a response to changes in ambient or environmental, temperature not to changes in body temperature within the physiological range (Fig. 8.1). If cold temperatures are too extreme, the animal will not be able to compensate for the energy lost by increasing its intake and will lose weight.

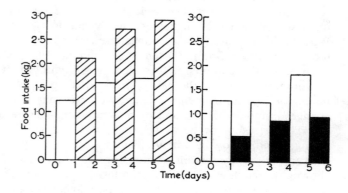

Fig. 8.1. Daily food intake of pigs fed ad libitum. Pig on left was subjected on alternative days to temperatures of 25°C (*white*) and 10°C (*hatched*). The pig on the right was subjected on alternative days to temperatures of 25°C (*white*) and 25°C (*black*) (Ingram and Legge 1974, copyright © 1974, with permission of Pergamon Press).

Decrease in the Rate of Glucose Utilization. Animals increase their food intake when the rate of glucose utilization in the brain falls. Experimentally, this phenomenon can be demonstrated by administering a competitive inhibitor of glucose, 2-deoxy-glucose, which decreases glucose uptake in the brain (Stephens 1980). Similarly, food intake can be increased by administering an agent that markedly lowers plasma glucose (Fig. 8.2), such as insulin. Although eating in response to a lack of utilizable glucose (glucoprivation) can be readily demonstrated in a variety of species, it is probably not a practical method for stimulating intake because the dosage necessary to stimulate food intake is perilously close to the dosage that produces hypoglycemic convulsions. Eating in response to glucoprivation is an emergency mechanism that the animal uses when its endogenous energy supply is approaching exhaustion. It probably is not involved in initiation and termination of meals in free-feeding animals.

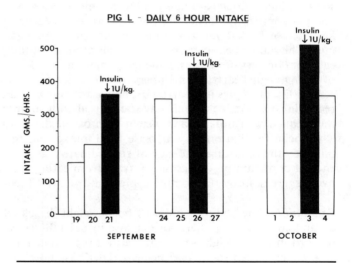

Fig. 8.2. Intake of a pig following injection of saline (*white*) or insulin (*black*).

Central Nervous System Depressants. Unlike insulin, central nervous system depressants, in particular the benzodiazepine tranquilizers, such as diazepam and elfazepam, can probably be used to stimulate food intake when given as food additives to pigs, cattle, and horses (Brown et al. 1976; McLaughlin et al. 1976). The action mechanism of these drugs is unknown; but, in general, drugs that stimulate the central nervous system, such as amphetamines, depress food intake, whereas those that depress the central nervous system, such as barbiturates or tranquilizers, stimulate food intake. These drugs may act on brain satiety mechanisms: the depressants depress satiety and, therefore, stimulate food intake; the stimulants stimulate satiety. Elfazepam, for example, has been shown to increase pigs' mean intake from a control level of 220 g/hr to 570 g/hr (McLaughlin et al. 1976). Opiates probably stimulate intake because administration of the opiate blocker naloxone suppresses intake (Baldwin and Parrott 1985).

Many compounds when applied directly to the central nervous system stimulate food intake: for example, calcium or magnesium ions or pentobarbital injected into the cerebral ventricles stimulate food intake for an hour after injection (Baldwin et al. 1975). Whether the effects of these ions, or of barbiturates and tranquilizers, can be used to economically and safely increase food intake of food-producing animals remains to be seen.

Social Facilitation. Two animals housed together usually eat more than the sum of their intakes when each is housed separately. This is true of most social animals. Social animals tend to do things as a group; therefore, when one pig goes to the feeder, all the pigs go to the feeder. Cole et al. (1967) have demonstrated that group-penned pigs eat more than separately housed pigs. The same phenomenon, social facilitation, leads all pigs to attempt to eat from one set of feeders while ignoring other feeders. Social facilitation of eating begins early; all members of a litter nurse together. Furthermore, several litters of pigs in a farrowing house will nurse at the same time. Tape recordings of nursing sounds have been used with some success to increase the frequency of nursing and weight gain of sucklings (Stone et al. 1974).

Social facilitation may increase food intake, but this tendency can be offset by the opposing tendency of subordinate pigs to eat less in the presence of dominant pigs. Pigs may refrain from eating even in the absence of overt aggression by the dominant pig (Baldwin and Meese 1979).

Palatability. The adult set point of body weight discussed above applies to an animal on a given diet. An increase in palatability can result in a shift upward in body weight set point. More simply put, animals eat more and gain weight when their food tastes good. Pigs show a marked preference for sweet substances (Kare et al. 1965; Kennedy and Baldwin 1972), consuming up to 17 liters of sucrose solution/per day. Advantage has been taken of this preference to increase intake (Aldinger et al. 1959). The intake and weight gain of pigs are not affected by the addition of a substance that tastes very bitter to humans, indicating a species difference in taste perception (Blair and FitzSimons 1970). Newly weaned pigs often show a drop in weight gain and sweet pig starters are used in an attempt to stimulate intake. Although suckling neonatal pigs show nearly as strong a preference for the various sugars as do adults (Houpt and Houpt 1976), these attempts are not always successful. A more rewarding approach may be to add to the sow's feed a flavor that will appear in her milk, and then to add the same flavor to the pig starter. Pigs will learn to associate a given flavor with a familiar food (in this case sow's milk) and they will more rapidly ingest that flavor in solid food rather than a strange flavor (Campbell 1976). As a result of testing 129 flavors, Mclaughlin et al. (1983) found that a cheesy flavor would increase the intake of newly weaned pigs.

Food Intake Depressants

Increase in Body Weight above Set Point. Animals that are force-fed will, as pointed out, decrease their food intake until their body weight reaches the pre-force-feeding level. Similarly, animals that have gained weight while being injected chronically with insulin will, when treatment ceases, decrease their intake and lose weight until their weight is at the preinjection set point.

High Environmental Temperature. Food intake is inversely related to environmental temperature; therefore, in hot weather animals eat less. The classical explanation is that "animals eat to keep warm and stop eating to prevent hyperthermia" (Brobeck 1955). Certainly inhibition of food intake in hot weather reduces specific dynamic action and other metabolic heat as further heat loads to the animal. Under normal conditions, food intake increases or decreases in response to environmental rather than body temperature; but when body temperature rises to pathological levels, as in fever, food intake also decreases.

Estrogen Levels. When female pigs are in estrus, food intake is depressed and activity levels rise. Gilts eat 4 kg less during a week in which they are in estrus than in weeks when they are in another stage of the reproductive cycle (Friend 1973). The increase in activity has been quantified: sows in heat walk 14,000 steps per day, whereas sows that are not in heat walk 5000 steps per day (Altmann 1941). Observation of the food intake and general activity of a sow can help the caretaker identify a female in estrus.

Central Nervous System Stimulants. Central nervous system stimulants, in particular amphetamines, have been used extensively in human medicine to treat obesity. Less stimulating and less abused but less effective drugs, like phenylpropanolamine, are also used as over-the-counter aids to dieting. Chemical control of obesity is rarely used in domestic animals. Domestic animals' weight is controlled by the livestock producer, and in rapidly growing meat-producing or lactating animals, depression of food intake is generally avoided. Amphetamines apparently affect neurons or neurotransmitters that mediate satiety or determine set point. Stimulants increase satiety and lower body weight set point.

Imbalance of Dietary Amino Acids. Animals generally show nutritional wisdom in that they select an adequate amount of protein. For example, growing pigs given a choice of a protein-free or adequate protein diet ingested sufficient protein for maximal weight gain (Devilat et al. 1970). When the protein contains an imbalance of essential amino acids, however, nutritional wisdom is not shown, and pigs will choose a nonprotein diet over a protein diet (Robinson 1975). The explanation for the marked depression of food intake seen when the imbalanced diet is the only one offered, or for the selection of no protein rather than an imbalanced protein, is unknown. It is speculated that an imbalance in essential amino acids leads to an imbalance in central nervous system neurotransmitters. Accumulation of a neurotransmitter might depress intake. The area of the brain involved in this behavior may be the cerebral cortex rather than the hypothalamus (Noda and Chikamori 1976).

Gastrointestinal Factors. None of the many factors that stimulate or depress intake has been involved in the control of normal meal initiation and termination. Glucoprivic eating is an emergency mechanism. Body weight set point is maintained, but over a period of days, not hours. Environmental temperature also influences food intake over a period of days. The most likely candidates for the meal-to-meal controllers of intake are those factors that are closely associated with the act of eating. Following a meal there are changes in the gastrointestinal and plasma level of various constituents, any one, or a variety, of which might influence food intake. Food intake ceases before intestinal absorption is complete. If it did not, animals would consistently overeat because food would continue to be ingested during the lag between ingestion of food and its absorption. Gastrointestinal hormones are released as soon as food is present in the upper gastrointestinal tract and may be satiety signals. Of the many gastrointestinal hormones, such as gastrin, secretin, and enterogastrone, cholecystokinin-pancreozymin (CCK) holds the most promise as a satiety agent (Gibbs et al. 1973). The hormone is released from the mucosa of the upper gastrointestinal tract. Cholecystokinin-pancreozymin has several physiological actions: it stimulates contraction of the gallbladder; it stimulates release of pancreatic enzymes; and it has been shown to inhibit food intake in hungry animals, including pigs (Houpt et al. 1979b; Houpt et al. 1979c). Cholecystokinin acts on the stomach, presumably by slowing gastric emptying, to produce feelings of satiety.

Satiety after meals high in protein or fat may be mediated through the release of cholecystokinin-pancreozymin, but another satiety mechanism may exist for high-carbohydrate foods. Suckling pigs depress food intake following gastric loads of isotonic glucose solution but not following gastric loads of isotonic sodium chloride solutions, which indicates that glucoreceptors may be in the gastrointestinal tract (Stephens 1975; Houpt et al. 1977) producing satiety. Hypertonic solutions of either glucose or sodium chloride depress intakes of both suckling and more mature pigs (Houpt et al. 1977; Houpt et al. 1983a,b). As food is being digested, the osmotic pressure in the intestine rises and stimulation of osmoreceptors in the intestine may be part of the basis of preabsorptive satiety (Figs. 8.3 and 8.4).

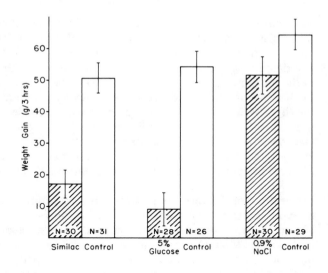

Fig. 8.3. Effect of various gastric loads on 3-hour intake of suckling pigs. Note that milk (Similac®) and isotonic glucose, but not isotonic saline, depressed intake (Houpt and Houpt 1976, copyright © 1976, with permission of Oxford University Press).

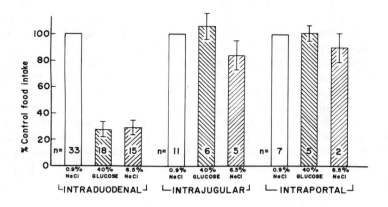

Fig. 8.4. Effect of hypertonic injections administered into the duodenum *(left)*, the jugular vein *(center)*, or the portal vein *(right)* on subsequent food intake of pigs. Only intraduodenal injections depressed intake relative to saline-injected controls.

Meal Patterns. Pigs are essentially diurnal animals; therefore, most feeding takes place during the day (Auffray and Marcilloux 1983). Feeding behavior is a circadian rhythm modified by environmental temperature. Pigs avoid eating when daily temperatures are highest, and, therefore, eat early in the morning and late in the evening (Feddes et al. 1989). Although social facilitation of feeding occurs in pigs, they eat simultaneously only 30% of the time. When meals of individual pigs were recorded, the pigs were found to eat 8 to 12 meals per day, the number of meals decreasing as the pig grew larger. Many "snacks" of less than 50 gm are also eaten (Bigelow and Houpt 1988). In contrast to anosmic rodents, anosmic pigs ate with the same meal patterns as intact pigs (Baldwin and Cooper 1979).

Integration of Factors That Stimulate and Inhibit Intake in the Central Nervous System. The controls of feeding in pigs have been reviewed by Houpt (1984). All the factors discussed — environmental temperature, rate of glucose utilization, palatability, social factors, estrus, gastrointestinal hormones, and the presence of glucose or hypertonic substances in the gastrointestinal tract — are operating at the same time. The information is integrated, presumably in the brain, and the animal eats more or stops eating accordingly. The role of the various neurotransmitters remains unclear (Jackson and Robinson 1971). Such factors as social facilitation and avoidance of protein-imbalanced diets are mediated through the cerebral cortex. Changes in food intake in response to temperature are mediated through the anterior hypothalamus.

Two areas have been identified that may integrate at least some of the information carried to the central nervous system from the rest of the body: the lateral and ventromedial hypothalamus. Stimulation of the lateral hypothalamus causes animals to eat. Lesions in the lateral hypothalamus of pigs cause a decrease in food intake and body weight (Khalaf 1969) whereas destruction causes animals to stop eating, even to die of voluntary starvation. The ventromedial hypothalamus is involved in determination of body weight set point. Destruction of that area results in hyperphagia and a rapid gain of weight until a new set point is reached, at which time food intake drops and body weight is maintained at the new, much higher levels. Increase in food intake and body weight has been observed in pigs following ventromedial hypothalamic lesions (Khalaf 1969). Use of brain lesions in meat-producing animals is probably not practical because the increase in weight is nearly all due to an increase in fat (Auffray 1969) (Fig. 8.5).

Clinical Problems

Polydipsia. Scheduled induced polydipsia has been produced experimentally in pigs (Stephens et al. 1983). Sows in a farrowing crate may show polydipsia, perhaps because the waterer may be the only thing available for them to manipulate (Rushen 1984). Pigs that are fasted or on a severely

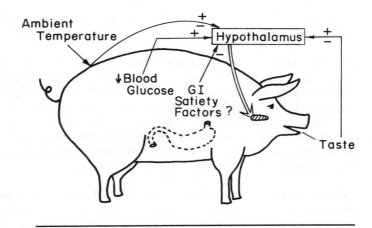

Fig. 8.5. Integration of factors that stimulate and depress food intake in the brain.

limited ration may also increase their water intake (i.e., show polydipsia) (Yang et al. 1981, 1984). Notice that in the free feeding pig or one that is only food deprived for a few hours, drinking accompanies eating, but in a very hungry one drinking occurs as a compensation for lack of food.

CONTROL OF FOOD INTAKE IN DOGS
Food Intake Stimulants
Defense of Body Weight. Dogs can increase their volume of intake when their diet is diluted (Janowitz and Grossman 1949b). Dogs deprived of food will eat when food is again available and, thereby, defend their body weight. Different breeds of dogs defend different body weights and different degrees of adiposity. A casual comparison of bulldogs and salukis makes this clear. The problem of overeating and obesity in dogs will be discussed in the section on abnormalities of ingestive behavior. Dogs can also regulate their protein and energy intake when offered diets differing in protein content. They chose a 30% protein diet (Romsos and Ferguson 1983).

Low Environmental Temperature. Food intake is believed to be controlled, in part, by the regulation of body temperature. Consequently, animals tend to eat more in the cold and less in the heat. This response has been demonstrated in environments as diverse as Alaska and Florida and with breeds as different as huskies and beagles. Food intake doubled in the winter as temperatures fell from a summer high of 20°C to a winter low of −17°C (Durrer and Hannon 1962). Indoor dogs eat less than those housed outside.

This reduction in intake is probably owing to the warmer indoor environment (Kuhn and Hardegg 1988).

Decrease in Glucose Utilization. Glucoprivic eating is seen in dogs. Dogs treated chronically with insulin eat more and gain weight (Grossman et al. 1947), and eating in response to glucoprivation produced by an inhibitor of glucose utilization has been demonstrated in dogs (Houpt and Hance 1969). A lack of glucose stimulates intake, but an excessive dose does not depress intake. Neither glucose injected directly into the portal vein of dogs nor a similar dose given via the jugular vein depresses intake (Janowitz et al. 1949; Bellinger and Williams 1989).

Central Nervous System Depressants. The effect of central nervous system depressants on food intake in dogs has not been systematically studied, but there is anecdotal evidence that dogs recovering from anesthesia will eat ravenously before they are fully conscious. Because of the danger of choking in dogs recovering form anesthesia, food should not be available to them. There is also anecdotal evidence that treatment with progestational agents increases food intake. This side effect of progestins used to prevent ovulation should be further investigated.

Social Facilitation. Many animal owners have noted anecdotally that the addition of another animal to the household increased the original pet's interest in food. At times the increase can be a pathological hyperphagia or increased food intake (Fox 1968). Social facilitation of food intake has been quantitatively measured in puppies (James and Gilbert 1955).

Palatability. Taste preferences must be determined for each species; one cannot assume that because something tastes good to humans, it tastes good to animals. Dogs prefer meat to a high-protein, nonmeat diet, and they prefer one meat over another. These preferences, in order, are beef, pork, lamb, chicken, and horsemeat (Lohse 1974; Houpt et al. 1978b). Not only flavor but also the form in which the food is offered is important in palatability. Dogs prefer canned or semimoist food to dry food (Kitchell 1972). They prefer canned meat to the same meat freshly cooked and prefer cooked to raw meat (Lohse 1974). Dogs do not prefer familiar food; in fact, they prefer a novel flavor of canned food. Puppies also tend to prefer novel food, but palatability and maternal effects may outweigh these preferences (Mugford 1977; Ferrell 1984) (Fig. 8.6). Dogs familiar with semimoist food eat only a little canned food at first, when both are available, but soon prefer the canned. Dogs also have an innate preference for sucrose in liquids or solid food (Grace and Russek 1969). Many semimoist dog foods contain large (25%) percentages of sucrose, which serves not only as a preservative, but to increase a palatability. They are unusual in preferring

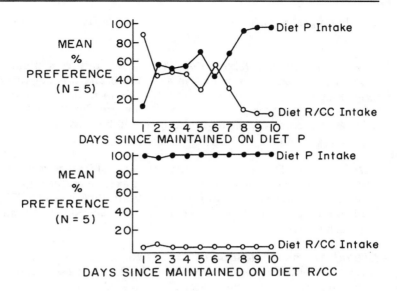

Fig. 8.6. Effect of novelty on food preferences. Dogs normally prefer diet P to diet R/CC but will show preference for either R/CC or P if it is a novel diet. Preference for P is sustained *(lower graph)*, but that for R/CC is reversed in a week *(upper graph)* (Mugford 1977, copyright © 1977, with permission of Monell Chemical Senses Center and Academic Press).

fructose to sucrose (humans and pigs prefer sucrose). Lactose, but not maltose, is also preferred.

Palatability does not depend on taste alone because elimination of olfaction (anosmia) eliminates preference for one meat over another, although anosmic dogs still preferred meat or a sucrose diet over a bland cereal diet (Houpt et al. 1978b). Odor is important, but only for detecting minor differences between food, as in distinguishing lamb from pork. Such major preferences as that for meat over nonmeat and for sucrose are not affected. Odor is most important for locating food (Rogers et al. 1967); taste is most important for identifying food.

Food Intake Depressants

Increase in Body Weight above Set Point. When dogs were given additional food each day through gastric fistulas they decreased their oral intake but not enough to prevent weight gain (Share et al. 1952). The lack of caloric compensation may be caused by the intragastric route of force feeding:

oropharyngeal signals were not elicited by the food, so satiety was not achieved.

High Environmental Temperature. Low temperatures stimulate food intake; high temperatures depress intake. Therefore, it is not surprising that dogs may be anorexic in warm weather. Temperature also affects feeding when a dog is transported from a cold to a warm climate. Unless other signs of disease accompany the anorexia, it should be considered part of normal thermoregulation.

Estrogen Level. During estrus, bitches tend to eat less; conversely, removing the source of estrogen stimulates food intake. Therefore, one of the side effects of ovariohysterectomy is a tendency for the spayed bitch to eat more and gain weight (Fig. 8.7). Presumably, this is caused by the removal of the estrogen source.

Central Nervous System Stimulants. High dosages of amphetamines depress food intake in dogs (Corson et al. 1972). Their use as a treatment for canine obesity has not been carefully evaluated.

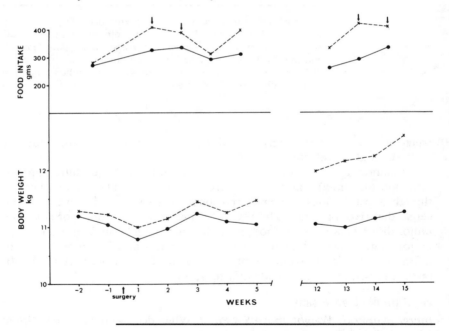

Fig. 8.7. Effect of ovariohysterectomy on food intake and body weight of beagle bitches. Solid lines = intact. Dashes = spayed bitches (Houpt et al., 1979, Effect of sex, copyright © 1979, with permission of *J. Am. Vet. Med. Assoc.*).

Gastrointestinal Factors. Oropharyngeal factors alone are not sufficient to induce satiety. Dogs surgically prepared with esophageal fistulas can swallow food, but if the fistula is open, the food will drop from the esophagus and not reach the stomach (sham feeding). In this case, dogs will sham feed for hours, stopping only to rest. It is interesting that placing food into the stomach at the same time that the dog is eating and swallowing food does inhibit further sham feeding (Janowitz and Grossman 1949a). Apparently, in dogs at least, both oropharyngeal and gastric stimuli are necessary to induce satiety. It has also been found that gastric loading immediately before a meal does not markedly inhibit food intake unless more than half of the normal meal is placed in the stomach.

Food placed in the stomach just prior to a meal does not inhibit food intake, but a gastric load given 20 minutes before a meal does (Janowitz and Grossman 1949a). The delayed effect of a gastric load indicates that food must be absorbed or that some humoral factor must be released by the presence of the food in the upper gastrointestinal tract to inhibit food intake. There is evidence in various domestic animals that both the products of digestion and humoral factors inhibit food intake. Therefore, glucoreceptors in the liver or portal system may be involved in satiety.

Food intake of dogs is suppressed by CCK and glucagon and markedly depressed by naloxone (Levine et al. 1984). Neither intraduodenal fat nor peripherally administered cholecystokinin suppresses sham feeding in dogs, although intraventricularly administered CCK does (Pappas et al. 1985). The difference in the effect of CCK on sham feeding and normally feeding dogs indicates the importance of integration of gastrointestinal and humoral events for satiety. The effect of naloxone indicates opiates stimulate intake.

Meal Patterns. Dogs that have free access to food 24 hours a day eat many small meals a day, mainly during daylight hours (Mugford 1977). This meal frequency indicates that dogs are diurnal rather than nocturnal; and that once-a-day feeding is not "natural," or at least not preferred, by dogs. Beagles living in individual kennels with dry food freely available tend to eat three times a day: at dawn, at dusk and whenever fresh food is given (Rashotte et al. 1984).

Integration of Factors That Stimulate and Inhibit Intake in the Central Nervous System. Dogs lesioned in the ventromedial hypothalamus show hyperphagia and weight gain (Rozkowska and Fonberg 1973). They also show hypersecretion of gastric acid (Nagamachi 1972). The excess production of hydrochloric acid may or may not be related to the increase in food intake that has been noted in laboratory rats and dogs with lateral hypothalamic or dorsomedial amygdalar lesions (Fonberg 1976).

Clinical Problems

Obesity. Obesity is the most common behavioral problem involving ingestion. The cause is simply an intake of energy that exceeds the output of
energy. Therefore, obesity is most often observed in a nonworking animal
fed a highly palatable diet.

Clinical reports (Mason 1970; Anderson 1974; Darke 1978) indicate
evidence of obesity in 20% to 30% of the dogs and 10% of the cats brought
to these clinics. These studies and a considerable body of anecdotal information indicate that twice as many spayed as unspayed bitches are obese.
The reasons for this were discussed in the section on controls of feeding in
dogs.

There are differences in breed susceptibility to obesity. When a palatable food is made freely available to beagles, some get very obese; in a
similar situation, terriers do not become obese (Mugford 1977) (Fig. 8.8).

Fortunately, obesity in animals is easily controlled by food restriction
and use of commercially available low-caloric dog foods, and most owners
find it easier to control their pets' weight than to control their own.

Anorexia. A much more difficult clinical problem is anorexia in dogs. An
example was a Siberian husky that would not eat. The dog's owner cooked
steak, eggs, stews, almost anything, but the dog was chronically under-
weight and ate intermittently. The dog was examined thoroughly, and no
organic cause for the ingestion problem was found. Discussion with the
owner revealed that she had simply taught the dog not to eat and that this
was the principal way in which the dog secured the attention of its owner,
who led a very busy life. This was the finicky eater gone to the extreme.
When the dog was 8 weeks old, the owner adopted the habit of sitting with
the dog while it ate and coaxing it along. The owner, a nursing home
nutritionist, was very concerned with eating. The first time the pup did not
eat it was whisked off to the veterinarian. The owner did not believe the
advice she got from this veterinarian, or several others, and instead of just
waiting for the pup to get hungry, went home and cooked a large meal of
eggs for it. This pattern persisted. Finally, not even steak excited the dog,
and it lived on dog biscuits that were always available and ice cream once a
day while holding out for novel foods. The treatment was to put the dog on
regularly scheduled meals with no treats and to wait. Within 3 weeks and
after a 5-day stretch of not eating, the dog was eating one good meal a day
and gaining weight.

Pica. Another abnormal ingestive behavior is pica, eating materials that are
not normally food. Puppies are notorious swallowers of inappropriate objects that must be surgically removed from the gastrointestinal tract. Occasionally pica may be the sign of a deficiency, but more frequently it is
simply an extreme form of oral exploration. One of the most common and

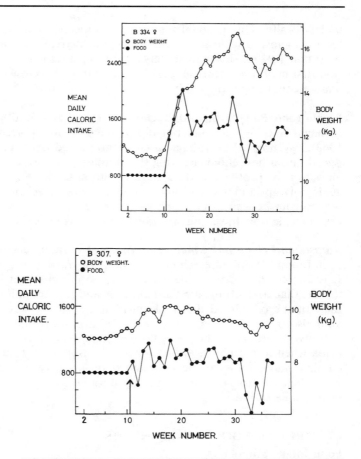

Fig. 8.8. Food intake and body weight of two beagles. Ad libitum feeding began at arrow. *Lower graph* is of a dog whose intake and body weight remained normal. *Upper graph* is of a dog that over-ate and became obese (Mugford 1977, copyright © 1977, with permission of Monel Chemical Senses Center and Academic Press).

most disturbing forms of pica is coprophagia in dogs. Unless the feces contain viable parasite ova, coprophagia affects the owner's aesthetic values more than the dog's health. The first approach should be a simple change in diet. Sprinkling pepper or some other noxious substance on feces may inhibit the vice. A better approach is to inject hot sauce into the center of several fecal masses so that the dog cannot tell that they have been adulterated. A much more effective treatment is to administer pancreatic enzyme tablets orally. This probably imparts an objectionable taste to the dog's feces. A more drastic treatment would be to treat the animal with apomorphine (.04 mg/lb) immediately after it ingests the feces. The taste aversion that may result from associating the eating of feces with vomiting and nausea could break the habit (Gustavson et al. 1974).

Psychogenic Polyphagia and Psychogenic Polydipsia. Both psychogenic polyphagia and psychogenic polydipsia have been reported in dogs (Fox 1968). Psychogenic polyphagia is frequently noted when a rival pet is introduced into a household. Feeding the animals in separate places sometimes alleviates the problem as does extra attention to the original pet. Psychogenic polydipsia may occur, but care must be taken to differentiate it from the polydipsia secondary to polyuria. Water intake may be restricted in psychogenic polydipsia but would be dangerous in pathogenic polydipsia.

Adipsia and Hypernatremia. This syndrome is rarely encountered but has been found in miniature Schnauzer dogs (Crawford et al. 1984) and in cats. It is interesting because it demonstrates two principles of ingestive behavior: (1) that controls of thirst are multiple and (2) that animals will increase their intake of diluted diet to keep their caloric intakes constant. These animals do not drink in response to the osmotic stimuli, so they present with hypernatremia. They will, however, respond to hypovolemia, so they drink when furosemide is administered. To insure that the animals consume enough water, they can be maintained on a slurry of canned food and water. The controls of food intake are normal, so they inadvertently ingest water while eating.

CONTROL OF FOOD INTAKE IN CATS
Food Intake Stimulants

Defense of Body Weight. The most important determinant of body weight in cats appears to be cyclical. Cats lose and gain body weight in cycles of several months' duration (Randall and Lakso 1968). For this reason, it has been difficult to demonstrate defense of body weight. In two studies, when their diets were diluted, cats did not eat more and, therefore, lost weight (Kanarek 1975; Hirsch et al. 1978). In both studies, the diet, a dry food, was diluted with a dry diluent, either kaolin or cellulose. The cats actually ate a smaller volume than they ate of the undiluted food, which indicated that it was unpalatable. In contrast, when cats' food is diluted with water, they do compensate by increasing the volume of intake and maintaining constant caloric intake (Mugford 1977; Castonguay 1981). Because the cats were consuming more of a diet when it was diluted with water, their water intake was increased. A watery cat food may be used to increase water intake when clinically indicated, that is, in cats with urolithiasis or hypernatremia (Carver and Waterhouse 1962).

Low Environmental Temperature. There has not been a complete study of the effect of temperature on food intake in cats, but one demonstration showed that a cold environmental temperature, rather than changes in brain or body temperature, affects feeding. This study found that cats drinking milk show a decline in brain temperature but also cease to eat (Adams 1963). If brain temperature was the factor affecting feeding, one would expect the cat to eat even more as its brain temperature fell.

Decrease in the Rate of Glucose Utilization. In contrast to most other species, cats have not been shown to respond to glucoprivation with an increase in intake. The glucose analogue, 2-deoxy-D-glucose, failed to stimulate feeding in cats (Jalowiec et al. 1973), but the failure to stimulate may have been a result of species differences in the effective dose of the drug rather than of species differences in the control of food intake.

Central Nervous System Depressants. Like dogs, cats increase their food intakes when treated with oral progestins. The same hormone might be used to stimulate feeding in anorexic cats, but there are dangerous side effects. The benzodiazepine tranquilizers also increase food intake in cats and may be used clinically for that purpose (Fratta et al. 1976). Direct application of depressants to the central nervous system has also stimulated feeding in cats (Feldberg 1959).

Social Facilitation. Cats do not increase their food intakes when housed in groups rather than individually, nor does the sight of one cat eating stimulate other cats to eat when food is freely available. Nevertheless, some cats will eat only in the presence of the owner or only while being petted.

Palatability. Cats are notoriously finicky, which reflects the strong influence of palatability on food intake in cats. There is one striking peculiarity in cats: they do not prefer sucrose as most animals do. Cats do not show a sucrose preference for aqueous solutions of sugar, but they do prefer sucrose in salt solutions. "Water" taste may block the expression of a sweet preference in cats (Bartoshuk et al. 1971). The feline preference for sucrose in saline solutions soon turns to aversion because vomiting and diarrhea result in the sucrase-deficient adult cat.

Cats will not eat diets containing medium chain triglycerides, or hydrogenated coconut oil. This aversion may be because they break the triglycerides down to fatty acids in their mouths (cats possess lingual lipase) and are peculiarly sensitive to bitter tastes. They will avoid as little as 0.1% caprylic acid or .000005 M quinine. In comparison rabbits and hamsters avoid quinine only at .002 M.

Cats prefer fish to meat and, unlike rodents, prefer novel diets to familiar ones (Hegsted et al. 1956). Cat food manufacturers, no doubt, take advantage of both of these feline preferences. If the new diet is not more palatable than the familiar one, the cat will, after a few days, begin to choose the familiar food (Mugford 1977). While cats are drinking milk their EEGs are synchronized as if they were drowsy (Cervantes et al. 1983).

Food Intake Depressants. Few studies have been done on the effects of environmental temperature, estrogen levels, or imbalances of amino acids on feline food intake. Ovariohysterectomized cats may become obese, but the low rate of obesity among cats (Anderson 1974) is evidence that this does not often occur. Castration of male cats may also lead to an increase in body fat content, but this has not been objectively demonstrated either. Food intake may be depressed when the set point of body weight of the cat falls during that portion of its cycle (Randall et al. 1978). Anorexia and weight loss of several hundred grams may cause the owner distress but are normal unless the anorexia is persistent or complete or other signs of disease are present.

Gastrointestinal depressants of feeding have been studied to some extent. Glucoreceptors that suppress feeding in cats appear to be in the liver (Russek and Morgane 1963). CCK and bombesin, a peptide related to CCK, suppress food intake of cats (Bado et al. 1988; Bado et al. 1989).

Opiates. The role of endogenous opiates in feeding has not been well investigated. They are probably important because the opiate blocker naloxone suppresses intake in most domestic animals, including cats (Foster et al. 1981).

Meal Patterns. Cats, like dogs, eat many small meals (12) per day when given free access to food. One might argue that this intake pattern is not natural, yet the caloric intake per meal is approximately equal to that contained in one mouse. A feral cat with good hunting skills might easily catch 12 mice (or 3 rats) per day. Unlike dogs, cats eat both in the light and in the dark (Mugford 1977).

Integration of Factors That Stimulate and Inhibit Intake in the Central Nervous System. Cats have been the subject of many studies of the central nervous system. Stimulation of the hypothalamus electrically, for example, influences food intake. It has been shown that cats with electrolytic lesions in the ventromedial hypothalamus become obese (Anand et al. 1955), whereas cats with lesions in the lateral hypothalamus are both aphagic and adipsic (Anand and Brobeck 1951). Lesions in the preoptic area may also lead to an increase in food intake (Wyrwicka 1978) as well as depress sexual behavior and scent marking (Hart and Voith 1978).

Clinical Problems

Anorexia. Anorexia is seen most often in cats. A moderately sick cat may further compromise its health by refusing to eat.

Although an otherwise healthy cat may fast for several days and will then be willing to eat large amounts or even hunt prey (Adamec 1976), anorexia can be a serious clinical problem. Anorexia is most commonly encountered in hospitalized cats or in originally healthy cats that are placed in a boarding kennel. Although an animal can be maintained by intragastric feeding, it is far more beneficial to the animal to reinstate voluntary eating because food taken by mouth stimulates gastric and intestinal secretions much more than does food given by stomach tube.

Advantage can be taken of the fact that depressants stimulate intake. The benzodiazepines, in particular diazepam, have been used to stimulate intake in anorexic patients. A common approach to this form of anorexia is treatment with benzodiazepine tranquilizers, which stimulate food intake in laboratory animals (Wise and Dawson 1974) and in horses (Brown et al. 1976).

Plant Eating. Cats frequently eat grass. This behavior is observed in free-ranging, prey-killing cats, as well as in those eating canned or dry diets. It is not surprising, therefore, that cats may eat house plants. Plant eating can have serious consequences to the cat because so many house plants are poisonous, but whether the plant or the cat is at risk, the behavior is undesirable. If owners observe their cat eating a plant, they can punish the cat or frighten it away, but the cat, like the destructive dog, is most apt to misbehave in the owners' absence. The best solution is to provide the green plants that the cat apparently needs and certainly desires. Plants that are safe for cats can be purchased at pet suppliers. While the cat is learning that it is permitted access to one plant, the decorative plants can be moved out of its reach. Later, the cat should be taught to discriminate the plants it must not eat from those that it may. A water gun can be used to punish the cat and aid in the discrimination process, but that approach depends on the owner's presence. The cat would have to be separated from the plants in the owner's absence. Another method is to spray the leaves of the plant with a hot pepper solution. Still another method that would prevent elimination in the pots of large plants as well as eating or climbing the plants is to place mouse traps in the pot upside down so that they will snap shut, and scare the cat, if the cat jostles the pot or the soil.

Wool Chewing. Wool chewing or sucking is a behavior problem that occurs with greater frequency in Siamese or Burmese cats than in other breeds. It should be differentiated from the nonnutritive suckling of many early weaned kittens. Wool chewing may or may not be related to early weaning. In comparison to the age at which free-ranging cats are weaned (6 months) most domestic kittens are weaned early. Wool sucking is usually not presented as a clinical problem until the cat is an adult. The behavior is characterized by chewing with the molars rather than incisors. The material chewed is usually wool, but in the absence of wool, the cat will generalize to other materials including upholstery. The cats do not chew on raw wool, but prefer knitted or loosely woven material. The behavior is sporadic, but large holes can be produced in a matter of minutes. The behavior seems to be related to feeding because (1) fasting stimulates the behavior and (2) access to plants or bones or even dry, as opposed to moist, food decreases the incidence of the behavior. There is no evidence of a nutritional deficiency, but it appears to be a craving for fiber or indigestible roughage. For that reason, treatment should be aimed at supplying fiber. A higher fiber diet should be fed and safe plants made available. If that does not eliminate the problem, the owner can give the cat wool, such as an old sweater or sock. To teach the cat to differentiate between things it is allowed to eat and those it is not, the owner can treat the wool object with cologne and a solution of hot pepper sauce. The principle is that the cat will learn to associate the smell of the cologne with the unpleasant taste and avoid objects that smell of the cologne. Then articles of clothing, blankets, and so on can be sprayed with the cologne alone to deter the cat. In severe cases in which the owner is considering euthanasia, feeding the cat one raw chicken wing per day should be recommended. The danger of injury to the gastrointestinal tract from the bones is less than the danger of euthanasia, particularly because many cats kill and eat birds with impunity.

CONTROL OF FOOD INTAKE IN HORSES

Food Intake Stimulants. The controls of feeding in horses have been reviewed by Ralston (1984). There have been several studies of the plants that ponies and horses choose to eat on pasture. Tyler (1972) found that the New Forest ponies of England eat eight species of plants but avoid the poisonous *Senecio*. These ponies will eat acorns, an overconsumption of which leads to acorn poisoning. Horses have been reported to become ill repeatedly after eating buttercups (*Ranunculus*) (Zahorik and Houpt 1977). Even in the laboratory, ponies find it difficult to learn to avoid a feed that poisons them; the ponies fail to learn if the illness does not occur immediately (Houpt et al. 1990; see Chap. 7). Archer (1973) found that of 29 species of grass, horses and ponies preferred timothy, white clover (but not red clover), and perennial rye grass. Dandelions were the most preferred herbs.

Central Nervous System Depressants. Brown et al. (1976) in a preliminary study found that the central nervous system depressant, diazepam, stimulates food intake in horses and that the commonly used tranquilizer, promazine, had a similar effect. Horses, in common with other more thoroughly studied species, increase their food intake when the brain, presumably the areas involved in satiety, is depressed (Fig. 8.9).

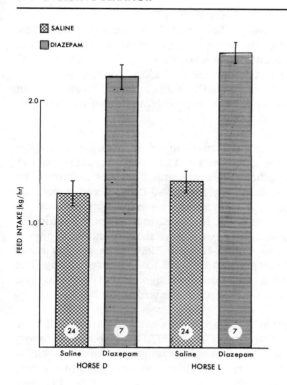

Fig. 8.9. Intake of two horses following intravenous saline (*hatched*) or diazepam (*lined*).

Palatability. Randall et al. (1978) have reported the basic taste preferences of immature horses. They show a strong preference for sucrose, but no preferences for sour (hydrochloric acid), bitter (quinine), or even salt solutions. At high concentrations, all the solutions except sucrose were rejected. In a study of adult ponies, nine of ten showed a strong preference for sucrose (Hawkes et al. 1985). Salt intake varies from 19 to 143 gm/day (Schryver et al. 1987). Presumably, a sodium-deficient horse would show a salt preference.

Grazing. Studies of grazing behavior indicate that horses are selective grazers. For example, in the Mediterranean climate of the Camargue, horses consume graminoids in the marshy areas, moving to the less preferred long grasses in the winter (Mayes and Duncan 1986). Similarly, meadow and shrubland were the vegetation types grazed by feral horses in the Great Basin of Nevada, but food preferences and nutritional needs had to be weighed against the dangers of exposure in the winter and the irritation of insects in the summer (Berger 1986).

Caloric Monitoring and Body Weight Set Point. Horses can regulate their energy balances by controlling the amount they ingest. In other words, horses do not eat a fixed amount or eat until they are ill, but increase or decrease the amount they eat to compensate for caloric dilution or enrichment (Laut et al. 1985). The pony gradually shifted to ad libitum feeding does not colic but does gain weight and is, therefore, in danger of laminitis, but the body weight reaches a plateau.

Meal Patterns. The normal feeding pattern of horses, as detailed in Chapter 3, is to graze continuously for several hours and to rest for longer or shorter periods depending on the weather conditions and distances that must be traveled to obtain water and sufficient forage. When offered a pelleted complete diet, a similar feeding pattern emerges — many long meals. It is interesting that whether the horse is grazing or eating pellets the statistical definition of a meal (feeding with breaks no more than 10 minutes long) is the same (Laut et al. 1985; Mayes and Duncan 1986). The grazing horse chews at a rate of 30 to 50 bites per minute for 8 to 12 hours per day. The frequency of oral vices in stalled horses on low roughage diets is probably a reflection of the oral activity seen in a natural environment.

Social Facilitation. Horses eat when other horses eat and eat more if they can see another horse eating (Sweeting et al. 1985). This response is important when creep feeding a foal and when encouraging an anorexic horse to eat. Horses appear to prefer to eat from the floor and from shallow buckets or mangers. This position enables the horse to see in all directions between its legs and may have evolved as an antipredator strategy. Grazing is, of course, from the ground rather than from the usual chest height of a manger. The problem with confined horses is to prevent ingestion of parasite ova while still encouraging intake.

Food Intake Depressants. The horse appears to depend more on pregastric signals to satiety than other species. When horses sham feed, that is, when most of what they have swallowed falls out of an esophageal fistula, they do not eat more than they do when the food reaches their stomachs (Ralston 1984). This response is in contrast to dogs that will eat very large meals under the same circumstances. Eventually the sham-fed horses compensate for the lost feed by eating sooner than normal, but apparently the signal for satiety is some form of oropharyngeal metering, that is, 25 swallows means enough has been eaten.

Another indication that chemoreceptors in the gastrointestinal tract are not important controls is that, when administered either intragastrically or intracecally, neither glucose nor cellulose — the main digestible constituent of grass, and its breakdown products, the volatile fatty acids solutions — depress intake until they have been absorbed. Intake is suppressed but only after a long latency. There is a mechanism for controlling caloric intake, but it appears to be postabsorptive (Ralston and Baile 1983; Ralston et al. 1983).

The first sign of colic in most horses is anorexia. This clinical observation as well as controlled studies indicate that pathological distention of the gastrointestinal tract inhibits feeding in the horse as in other species. This anorexia is probably mediated through pain receptors that travel in the sympathetic nerves. Analgesics intended for use in horses are often tested by determining whether a horse with a dilated cecum will eat when treated with the drug. In this case, the drug is probably working peripherally so that pain signals are not relayed to the brain, whereas diazepam works by inhibiting a centrally acting inhibitor of appetite.

Clinical Problems

Cribbing. Cribbing is an oral behavior in which the horse grasps a horizontal surface, such as the rim of a bucket or the rail of a fence, with its incisors, flexes its neck, and swallows air. Some horses swallow air without grasping an object, which is called aerophagia or windsucking; the latter term can also be used to refer to pneumovaginitis. Two and one-half percent of Thoroughbreds crib (McBane 1987; Vecchiotti and Galanti 1987). It is a clinical impression that cribbing occurs more frequently in confined horses but once established may persist even when the horse is on pasture. It may be a result, instead of or in addition to a cause, of gastrointestinal problems. Swallowing air may be a pleasurable sensation to an animal experiencing gastrointestinal discomfort. The one undeniable consequence of cribbing is excessive wear of the incisor teeth.

The simplest method used to prevent cribbing is to place a strap around the throat just behind the poll so that pressure is exerted when the horse arches its neck and even more pressure is exerted when the animal attempts to swallow. The horse is, in effect, punished for cribbing. If a plain strap does not suffice a spiked strap or collar can be used. A common observation of a horse wearing a cribbing strap is that it continues to grasp horizontal objects with its teeth, but does not swallow as much air. Many stables are designed or modified so that there are few horizontal surfaces available, but water and feed containers usually provide the horse some opportunity to crib. Shock has also been used to punish horses that crib. Apparently, release of endogenous opiates is involved in cribbing because blocking opiate receptors stops horses from cribbing but only while the drug is present, that is, the horse does not learn not to crib (Dodman et al. 1988).

There have been a variety of surgical treatments for cribbing: buccostomy, cutting the ventral branch of the spinal accessory nerve (9th cranial); myotomy of the ventral neck muscles; or a combination of partial myectomy of the omohyoideus, sternohyoideus, and sternothyroideus and neurectomy of the ventral branch of the spinal accessory nerve. The success rates of these treatments vary from 0% to 70% (Forssell 1926; Karlander et al. 1965; Hamm 1977; Firth 1980; Owen et al. 1980; Greet 1982; Turner et al. 1984).

Is it necessary to prevent cribbing? Unless the otherwise healthy horse is losing weight or suffering from colic as a result of swallowing air or flatulent colic, the vice is not interfering with the horse's well being. The noise of cribbing often annoys the owner, but that is not a good reason to subject an animal to the risk of surgery and the possible side effects of infection and disfigurement that may interfere much more with the animal's function than cribbing did. Cribbing is considered an unsoundness, but this may not be justified. It is believed that horses learn to crib by observing other horses cribbing, but this has never been proven. It is possible that the environment that causes one horse to crib causes the other horse in the same environment to do so. Young horses may be more likely to learn the habit from an adult than are other adults.

Wood Chewing. Both cribbing and wood-chewing horses grasp horizontal surfaces with their teeth, but the wood-chewing horse actually ingests the wood, whereas the only damage the cribbing horse does is to mark the wood with its incisors. In contrast to cribbing, wood chewing appears to have a definite cause—a lack of roughage in the diet. Several investigators have noted that high-concentrate diets or pelleted diets increase the incidence of wood chewing. Feral horses as well as well-fed pastured ponies have been observed to ingest trees and shrubs, so there is some need or appetite for wood even when grasses are freely available. Farm managers are well aware that trees, especially young trees, must be protected from horses on pasture. Horses cannot digest wood; nevertheless, there may be some role for indigestible roughage in equine digestion. Jackson et al. (1984) have found that wood chewing increases in cold, wet weather.

Eliminating edges, covering edges with metal or wire and painting the surface with taste repellents are the traditional methods for preventing horses from wood chewing but providing more roughage is a better practice both behaviorally and nutritionally. If roughage is provided, the horse's motivation to wood chew is reduced rather than thwarted. Boredom may be a factor; an increase in exercise may reduce the rate of wood chewing.

Polydipsia. Psychogenic polydipsia is also seen clinically in horses. The owner's main complaint is that the horse urinates frequently and copiously so that its stall is wet; in this case the polyuria is secondary to the polydipsia. In horses as in small animals the condition must be differentiated from diabetes insipidus and other causes of secondary polydipsia.

CONTROL OF FOOD INTAKE IN RUMINANTS. Ruminants will be considered as a group because their common anatomical and metabolic characteristics have profoundly influenced the controls of food intake.

Food Intake Stimulants

Defense of Body Weight. Cattle can, within limits, decrease their food intake as caloric density increases. Even young calves demonstrate this ability by decreasing their intake of milk replacer as the percentage of water in the replacer falls (Pettyjohn et al. 1963). The ability of cattle and other ruminants to respond to dietary dilution, that is, to increase their volume of intake as the nutrient density decreases, has been reviewed by Baile and Forbes (1974).

Low Environmental Temperature. Cold ambient temperature increases food intake in ruminants as in simple-stomached animals. This response can be observed not only when environmental temperature falls but also when the animals' heat loss is increased. Sheep, for example, eat more following shearing (Ternouth and Beattie 1970). At very low temperatures ($-10°C$), intake may be inhibited and the consequences of greater losses of energy of heat will be compounded by a decrease in energy intake.

Decrease in Glucose Utilization. Cross-circulation from hungry to satiated sheep stimulates intake of the satiated animal (and depresses intake of the hungry one), indicating the existence of humoral factors that influence feeding (Seoane et al. 1972). Glucose or insulin levels may be among the humoral factors. Because almost everything that the ruminant eats is exposed to rumen bacteria before reaching the intestine, very little dietary glucose becomes available to the animal. Instead, the ruminant depends on the volatile fatty acids for energy and produces glucose by the process of gluconeogenesis. Plasma glucose is low, approximately half that of simple-stomached animals. It was assumed, for these reasons, that ruminants were fairly independent of glucose utilization and would not be expected to eat in response to glucoprivation. However, it has been demonstrated in both sheep and goats that food intake can be markedly increased by insulin and the glucose analogue 2-deoxy-D-glucose (Houpt 1974).

Central Neural Mechanisms. The easiest generalization that can be made about central stimulants of intake is that factors that depress activity, such as anesthetics and opiates, increase feeding. For example, central nervous system depressants, such as calcium, barbiturates, and benzodiazepines, also stimulate intake in ruminants (Baumgardt and Peterson 1970; Martin et al., 1973a; Seoane and Baile 1973a,b; McLaughlin et al. 1976; Dinius and Baile 1977; Krabill et al. 1978).

As the number of known neurotransmitters and neuromodulators grows greater, the hope of completely understanding the pharmacology of central neural controls of feeding and other behaviors grows dimmer. Cholecystokinin, discussed as a peripheral satiety factor, also exists in the brain where it can also act to produce satiety in sheep (Baile et al. 1986). Immunizing lambs against endogenous CCK did not result in a significant increase in intake (Trout et al. 1989).

The opioid peptides are also involved in feeding behavior. The mu and kappa receptor agonists given intracranially appear to stimulate intake in sheep; gamma agonists depress intake (Baile et al. 1987).

Although sheep increase their food intake when norepinephrine is injected via the cerebral ventricles, cattle do not show the same response nor does a clear picture emerge when adrenergic agonists or antagonists are injected (Baile et al. 1972). Cerebrospinal fluid from hungry sheep stimulates intake of satiated sheep indicating that there are humoral factors within the brain (or the ventricles) that influence feeding (Martin et al. 1973a).

There has not yet been any practical application of these findings to increasing the food intake of fattening cattle or lactating dairy cattle. It is not clear whether these drugs, which stimulate intake over the short term, would, if administered chronically, produce a long-term increase in food intake or meat and milk production (Simpson et al. 1975). The problem of the effects of barbiturates and tranquilizers on human consumers also remains unknown. The latest approach is to stimulate production with bovine growth hormone or somatotropin and rely on the animal to increase intake to match the increased production (output).

Social Facilitation. Confined, isolated lambs eat less than grouped ones (Webster et al. 1972). Sheep eat less when in metabolism cages, which may be the result of a lack of social facilitation in a gregarious species, as well as a response to the stress of confinement (Foot and Russel 1978). Cattle will also eat more in groups than they will individually and first-calf heifers will eat more if grouped with older cows. Production may not be increased. These increases must be balanced against the negative effects of dominance disputes.

Palatability. Ruminants appear to have a definite sweet preference. Goats readily consume glucose solutions at concentrations as high as 40% (Bell 1959). Calves also show a glucose preference that is similar within a pair of twins (Bell and Williams 1959). Cattle reject 20% sucrose, but goats do not (Goatcher and Church 1970c). Sheep do not appear to have a very marked sweet preference. Nevertheless, it was only 5% sucrose and glucose that were preferred; higher concentrations produced no preferences or aversion. Lactose was rejected at a concentration of 2.5%. Sheep also show no preference for molasses for which cattle show a strong preference (Goatcher and Church 1970c).

Sheep are much more sensitive to unpleasant taste. They reject sour solutions of hydrochloric acid, acetic acid, lactic acid, or butyric acid at concentrations lower than those found in the rumen, yet the volatile fatty acids must surely be tasted when the sheep ruminates. A bitter solution of quinine (.004 M) was rejected, but not one of urea, even at concentrations high enough to produce fatal toxicity (Goatcher and Church 1970b). Experience also plays a role in that sheep that initially rejected quinine-treated hay, eventually accepted it even when nonbitter hay was available (Forbes 1986).

Although ruminants show a salt preference when sodium deficient (Denton 1967), only a very weak preference is shown by normal sheep, cattle, or goats (Goatcher and Church 1970c). Goats do not prefer bitter substances, but will accept quinine—adulterated water at nearly 10 times the concentration that a rat will (Bell 1959, 1963). Field studies confirm that goats have a high tolerance for bitter taste (Cory 1927). Cattle reject acetic acid at .08% and quinine at concentrations greater than 2.5% (Goatcher and Church 1970d). Twin cattle show similar rejection responses, as well as similar preference responses (Bell and Williams 1959). Of the more usual feed substances, lambs prefer soybean meal to commercial pellets, and those two were strongly preferred over fish meal, flocked maize, or whole oats. Rolled barley and sugar beet pulp were moderately well accepted (Davies et al. 1974). Other factors in addition to taste and odor must be considered. Texture and ease of prehension may explain why cattle prefer pelleted to unpelleted feed. Unchopped silage is preferred to chopped silage (Duckworth and Shirlaw 1958), in part because more bites are necessary to collect the chopped silage. Sheep prefer pelleted feeds to chopped feeds (Van Niekerk et al. 1973).

The preferences of ruminants for plant species vary and depend on a number of factors. Including (1) the growth stage of the plant—most ruminants prefer fast-growing succulent species; (2) the mixture of species—clover, for example, may not be eaten if it is growing in a pure stand but will be eaten if grasses are growing with it; (3) the season of the year—ruminants will consume species that are green during the winter although the same species will be rejected in the summer when other species are green; (4) the relative abundance of herbage. These factors are particularly important in determining when a ruminant will ingest a poisonous plant. Many poisonous plants are bitter in taste or harsh in texture and are, therefore, avoided by ruminants; but when the poisonous plant is the first green plant to appear in the spring (hemlock or skunk cabbage) or the only forage available on a pasture, it will be ingested. In addition to these factors, the location of the plant, in particular its distance from water (in arid regions) or shelter, will influence the amount of it consumed (Arnold and Dudzinski 1978).

Various methods have been devised for determining preferences of grazing animals. The most accurate and most direct is to collect the food eaten by sheep or cattle with open esophageal fistulas. Another method is to closely observe the animals. This method is more accurate if the forages are planted in narrow pure stands from which the ruminants may choose. A third method is to sample the plant species and heights of the plants before and after the pasture has been grazed. Studies using these methods have revealed that cattle graze selectively; that is, they do not necessarily eat plants in proportion to the number of each plant species in a pasture (Herbel and Nelson 1966b). For example, Cowlishaw and Alder (1960) found that cattle most preferred meadow fescue and timothy; perennial rye grass and cocksfoot were the next most preferred plants and *Agnotis* and red fescue the least preferred. Rapidly growing plants are the most palatable, if they are high enough in fiber (Garner 1963).

Sheep showed similar, but more variable, plant preferences and appear to be more selective than cattle (Meyer et al. 1957). They preferred cocksfoot more and meadow fescue less than cattle. In general, the preferences were correlated with the dry matter and carbohydrate content of the grasses. Sheep select a diet higher in protein and lower in fiber content than that obtained from clipped pasture samples, indicating the advantage to the animal of selective grazing (Weir and Torell 1959; Heady 1964; Strasia et al. 1970). A practical consideration in pasture management is that sheep, as well as horses, avoid fecally contaminated grasses, although they seem to prefer grass subjected to urine contamination (Marten and Donker 1964).

It is unclear what senses are involved in forage preferences. Anosmic sheep do not avoid fecally contaminated feed as normal sheep do (Tribe 1949). However, anosmic sheep and sheep with impaired vision (hooded with translucent eye coverings) showed preferences very similar to those of intact sheep, although the visually impaired sheep tended to graze at one level rather than grazing higher or lower on the plant to select the most succulent portions (Arnold 1966a). Even section of the gustatory nerves had little influence on forage preferences (Arnold 1966b).

Food Intake Depressants

Body Weight Set Point. When dry cows are allowed to eat ad libitum, intake reaches a plateau in Jerseys that indicates a set point of body weight. Holsteins under similar conditions continue to gain for a longer period, which probably indicates that a breed selected for high production has less inhibition of intake (Monteiro 1972). Blaxter et al. (1982) found that intake in ad libitum feeding sheep remained relatively constant from one year of age onward.

High Environmental Temperature. Cattle and sheep eat less when environmental temperatures are high. In addition to the direct effects of heat on intake, effects on plant growth may render them less palatable. One of the problems of raising cattle in the tropics or other hot environments is that food intake, and, therefore, production, falls. The use of tropical breeds and crosses with tropical breeds of ruminants helps alleviate the problem because these cattle eat more in the heat or eat more at night when the temperatures are lower. (See Baile and Forbes [1974] and Forbes [1986] for reviews of all aspects of the controls of food intake in ruminants including thermostatic factors.)

Estrogen Level. Food intake falls in estrous cattle (Hurnick et al. 1975) and goats (Forbes 1986). The decrease in intake can be used to identify cows that are ready for breeding. Stilbestrol is an estrogenic drug that has been used to fatten steers. In the doses given, it does not suppress intake and the anabolic effect of stilbestrol leads to weight gain (Heitzman 1978). After stilbesterol was banned, a mixture of estrogen and progesterone has been used to improve weight gain in feedlot cattle.

Rumen Fill and the Products of Rumen Fermentation. The anterior gastrointestinal tract of the ruminant (the rumen) is an anaerobic fermentation chamber in which bacteria-producing cellulase can release the energy that is used by the animal. This energy is otherwise unavailable to the animals and enables cattle, sheep, and other ruminants to survive on grass and roughage diets. In grazing ruminants, the time necessary to collect food and fill the rumen is probably the limiting factor in food intake. For example, when cattle are grazing on the open range they spend 56% of their time grazing and another 21% of the time ruminating (Hafez et al. 1969). In cattle fed diets high in grain for fattening and in lactating dairy cattle eating concentrates, other factors inhibit food intake because cattle fed a 75% concentrate diet spend only 2.5 hours a day eating (Hoffman and Self 1973).

The end products of bacterial digestion in the rumen are the volatile fatty acids: acetic, butyric, and proprionic. When acetic acid is added to the rumen, food intake is depressed; similar amounts injected via the jugular vein have no effect on food intake (Baile and Pfander 1966). Therefore, it seems likely that acetic acid may stimulate receptors located in the rumen epithelium or in the portal circulation. These receptors may, through central nervous system connections, inhibit food intake. On low-roughage, high-energy diets, production of volatile fatty acids is probably the most important satiety factor. On low energy, high-roughage diets, rumen fill or abdominal fill is probably a satiety factor (Fig. 8.10). Intake falls in pregnant or fat cattle, presumably because of the increase in abdominal fill by the fetus or fat (Bines et al. 1969; Jordan et al. 1973).

Parasitism. Gastrointestinal helminths stimulate the release of their host's gut hormones, including cholecystokinin. Therefore, the heavily parasitized ruminant loses weight not only because the parasites consume some of the food their host ingests but also because the host ingests less (Symons and Hennessy 1981).

Meal Patterns. The time that cattle and sheep spend grazing on pasture and eating in feedlots or loose housing has been reviewed in Chapter 3. Rates of

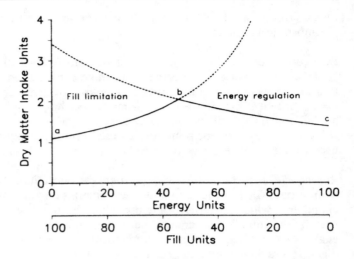

Fig. 8.10. Illustration of the reciprocal relationship between dietary characteristics and dry matter intake. Line *a* to *b* represents intake limited by the fill effect of the diet. Line *b* to *c* represents intake limited by the energy demand of the animal. Dashed lines represent unattained intake predicted by extrapolating the theoretical equations (Mertens 1987, with permission of *J. of Anim. Sci.*).

eating have also been measured. Cattle can eat 6 pounds of hay in an hour (Blaxter 1944). An important consideration for dairy farmers is the speed with which cows can consume concentrates because in many dairy parlors the cows are fed only while they are being milked. It takes a cow one to two minutes to eat a pound (0.5 kg) of grain; slightly less time is required if the grain is cubed (Jones et al. 1966). If a high-producing cow is milked in too short a time, she will not have time to consume the grain that she needs; another arrangement must be made to feed her. If cows are not fed grain in the milking parlor, however, they will not be as eager to enter. All these factors must be considered when planning a milking parlor and feeding system.

Olfactory bulbectomy influences meal patterns in rats but not in sheep (Baldwin et al. 1977). Sheep eat approximately a dozen meals a day in a laboratory. Cattle in a similar environment show similar patterns (Chase et al. 1976). Cattle show individual cyclic intake patterns (Stroup et al. 1987).

Rate of chewing in cattle has been studied by direct observation or by recording jaw movement using either a balloon under the jaw or a strain gauge on the halter (Deswysen et al. 1987; Beauchemin et al. 1989). These studies have led to the conclusion that cattle that eat fast, chew fast and eat more rather than swallowing relatively unmasticated food. Cows should be selected for rapid intake both to solve the problem of eating enough while being milked and to help reduce negative energy balance during early lactation.

Integration of Factors That Stimulate and Inhibit Intake in the Central Nervous System. The hypothalamus of ruminants appears to be as intimately involved in the control of food intake as the same brain area is in animals with simple stomachs. Electrical stimulation of the lateral hypothalamus elicits eating in goats, and stimulation of the ventromedial hypothalamus depresses intake in the same species. Lesions of the lateral hypothalamus produced hypophagia (Baile and Forbes 1974). Kendrick and Baldwin (1986a,b) have shown that cells in the zona incerta of sheep respond to the sight of food, especially a preferred food, and other cells respond when food and/or liquid is in the mouth. Cells in the hypothalamus respond to the sight of food, the approach of food, and the ingestion of food (Madison and Baldwin 1983). The factors influencing food intake in ruminants are shown in Figure 8.11.

Clinical Problems. The greatest problem of ruminant intake is that energy intake is less than energy output for the first few weeks of lactation. Cows, especially high-producing ones, frequently suffer from production diseases such as ketosis as a consequence. One of the most urgent needs in applied animal behavior is for a means to stimulate intake. Cattle on high-concentrate diets are probably not constrained by rumen fill or rate of passage but by a lack of appetite. Appetite could be stimulated either indirectly by blocking a satiety factor or directly by stimulating the relevant central neural structures. Unfortunately, it will be difficult to find a means of stimulating appetite that does not involve adding a chemical to the cow's system that could be secreted into the milk.

The Fat Cow Syndrome. An abnormality of feeding behavior that is being seen with increasing frequency is the fat cow syndrome (Morrow 1976; Morrow et al. 1979). Under some types of management, dairy cows may be fed in a group that includes both lactating and dry cows. The dry cows may overeat, gain weight — mostly fat — and later at calving or during lactation become ill or even die. The pathogenesis of the syndrome remains unclear, but the initial problem is that the cow overeats. High-producing dairy cattle have been bred to eat large amounts to supply the metabolic fuel for lactation. When lactation ceases, the increased food intake may persist and the fat cow syndrome may result.

Pica. Cattle frequently ingest indigestible items such as metal. These items may lodge in the reticulum and cause lesions there. Cattle will lick batteries, apparently attracted to the taste, but they may suffer lead poisoning as a consequence.

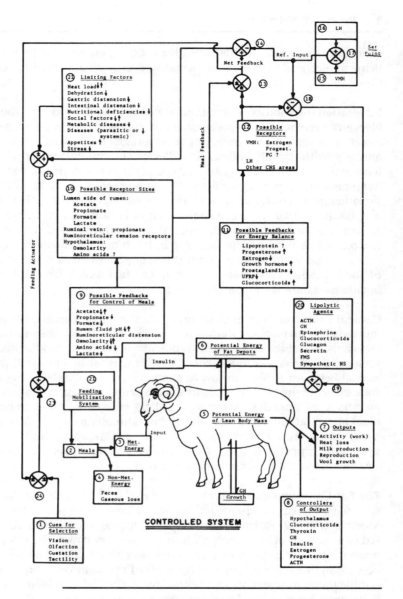

Fig. 8.11. Scheme illustrating some interrelationships of a regulatory system of energy balance with special emphasis on control of feed intake in ruminants. Factors for a hunger-satiety system probably differ greatly between species and within species under various environmental, physiological, and nutritional conditions. Boxes 9, 11, and 21 indicate each factor's probable effect on feeding behavior. Symbols denote comparators of signal inputs: FMS = fat-mobilizing substance; LH = lateral hypothalamus; VMH = ventromedial hypothalamus; UFRP = unidentified factor in ruminant plasma (Baile and Forbes 1974, copyright © 1974, with permission of *Physiol. Rev.*, Am. Physiol. Soc.).

WATER INTAKE. At least three types of stimuli elicit thirst: an increase in the osmotic pressure of the blood, a decrease in the blood volume, and a dry mouth.

Increase in Osmotic Pressure. Hypertonic sodium chloride infused intravenously causes thirst in pigs, horses, and dogs (Wood et al. 1977; Ingram and Stephens 1979; Sufit et al. 1985). An increase in the osmotic pressure of the blood is believed to stimulate osmoreceptors located in the anterior hypothalamus (Wolf 1950). Goats drink copiously when hypertonic saline solutions (2% NaCl) are injected into the anterior hypothalamus through a cannula (Andersson and McCann 1955). An increase in the osmotic pressure of the blood would be expected, for instance, in a cow that produces 80 pounds of milk (95% water) per day. A lactating cow drinks 100 pounds more water a day than a dry cow for this reason (Hafez et al. 1969). Similarly, a panting dog loses water alone and will exhibit thirst as a result of an increase in the osmotic pressure of its blood and consequent stimulation of osmoreceptors.

Decrease in Blood Volume. Decrease in blood volume, for instance as a result of hemorrhage or peritoneal dialysis (Cizek et al. 1951), may stimulate thirst. The drug furosemide, frequently administered to race horses shortly before they perform, causes a decrease in plasma volume and thirst in horses (Sufit et al. 1985) and sheep (Zimmerman and Stricker 1978). Water is lost when animals are heat stressed whether they cool themselves primarily by panting (dogs), saliva spreading (cats) or sweating (horses). Therefore, it is obvious that water intake increases with environmental temperature.

Dry Mouth. A desalivate animal takes frequent small draughts of water to swallow dry food. The importance of dryness of the mucous membrane to initiation of drinking is not known.

Angiotensin. Experimental interest has recently focused on still another type of drinking, that in response to the hormone angiotensin. Angiotensinogen is acted upon by the kidney hormone, renin, and then by a pulmonary converting enzyme to form the octapeptide angiotensin II, which has several actions. It releases another hormone, aldosterone, which is involved in sodium reabsorption by the kidney. As its name implies, high doses of angiotensin II can increase blood pressure. Most interesting is its action on water intake; angiotensin II is the most potent dipsogen known. Angiotensin II, or procedures known to release the hormone, stimulate water intake in a wide variety of animals, including dogs (Fitzsimons and Szczepanska-Sadowska 1974), cats (Sturgeon et al. 1973; Cooling and Day 1975), sheep (Abraham et al. 1975), goats (Andersson et al. 1970), pigs (Baldwin and Thornton 1986), and horses (Andersson et al. 1987).

The role that angiotensin release may play in normal thirst remains to be determined. In addition, the central nervous system site where angiotensin acts is still controversial. Situations in which blood supply to the kidneys is compromised stimulate renin release and, therefore, angiotensin release that could be expected to lead to thirst. Indeed, dogs in congestive heart failure show increased water intake (Ramsay et al. 1975). Although acetylcholine injected intracranially elicits drinking in rats, it does not seem to have a dipsogenic effect in any of the other species tested, including cats (Myers 1964).

Multiple Causes of Thirst. Just as most food intake appears to occur without a major change in body energy balance, most drinking occurs without a change in body fluid balance. The osmotic and volume depletion stimuli to drinking are emergency mechanisms to restore body water under life-threatening conditions.

Most drinking in domestic animals is prandial, that is, in association with meals. For example, 75% of water intake in pigs is in association with meals (Herring and Scapino 1973). Twenty percent of the drinking occurs after feeding (within 10 minutes), 30% during pauses in feeding, and 25% preceding meals. The drinking occurs before any changes in blood volume or osmolality. The animals appear to be drinking in anticipation of their needs and the physiological mechanisms are the same as that which controls the cephalic phase of gastric secretion (Houpt et al. 1986).

Thirst produced by overnight water deprivation is a combination of osmotic thirst and hypovolemic thirst. In contrast to the osmoreceptors that stimulate vasopressin secretion, those that stimulate thirst in the dog appear to be located on the blood side of the blood brain barrier (Thrasher et al. 1980). Water intake of thirsty dogs is reduced 80% if water is injected intracarotidly so that the osmoreceptors of the brain are no longer stimulated. Intravenous administration of an equal volume of water has no effect. If thirsty dogs are injected with isotonic saline so that peripheral blood volume is returned to normal, their water intake is reduced 20%. If both intravenous isotonic saline and intracarotid water is administered, water-deprived dogs do not drink (Ramsay et al. 1977). Despite this evidence that thirst is inhibited when blood volume and osmotic pressure return to normal, dogs cease drinking before these parameters return to normal (Wood et al. 1977). Signals from the gastrointestinal tract may inhibit water intake as signals appear to inhibit food intake. The stomach has been implicated (Towbin 1949), but gastric fill alone does not suppress drinking in dogs (Sobocinska 1978); duodenal osmoreceptors are more likely candidates for the source of inhibition.

SPECIFIC HUNGERS AND SALT APPETITE. The concept of nutritional wisdom in animals has not been validated. Animals apparently cannot innately choose a diet containing a vitamin or other nutrient in which the animal is deficient (Rozin 1967). Horses fed a diet deficient in calcium will not eat more of a calcium supplement to correct the deficiency (Schryver et al. 1978). There is better evidence for a phosphorus specific hunger. Cattle and other ruminants chew on bones and some will consume phosphorus supplements. What animals, at least laboratory animals, can do is to learn to associate a particular diet with improvement in health. Conversely, animals can learn to avoid a diet that they associate with a feeling of malaise. Given proper clues, such as a distinctive flavor, most domestic animals could probably learn to choose a diet that would correct a deficiency rather than choose a deficient diet.

There is one nutrient that nearly all species select innately when they are deficient: sodium (Holmes and Cizek 1951). Salt hunger is a well-recognized phenomenon, especially in herbivores whose diet tends to be low in sodium. In most species removal of the adrenal glands and the consequent hyponatremia is followed by life-saving ingestion of sodium chloride. Pigs, for instance, drink sodium chloride solutions and survive, following experimental adrenalectomy (Lustgarten et al. 1973). Ruminants can be made experimentally sodium deficient by creating a salivary fistula. The large loss of sodium bicarbonate in saliva produces a deficiency that sheep can correct by drinking precisely the quantity of sodium solution they need to bring their sodium levels up to normal. Treatment with a diuretic also stimulates sodium appetite in sheep (Zimmerman and Stricker 1978). Sodium-deficient cattle and sheep are able to learn an operant response to obtain a sodium reward (Abraham et al. 1973; Sly and Bell 1979). Sodium status is apparently assessed in the hypothalamus where changes in the concentration of intracellular sodium initiate transcription and translation processes. The protein synthesized alters the ionic or membrane characteristics or increases the neurotransmitter capacity of the neurons (Denton 1982). The hormones that mediate salt hunger are angiotensin and aldosterone, acting in synchrony. Because of the hormonal involvement, salt appetite does not appear until hours or days after a deficiency exists.

SUMMARY. A variety of factors influence feeding in animals; some of these factors act in the short term, others act in the long term, and still others only operate in emergency situations. Taste can either stimulate or inhibit intake. Taste factors have been used to increase intake of newly weaned pigs. Gastric factors, particularly gastric fill, can suppress feeding. Changes within the intestinal tract, such as an increase in osmotic pressure and release of the hormone cholecystokinin, can bring a meal to an end, thus inhibiting intake. These are all short-term factors. The long-term controls of feeding in which intake is controlled as part of the regulation of body weight or body fats are obviously operational in pigs and other domestic animals. This regulation is indicated by their responses to manipulation of the caloric density of their diets, but the mechanisms underlying long-term controls remain unknown. Eating in response to a lack of metabolizable glucose is an emergency mechanism that is probably not involved in meal to meal initiation of eating. Increased blood levels of glucose do not suppress feeding. When food is freely available, animals, including carnivores, eat many meals a day. Drinking is associated with eating. Under these circumstances, initiation of a meal is probably a response to waning of the satiety signals. These signals are integrated in the brain where a variety of neurotransmitters and anatomical sites are involved in feeding. The finding that depressants introduced into the brain stimulate feeding reinforces the concept that hunger occurs when satiety signals weaken.

9

Miscellaneous
Behavioral Disorders

This chapter contains descripitons of common and uncommon behavior problems of domestic animals that do not fit into the classifications of aggression, maternal behavior, sexual behavior, or other problems covered in previous chapters. In addition, samples of behavioral history forms will be given.

THE INTERVIEW AND BEHAVIORAL HISTORY. The interview should be a pleasant experience for owner, clinician, and patient. Owners should be put at ease with a handshake, eye contact, and a compliment about the animal. Interviews should be conducted in an office or consulting room rather than in an examination room, to put both the owner and the animal at ease. The clinician should schedule plenty of time. The average case of an aggressive dog takes two hours for history and instruction. It is very important to be nonjudgmental in asking questions. The clinician should ask where the dog sleeps rather than say, "You don't let the dog sleep on the bed, do you?" Later, the clinician can remind the owners of how bad they will feel if their aggressive dog bites a child rather than demand that they never allow their dog to run free.

The type of history form depends on the practitioner's personal preference. In a university clinic where many people, including clinicians and owners, will be using the forms or when information is to be statistically analyzed, a computerized or multiple-choice form is best, whereas those who interview all their own patients may prefer an essay-type form like the one shown in Figure 9.5. Owners will be daunted when asked to fill in such a long form and will answer each question with a word or two, whereas they will be more than willing to answer the same questions at length verbally.

Behavioral History. Before instituting treatment of any behavior problem, such as a destructive dog, a detailed history should be carefully taken (Fig. 9.1). It is important to determine just when and where the animal is destructive. The owners have usually attempted to deal with the problem themselves, and the methods they have used should be recorded. It is always important to maintain a neutral attitude when taking a history because any criticism of the owners' methods at this time will probably inhibit the owners from volunteering additional information.

Environment. The dog's environment should be fully described. In particular, the schedule of the entire family should be noted. The usual location of the dog both when people are at home and when it is alone should be determined. The amount of exercise or opportunity to exercise given to the dog should be recorded. Perhaps the most important information in the dog's history is the owners' usual behavior just before they leave the dog and just after they return home. Excessively emotional or stimulating leave-takings and greetings by the owner have been implicated in canine destructive behavior. The actions of the owners when they discover that the dog has been destructive, in contrast to their actions when it has not been destructive, should be ascertained.

Early History. The dog's early history often gives a clue to the cause of misbehavior. For example, if the dog was obtained from a kennel at 6 months of age, it may never have been properly socialized to people during the socialization period from 4 to 14 weeks of age. If the dog was obtained as a 3-week-old puppy or was hand-raised, it may be too dependent upon people and not properly socialized to dogs. If the dog was obtained from a pound or an animal shelter, it may have been placed there because it was destructive in its original home. The owners should be able to supply this information and to indicate why the dog was obtained by them. For instance, if the dog was obtained to be a companion for a child during the summer, the dog may exhibit destructive tendencies when the child goes off to school (and the parents to work) in the fall.

Training. Along with the dog's early history, any training that the dog received should be noted. The relationship of the dog to the owner can best be elucidated by the type of training, or the absence of training, given to the dog. If the dog could not be trained by the owner, the dog may either be dominant over the owner or it may be genuinely untrainable. It is not unusual to encounter cases in which dogs both are destructive and eliminate in the house. Occasionally, the dog has never been completely house

Date _____

BEHAVIORAL HISTORY FORM FOR DOGS

Owner's
   Name _____

Veterinarian's
   Name _____

Diagnosis _____

Address _____

Address _____

Breed _____

_____

_____

Coat Color _____

_____

_____

Sex _____

Telephone _____

Telephone _____

Date of Birth _____

Name _____

BEHAVIOR PROBLEMS

1. What is the main behavior problem or chief complaint? Describe the last three examples of the behavior problem on the back of this page.

2. When did problem begin?

3. When does animal misbehave? How often and under what circumstances?

4. Has there been a change in frequency or appearance of the problem?

5. What has been done so far to correct the problem?

6. How is the dog disciplined for this and other misbehaviors?

7. Are there any other behavior problems?

ANIMAL'S ENVIRONMENT

8. What persons are in the animal's environment? What are their schedules?

9. What other animals are in the house or on the premises?

10. Where is animal kept during day? At night? When owner is away? When guests come?

11. How is the animal exercised? Does it run free? How and when do you play with it?

12. Describe a typical day in the animal's life.

EARLY HISTORY

13. Why was animal obtained?

14. Source of animal?

15. Age at weaning?

16. Age when obtained by present owner?

Fig. 9.1. Behavioral history form for dogs.

EDUCATION

17. Method of house breaking? Age when house broken?

18. Does the animal ever eliminate in the house now?

19. Who trained the animal? How well does the animal obey for each person? (Please use chart: use good, fair, poor)

PERSON	SIT	DOWN	COME	STAY	HEEL

20. Does the animal obey better in certain places?

21. Other obedience training (CD, CDX, etc.)?

22. Any tricks such as fetch, shake hands?

23. Hunting or harness training?

FEEDING BEHAVIOR

24. What is the animal fed and when is it fed?

25. Who feeds the animal? Can you take the food away from the animal? bones?

26. Does it have a good appetite? Does it like treats? Can you take treats away?

SEXUAL AND MATERNAL BEHAVIOR

27. Has the animal shown mounting behavior (male or female), or been in heat (female)? If the dog mounts, does he/she mount dogs (sex) or people?

28. Has the animal been bred or used for breeding? Was it a good mother? Does the animal ever "mother" toys or other animals? Can you take toys away?

GROOMING BEHAVIOR

29. Are there any areas the dog licks or chews excessively?

30. Does the animal tolerate brushing or enjoy it?
Does the animal tolerate nail trimming?
Does the animal tolerate bathing?

SOCIAL BEHAVIOR

31. Is the animal aggressive or timid with other animals of the same species? With other species?

32. How does the animal act with:
    Friends -
    Children -
    Strangers - At door?
                People on street while dog is inside?
                People on street while dog is outside?
                People outside car?
    Veterinarian -

33. When does the dog bark? Whine?

LEARNING

34. Would you describe your pet as a good, fair, or poor learner?

SLEEPING BEHAVIOR

35. Does the animal sleep through the night? Is he restless at night? Where does the animal sleep at night?

36. Can you get him off bed or furniture? How does he react?

BRIEF MEDICAL HISTORY

trained, but, in most cases, the dog is trained and will not urinate or defecate in the house — if the owners are home. It is only in their absence that it eliminates in the middle of the living room, for instance. Many owners regard this as spiteful behavior, but it is much more likely to be a sign of emotional stress. Dogs may also eliminate in the house because they are scent marking with feces and/or urine. At least some "accidents," especially those that occur on beds or couches, may be scent marking in response to strong human odors.

Other Behavioral Problems. Finally, the clinician should carefully question the owner about any other behavior problem exhibited by the dog. Most dogs that exhibit one problem exhibit many, possibly in response to one cause. It often helps to inquire specifically about behavior that may be abnormal, such as feeding or sexual behavior. The dog may be a finicky eater or it may overeat and be obese. It may bark or mount people's legs. The veterinarian should be aware of other behavior problems for two reasons: a complete picture of the dog's behavior will emerge, and the other misbehaviors may become worse as the animal becomes less destructive unless the cause of the destructive behavior is removed.

Behavior history forms for cats (Fig. 9.2) vary somewhat from those for dogs because cats are rarely trained and do not require the attention that dogs do. Equine histories are often incomplete because horses, particularly those with behavior problems, change hands many times. Race horses have a variable environment as they move from home farm to trainer to race track, and the personnel who handle them are rarely the owners. The veterinarians will be different in each location, so a complete medical history will be difficult to obtain. A history form for general equine behavior problems (Fig. 9.3) and ones for specific problems such as foal rejection (Fig. 9.4) and trailering (Fig. 9.5) are included.

Instructions. Instructions are the most important part of treating a behavior case because owner-animal interactions rather than drugs or surgery alone determine the outcome. If possible, the clinician should demonstrate the treatment techniques and always give written instructions both to remind the owner and to inform any absent member of the family, then send a copy to the referring veterinarian. The clnician should ask the clients what they are going to do to be sure they understand and are willing to comply.

DOGS

Destructiveness. Dog owners complain about two major types of behavior problems: aggressiveness and destructiveness. The first type is a serious threat to public safety and has been discussed in Chapter 2. The second type occurs with greater frequency and is less amenable to treatment than is

BEHAVIORAL HISTORY FORM FOR CATS

Date _____

Owner's
  Name _____

Veterinarian's
  Name _____

Diagnosis _____

Address _____

Address _____

Breed _____

_____

_____

Coat Color _____

_____

_____

Sex _____

Telephone _____

Telephone _____

Date of Birth _____

Name _____

BEHAVIOR PROBLEMS

1. What is the main behavior problem or chief complaint? Describe the last three examples of the behavior problem on the back of this page.

2. When did problem begin?

3. When does animal misbehave? How often and under what circumstances?

4. Has there been a change in frequency or appearance of the problem?

5. What has been done so far to correct the problem?

6. How is the cat disciplined for this and other misbehaviors?

7. Are there any other behavior problems?

ANIMAL'S ENVIRONMENT

8. What persons are in the animal's environment? What are their schedules? Has there been a change?

9. What other cats or animals are in the house or on the premises? What is the relationship between the animals (friendly, neutral, hostile)?

10. Where is cat during day? At night? When owner is away? When guests come?

11. Does cat run free? How does it signal to go outside?

12. Where can the cat usually be found?

EARLY HISTORY

13. Why was animal obtained? Has the cat met the goal for which it was obtained?

14. Source of cat?

Fig. 9.2. Behavioral history form for cats.

15. Age at weaning?

16. Age when obtained by present owner?

EDUCATION

17. Method of house breaking? Age when house broken?

18. Does the animal ever eliminate in the house now?

19. Is kitty litter used?
    What brand?   Shape and size of pan?
    Has the brand been changed recently?   Covered?
    How often is fecal matter removed from litter?   Liner?
    How often is pan completely cleaned?   Additives used?
    Where is the pan located?   Does the cat cover?

20. Does owner play with cat? If so, when and how?

21. Does the cat come when called?

22. Any tricks?

23. Is the cat leash trained?

FEEDING BEHAVIOR

24. What is the animal fed? When and where is it fed?

25. Who feeds the animal?

26. Does it have a good appetite? Does it like treats?

SEXUAL AND MATERNAL BEHAVIOR

27. Has the animal shown mounting behavior (male or female), or been in heat (female)?
    If the cat mounts, does he/she mount cats (sex), people, or objects?

28. Has the animal been bred or used for breeding? Was it a good mother? Does the
    animal ever "mother" toys or other animals?

GROOMING BEHAVIOR

29. Does the animal keep its coat in good condition? Are there any areas which are
    licked excessively?

30. Does the animal tolerate brushing or enjoy it?

SOCIAL BEHAVIOR

31. Is the cat aggressive or timid with other cats? With other species?

32. How does the animal act with:
    Friends?
    Children?
    Strangers?
    Veterinarian?

33. When does the cat meow, growl, purr?

SLEEPING BEHAVIOR

34. Does the cat sleep through the night? Is he restless at night? Where does the cat sleep at night?

MEDICAL HISTORY

35. Is the cat declawed? If so, when and what was aftercare? Is there a scratching post? What type and brand? Does the cat use it?

36. Is the cat on any medication now?

37. Brief medical history.

BEHAVIORAL HISTORY FORM FOR HORSES

Owner's Name _____	Veterinarian's Name _____	Horse's Name _____
Address _____	Address _____	Breed _____
_____	_____	Sex, Age _____
Telephone _____	Telephone _____	Color _____

MAIN BEHAVIOR PROBLEM

1. Chief complaint?

2. When did problem begin?

3. When does horse misbehave? How often and under what circumstances?

4. Has there been a change in the frequency of appearance of the problem?

5. What has been done so far to correct the problem?

HORSE'S ENVIRONMENT

6. Type of housing (stall, pasture, run-out shed)?

7. Diet?

8. Exercise (hrs per wk ridden, hrs per wk in paddock, type of bit used, martingale)?

9. Other horses in environment and relations between horses (friendly, aggressive, neutral)?

10. Other animals in environment?

EARLY HISTORY

11. Why was horse obtained? Is it still used for this purpose?

12. Source of animal?

13. Age at weaning?

14. Age when obtained by present owner?

15. Were there previous owners?

16. Do related horses have similar problems?

Fig. 9.3. Behavioral history form for horses.

EDUCATION

17. Age at halter breaking?

18. Method of training to saddle or harness; age when training began?

19. Other types of training methods?   Driving   jumping   dressage   games
    trail riding   cutting

OTHER BEHAVIOR PROBLEMS

20. Shying, how often and at what? Any other phobias?

21. Head shy?

22. Resentful of grooming?

23. Aggression towards humans or animals (dogs, cows, etc.)?

24. Aggression toward other horses (threatens, strikes, bites, kicks, chases)?

25. Misbehavior under saddle (circle appropriate behavior):   Moves while rider mounts
    backs in harness   bucks   rears   wants to lead or will only follow other horses
    runs away   slow to leave and quick to return to barn   hard to keep on right or
    left   other

26. Barn vices (circle appropriate one): Cribs   chews wood   paws   kicks stall

27. Sexual behavior: Excessive   inadequate   abnormal?

28. Maternal behavior: Excessive   inadequate   abnormal?

PHYSICAL HISTORY

29. Present medical problems?

30. Past medical problems?

31. Drug history?

32. Results of diagnostic tests?

STALLION BEHAVIOR

33. Was he raised alone or in the company of other foals?

34. Was he ever punished for acting like a stallion (studdish)?

35. Have stallion rings or a brush to prevent masturbation been used in the past or
    present?

36. How many mares has he bred? How many have you attempted to have bred?

37. Has he ever been allowed to breed an unrestrained mare?

38. Has he watched other stallions breed?

39. How many mares have been used to "tease" him?

40. Does he flehmen (horse laugh)? Lick the mare's legs or hindquarters? Attempt to mount? Mount? Intromit? Ejaculate?

41. Has he ever been used as a "teaser"?

42. Does he attempt to bite or kick mare before or after breeding?

Date _____

HISTORY FORM FOR FOAL-REJECTING MARES

Owner's
Name _____

Veterinarian's
Name _____

Name _____

Address _____

Address _____

Breed _____

_____

_____

Sex, Age _____

_____

_____

Color _____

Telephone _____

Telephone _____

1. Is this the mare's first foal? If not, how long has it been since she last foaled? How many foals has she had? How has she behaved towards previous foals (has she rejected a foal before)?

2. Did her dam show normal maternal behavior? Toward the mare? Toward other foals?

3. Date and hour of foal's birth?

4. Was the birth observed?

5. Was the veterinarian present?

6. Was the birth assisted (i.e., was the foal pulled)?

7. Were there any complications during parturition?

8. Where did foaling occur? If outside, were other horses present?

9. If in a stall, was the stall rebedded after parturition? If yes, how long after the foal's birth?

10. Were the placenta and fetal membranes (afterbirth) removed?

11. How soon after the foal's birth did anyone enter the stall and for how long?

12. Was the foal's umbilicus treated with iodine? If so, when?

13. Was an enema administered? If so, when?

14. How long before the mare stood after parturition?

15. How long until the foal stood after parturition?

16. How long until the foal first nursed?

17. Was the nursing assisted?

18. Were there other horses in the barn? If so, were they visible to the mare?

19. How many people visited the mare and foal during the first 24 hours?

Fig. 9.4. History form for foal-rejecting mares.

20. How many people entered the stall and how often was the stall entered during parturition? During the first 24 hours?

21. How did the mare react toward humans during parturition? During the first 24 hours?

22. How does the mare react to dogs, other foals?

23. Describe in detail the mare's behavior toward the foal.

24. What treatment has been attempted (twitching, tranquilizers)?

```
                                                    Date _____

             BEHAVIORAL HISTORY FORM FOR HORSES WITH TRAILER PROBLEMS

Owner's                    Veterinarian's            Horse's
   Name _____       Name _____      Name _____

Address _____    Address _____   Breed _____

_____    _____   Sex, Age _____

Telephone _____    Telephone _____   Color _____
```

1. What is main problem?

2. Why is horse trailered?
   a. to show
   b. to trail ride
   c. to race
   d. to hunt
   e. other

3. When did problem begin?

4. Did horse previously trailer well?

5. When does problem occur?
   a. loading?
   b. traveling?
      1) on curves
      2) when braking
      3) as soon as trailer moves
      4) after traveling _____ minutes or _____ miles
   c. unloading?
   d. standing in stationary trailer?
   e. other?

6. Does the problem occur:
   a. on way to event?
   b. on way home?
   c. both?
   d. other?

7. Does change of driver or trailer have any effect?

8. Type of trailer:

   Brand _____ Model _____ Year _____ Height, Width _____

   a. one horse   1) can partition be removed?
   b. two horse   2) does partition reach floor?
   c. goose-neck   3) can horse turn neck to face out?
   d. van
   e. livestock
   f. other

Fig. 9.5. Behavioral history form for horses with trailer problems.

9. Type of entrance-exit:
   a. ramp
   b. walk through
   c. step down
   d. other

10. Type of flooring?

11. Does horse face:
    a. direction of travel?
    b. opposite to direction of travel?
    c. sideways?
    d. other?

12. Does trailer have doors:
    a. in front?
    b. one or both sides?

13. Is trailer lighted inside?

14. Are there side or front windows?

15. Has the horse been injured while riding, in loading or unloading?

16. Has there been an accident while horse was in trailer?

17. History of problem. Describe last 3 instances in detail.

18. What treatments have been used?
    a. tranquilizer
    b. cattle prod
    c. feeding in trailer
    d. other

19. Did sire, dam and siblings:
    a. travel well?
    b. have similar problems?
    c. unknown?

aggression. A dog quickly ceases to be man's (or woman's) best friend when it has scratched the new rug to shreds, gnawed the antique desk, and strewn the contents of the garbage pail through the house. Destructiveness appears to be increasing in frequency, although a careful survey has not been performed. Two recent sociological trends may help to explain the greater incidence of destructiveness: more people, especially young people, own dogs; fewer spouses remain at home during the day. The net result is that many more dogs spend 8 hours a day or more alone, and, therefore, the opportunity for destructive behavior is present.

Clinical Cases

CASE 1. A beagle that was well trained as an obedience dog was sometimes destructive when left alone. Because it was owned by a veterinary student, it was left alone for 10 or more hours a day. The dog would chew up anything left on the floor, but it would not pull books off tables or chew the furniture itself. The dog also barked in the owner's absence. As is often the case, it was usually the roommate's books and shoes that were destroyed rather than the owner's. An animal's behavior problem is often the focus of disputes between the people who share the animal's environment. In this case, the roommate moved out, and the owner attempted to treat both the destructiveness and the barking by the behavior modification method described in the section on treatment. The behavior modification was only partly successful; destructiveness decreased but did not cease. This dog was a well-trained animal, an obedience champion, which indicates that even a dog that is well behaved in the owner's presence can be destructive when alone or, in this case, with only a cat for company.

CASE 2. This case was much more dramatic. The dog was also a beagle, a breed that seems predisposed to destructiveness. The dog had a history of convulsions and was maintained on an anticonvulsant. The convulsions increased in frequency, and, at the same time, the dog became very destructive in the owner's absence. She ate through several wooden cupboard doors and finally was taken to the veterinary clinic when she ate all the Christmas presents. In the clinic the dog responded well to a different anticonvulsant and upon discharge was no longer destructive. The response to medication indicates that organic disease was present, but the dog showed the typical destructive behavior pattern of chewing only in the owner's absence.

CASE 3. An adult female German shepherd had been accustomed to being alone and had been well behaved until she had a litter of puppies. When she was temporarily separated from the puppies by a barricade, she began to chew, not the barricade but the furniture. It is interesting to note that the owner separated her from the puppies because he did not think that the bitch should be eating the puppies' feces, a behavior that is a normal component of canine maternal behavior. The destructiveness ceased when she was again allowed free access to the puppies.

In this case, the destructiveness has recurred sporadically. It has usually been controlled by painting the chewed object with a noxious "hot" sauce, but only after the large dog and two of her now fully grown offspring did considerable damage. They have, in separate incidents, ruined an antique table, spread plants and their soil over the house, and eaten several rungs of a banister. Although the bitch may have "missed" her puppies in the initial episode, each dog had two other dogs for companionship in the subsequent episodes. It has never been determined if all three dogs or only one or two are actually doing the damage.

CASE 4. An adult spayed German shepherd had been given to the present owner because it could not be restrained in a pen or a cage. Its early history was unknown. The present owner tried to keep the dog in a fenced yard, but the dog jumped the fence. When the fence was made higher, the dog clawed at the back door or jumped against the windows of the house. If it was tied, it struggled until it was in danger of choking itself and howled. If it was kept inside, it would claw the door, jump through the window, and try to get out as desperately as it tried to get in. It could tolerate no restraint of any kind. It behaved in a similar fashion when it was restrained in a room with other dogs.

The owner finally put the dog in the basement while she was at work. The dog tore up the rug on the basement stairs and then began to eat the wood of the stairs themselves. It would also defecate all over the basement floor rather than on the papers provided. When a young dog was put in the basement with it, the young dog began to eliminate everywhere also, and the older dog continued to be destructive. Neither tranquilizers nor anticonvulsants attenuated the behavior. The owner had trained the dog fairly well and it was obedient, but somewhat aggressive, when the owner was present.

The owner tried behavior modification for a few weeks, but there was no improvement, and the dog was euthanized. No abnormalities were evident on either gross or microscopic examination of the tissues, including the central nervous system.

Etiology of Destructiveness. What are the causes of destructive behavior? Unfortunately, the definitive answer to this question has not been found. Some intelligent guesses can be made on the basis of the situations that lead to destructiveness, but the inadequacy of current knowledge is reflected in the lack of success of the treatments proposed for destructiveness. Once the causes of destructiveness are determined, it should be possible to find effective treatments. Different types of destructiveness may have different causes.

ABSENCE OF THE OWNER FROM THE HOME. An examination of the environment in which destructiveness occurs reveals that the behavior almost always takes place in the owner's absence. It is not just that the owners punish the dog when they see the dog chew or scratch and that the dog has, consequently, learned not to do so in their presence; the owners have seldom seen the dog misbehaving in that way. This circumstance and the fact that the rise in incidence of destructive behavior appears to parallel the rise in the numbers of working couples indicate that being alone must have something to do with the dog's misbehavior.

It is very difficult to determine just when a dog is destructive and what its behavior is just before and just after it chews and scratches because the behavior occurs only when no one is home. One way to observe a dog by itself is to use closed-circuit television. A small closed-circuit system, similar to those used in stores to prevent shoplifting, would be a good investment for practitioners genuinely interested in treating behavior problems. Those who have used this method have found that most dogs appear to become destructive shortly after the owner leaves, but a few become anxious at approximately the time that the owner is expected home. The dog then begins to chew. Dogs appear to have a fairly accurate sense of time and can, therefore, predict arrival times, if the owners have definite schedules. Other dogs simply cannot tolerate being left alone. This intolerance is usually expressed by destructiveness, sometimes by barking, and sometimes by both.

LACK OF ENVIRONMENTAL STIMULATION. Boredom may be a cause or a contributing cause of destructive behavior especially if the dog is a large and/or young animal that does not receive adequate exercise. Animals appear to need environmental stimulation. Dogs will work to see other dogs by pushing a panel with their muzzles; and other animals will work for light, to obtain access to another environment, or for brain stimulation. Dogs, therefore, may find chewing and scratching less boring than just lying around. In more scientific terms, they find such activity rewarding.

BARRIER FRUSTRATION. Another type of destructive dog cannot tolerate being enclosed and exhibits barrier frustration. It may result if the dog has been punished by being put into a closed room or into a fenced yard, but sometimes it occurs in the absence of such treatment. Barrier frustration is also caused by the presence of something very desirable on the other side of the barrier. An older dog may suddenly begin to chew the window sill after years of remaining quietly at home alone. Careful observation may reveal that a dog or cat parades past the window frequently; the dog inside cannot give chase, but can only try to remove the barrier.

Another dog showed extreme barrier frustration suddenly but probably from different etiology. This dog had previously lived in a house with other dogs, but when its owners, both of whom worked, got an apartment to themselves, the dog tried to accompany them to work. It tore at the door or jumped through a closed window and ran to the restaurant where the husband worked. The dog would act similarly if tied in the back of a pickup truck but would be content to sit alone in the cab of the same truck. Many dogs are more content in cars than in houses, possibly because the car is denlike and, no doubt, smells of both the owners and the dog. The German shepherd of case 4 showed barrier frustration because it tried as hard to get into the house when outside as to get out when it was inside.

The three environmental causes of destructiveness appear to be the absence of the owner from the home, lack of environmental stimulation, and barrier frustration.

Breed Incidence. In addition to environmental causes of destructive behavior, breed incidence may be a factor as well. In our experience, beagles, German shepherds, and malamutes seem to be particularly prone to this behavior problem, but this may be a function of the popularity of the three breeds. A survey of the incidence has not been conducted.

The potential dog buyer should be warned of the possibility of destructiveness as a problem. According to a survey of veterinarians and obedience judges, West Highland white terriers, Irish setters, Airedale terriers, German shepherds, Siberian huskies, and fox terriers are the breeds most apt to be destructive (Hart and Hart 1988). Working couples, students, and apartment dwellers should be discouraged from selecting those breeds. The peak of destructiveness occurs when the dog is 18 months. Dogs appear to "grow out of," this problem, but the behavior can become chronic if mistreated or mishandled. These people should not be denied the pleasures of owning a dog, but they should choose their breeds carefully. There have been facetious suggestions that urban dogs should be of a new breed that needs no exercise and can be trained to eliminate in a litter pan. Another characteristic should be added—ability to tolerate owner absences for 10 hours or so. Until such a dog appears, potential dog owners would do well to question the breeder about the sire's and dam's behavior. One can ask where the dog is kept during the day or while the owner is away to elicit the information. Only exceptionally honest breeders would volunteer that their dogs have behavior problems.

Prevention. Prevention of the development of misbehavior is always easier than eliminating established misbehavior, but because the exact cause of destructive behavior remains unknown, preventative techniques have not been established. There are, however, several things that the puppy owner should avoid, or destructiveness may develop. Puppies should not be given old shoes or a piece of rug to chew upon because they will not be able to differentiate between the castoff shoe and the new $40 pair of shoes nor between an old scrap of rug and the living room carpet. Toys can be provided, but they should be of a type and texture that the dog can easily distinguish from forbidden objects. One toy given only when the owner leaves is better than many toys that are always available. Leave-takings and greetings should not be emotional. Playing tug-of-war games may also increase the puppy's tendencies to chew in play or self-stimulation.

Puppies should be taught to stay in a crate while the owner is absent. The time that the puppy is left in the crate should be gradually increased from an hour or two to a working day if the dog has adequate bowel and bladder control.

Treatment. The first approach to destructive behavior is to determine whether it is a result of lack of environmental stimulation, barrier frustration, or separation anxiety. The young dog that is chewing furniture but not doorways is most likely to need more environmental stimulation. Increasing its exercise, leaving it with other young dogs, providing a chew toy that it only has when the owner is away are the treatments of choice. Barrier frustration and separation anxiety are more serious. The first step should be to reduce the animal's anxiety. The owner should be told not to punish the dog unless it is caught in the act of chewing or scratching. Drug therapy is also indicated. Mild and early cases may be relieved with acepromazine. For more difficult cases, the mood elevator, amitriptyline (0.5-1 mg/kg) has proven successful.

SUBSTITUTION OF REWARDS. A behavior has to have some reward value to an animal for it to persist. It is easy enough to see the reward value of digging in the garbage; the reward is food. It is more difficult to imagine what the reward is for a dog that scratches, but escape from confinement is one possibility, especially if the scratching is directed at doors or windows. Chewing can have several rewards: escape or a form of play or self-stimulation. Some dogs run in order to play; others chew. Dogs that chew for stimulation should be able to transfer their chewing behavior to a more suitable object, such as a rawhide or rubber toy. If care is used in selecting a bone that cannot be splintered, bones can be provided because it is much easier to transfer chewing behavior to a naturally preferred object like a bone than to an artificial toy. Provision of hard chewable dog biscuits or chow should also help. If a dog chews just before the owner comes home, the owners should teach the dog to fetch and chew a bone or chew toy whenever they come home. The dog may then transfer its chewing from furniture to the toy.

Meanwhile, the objects that the dog should not chew should be made as undesirable to the dog as possible. This attempts techniques to substitute a negative consequence for a positive reward value to the dog. As long as the negative aspects of chewing are greater than the positive rewards the dog will refrain from chewing. Most owners have thought of this solution, and in some instances (e.g., case 3) painting the chewed object, such as a table leg or a plant, with a hot sauce of chili and cayenne pepper is effective.

In a variation of this technique, Campbell (1975) suggests spraying at the dog with an aerosol deodorant while holding the spray can near the object chewed. The spray should not be directed at the dog's eyes, but at its muzzle. The object chewed should be sprayed also. Most dogs dislike aerosol sprays and will avoid the smell of an object when the smell is associated with the spray. Several commercial dog repellant sprays are available, but the efficacy of these products, particularly over a long period, is questionable.

The main disadvantage of making any one object or even several objects unpleasant to taste or unpleasant to smell is that the dog will simply transfer its chewing to another site. Another disadvantage is that some objects, curtains or rugs, for example, can not be spread with a noxious concoction. These negative techniques are most effective for dogs that have just begun to be destructive. Like most bad habits, human or canine, it is easier to break earlier in its course.

COMPANIONSHIP. Loneliness has been mentioned as a cause of destructive behavior, and in some cases, relieving the loneliness relieves the dog of its destructiveness (Fox 1968). Students often find that their dogs can safely be left with their parents during holidays because the parents also have dogs. A destructive dog may stop misbehaving if a cat or dog or even a turtle is provided for companionship. So far, no one has been able to predict which dogs would profit from having a companion. It rarely helps a dog that shows barrier frustration (case 4). The one companion that does appear to inhibit destructive behavior in nearly all cases is the owner. Unfortunately, few owners can change their life-styles so that the dog is not left alone for long periods. If the dog is a desirable pet except for its destructive tendencies, the veterinarian might suggest that a home be found in which the dog would not be left alone. A family with small children and a spouse who does not work would be suitable, but any prospective owner must be warned that the dog can never be left alone in a place where it could do any damage.

BEHAVIOR MODIFICATION. The method that is most effective in treating destructiveness is a form of behavior modification. Instead of punishing the dog for being destructive, it is rewarded for good behavior. The first step of the behavior modification program is for owners to determine the longest possible period that the dog can be left alone, which may be as short as 15 minutes. While the owners are determining that period, they should also keep a behavioral diary in which the dog's actions, as well as the owners' in relation to the dog, are all recorded. The behavioral diary may contain instances of abnormal behavior by the dog or counterproductive treatment of the dog by the owner. When the longest period of time that the dog can be left alone has been determined, the behavior modification procedure can begin.

The dog is given a new toy and perhaps a new rug to lie on. The toy is only given to the dog when the owners leave. The owners should handle the toy so that it has the owners' scents. The radio or, if possible, the television set is left on while the dog is alone. All these things—the toy, bed, and sounds—are to increase the stimulation provided by the dog's restricted environment. They also cue the dog that things are going to be different—in particular its behavior is going to be different in the new environment. On the first and on all subsequent days of training, the owners make all the usual preparations for leaving for the day. Coats are put on, keys collected, briefcases and handbags picked up, and doors locked. The owners then leave. They drive off in the car, if that is their usual means of departure. They return in 15 minutes, and if the dog has not been destructive, they praise it mildly, put away their coats, and so on. The next day the procedure is repeated except that the dog is left alone for half an hour. Each day the length of time the dog is left alone is increased. The increase should be gradual. Many owners will be so elated by 1 hour of good behavior in their absence that they leave the dog for 6 hours and return to find a torn-up cushion or worse. The veterinarian would do well to give the client a schedule to follow to avoid misunderstanding or mistakes.

The scheduled absences should be 15 minutes, 30 minutes, 45 minutes, 1 hour, 90 minutes, 2 hours, 3 hours, and so on. Obviously, the behavior modification method takes a long time. Unless the owners are willing to devote their month's vacation to the procedure, it will be exceedingly difficult for them to engage in their usual occupations and still modify the dog's behavior. Two solutions have been used by owners who can only use the behavior modification schedule on weekends. One solution is to leave the dog in the car during the day if the weather is cool enough. In the summer, temperatures in a parked car can be lethally high. Another solution is to leave the dog somewhere outside the home or at least outside the room or rooms where it is left in the owners' absence. The dog can be taken to friends who stay home or have other dogs to keep the problem dog company. If there is no other solution, the dog can be kept in a cage when the owners are away from home.

It is important that the dog get enough exercise to help attenuate its destructive tendencies, and exercise is particularly important if the dog must be closely confined during most of the day. Regular exercise periods and obedience training should be instituted, preferably in the morning and evening. These exercise periods serve two functions: the obvious one of ensuring that the muscles of the dog are active and providing a time when the dog will regularly receive attention from the owner. If the dog is obedience trained, one daily session of reviewing its training will provide another opportunity for the dog to be rewarded for good behavior.

The owners should be warned that the dog may be destructive during the behavior modification training period. In that event, the owners should not punish the dog. The dog should not be rewarded with any attention at all but should be ignored. Any mess that the dog made should be cleaned up while the dog is not present. The next day the period of absence should be shortened and the training resumed.

The theory behind behavior modification of destructive behavior is that the dog learns that (1) the environment has changed (more stimulating with toys and music); (2) non-destructive, not destructive, behavior will be rewarded; (3) it will receive attention from the owners without having to resort to destructiveness. The behavior modification method is theoretically the best, but it is only 50% successful. Few owners have the patience to continue the training. A modification of the method in which many short absences are used, rather than absences of gradually increasing length, may be effective. Long-acting progestins, which tend to calm the dog, may also be used for short periods to curb destructiveness while the owners begin obedience work and behavior modification.

DENNING. It is sometimes possible to reduce destructiveness in the owners' absence by a method called "denning," which depends on gradual reduction of confinement. The theory is that the dog will become used to the crate and treat it like a den into which it can retreat when it is anxious. Alternatively, the contrast between confinement and the relative freedom of the house may make the dog more content in the house. This technique can only be used on dogs that do not injure themselves or soil in a crate.

An airline crate is much more effective than a wire cage because it is darker and because the dog can not catch its feet in the wire. If an open wire cage must be used, cover it with a blanket or towels so that light can enter only from the front. Put food and water and an attractive toy in the cage. Put the dog in the cage after a play session with the toy or for its dinner the first few times. Try to associate the cage with pleasant things.

The whole denning process takes 6 weeks. The first few days are the most important because that is when the dog may bark; therefore, the process should begin on a weekend or some time when the owners can devote all their time to it. The owners put the dog in the cage. If it barks or howls they punish it by throwing a light chain (the collar) against the cage or by throwing water on the dog. Such methods will not hurt the dog but will startle it. If the owners let the dog out when it barks, it will learn to bark many times in hopes of the reward of freedom or attention. The dog should live in the cage for the first 2 weeks and should only be taken out to eliminate (3 or 4 walks on a leash per day) and for obedience training (at least 20 minutes per day, preferably in several short sessions). During the second 2 weeks, the dog should be allowed out of the cage when the owners are at home and awake. The owners should leave the cage door open when the dog is outside the cage so it can return when it wishes. Leash walks and obedience sessions should continue. During the third 2-week period, the cage should be open and the dog free to go in and out of it as long as the owners are home, whether they are awake or asleep. Training continues during this period. After 6 weeks, the dog should no longer be confined in the cage at all, but the cage should remain with the door open for it to use.

DOG DOORS. As noted previously, many dogs show a particular kind of destructive behavior, barrier frustration. When possible, these dogs should be provided with a "dog door" so that they can go in and out at will. They should not run free but may be content to choose whether they are inside or outside.

PHARMACOLOGICAL TREATMENTS. The rationale for amphentamine use is that destructive dogs are animal models of hyperactive children, on whom amphetamines have a calming effect. Amitriptyline (Elavil 1–2 mg/kg) and chlorazepate Tranzene 0.5–1mg/kg have proven successful in some cases of destructiveness.

OPERANT CONDITIONING. Another solution is to provide the animal with a task that will command its attention and give it the opportunity to use its paws and/or muzzle. The solution is to operantly condition the dog to obtain its food by pressing a panel with its muzzle or a lever with its paws. The dog can be easily trained, as many laboratory dogs have been, to press a panel a number of times for a few pieces of dog chow. It will have to work several hours a day to obtain its nutritional requirements. The apparatus must be constructed to fit the dog, but it can be made of rugged parts that a dog cannot destroy. Operant conditioning has been used with considerable success to treat and prevent stereotyped behavior in zoo animals (Schmidt and Markowitz 1977).

Self-mutilation. Self-mutilation can vary from excessive grooming to the commonly encountered lick granuloma or even to life-threatening, self-inflicted wounds. Many dogs lick their paws until they are hyperemic, and the fur is discolored. Whenever dogs bite or scratch at their coats, a physical basis, such as impacted anal sacs, flea bite allergy, or dermatitis should be considered.

Occasionally, however, there is a behavioral basis for such actions. A German shepherd developed a granulomatous lesion on his foreleg. The dog had begun to lick the affected paw when the little boy with whom the dog normally spent much time was in the hospital for a few days. The abnormal behavior ceased and the lesion healed with application of topical corticosteroids and the return of the child. The behavior and the lesion recurred when the child started school. Again, corticosteroid medication and adjustment of the dog to the child's new schedule were followed by healing of the lesion.

Endogenous opiates may be involved in self-mutiliation; the opiate blocker naloxone has been used successfully to treat lick granulomas (Dodman et al. 1988; White 1990).

Shock collars have been used to punish dogs when they lick a granulomatous lesion. This method might also be used for the more acute and more common problem of the dog chewing out its stitches or eating its cast. Postsurgical self-mutilation is not a strongly entrenched behavior pattern and should be more easily suppressed than long-standing licking problems. Extreme cases of self-mutilation during which the dog barks or screams are usually accompanied by abnormal electroencephalograms indicating that antiepileptic drugs may attenuate the problem.

Tail Chasing. Tail chasing can be a form of play or a sign of gastrointestinal parasites in puppies but is a behavior problem in adult dogs. It is most common in bull terriers. The cause is unknown. Trauma to the tail caused by self-mutilation makes it difficult to determine whether a lesion in the area preceded the behavior. Restraint seems to exacerbate the condition. This condition has been noted in Scottish terriers (Thompson et al. 1956). In mild cases, eliminating cage confinement and counterconditioning the dog to obey a command have been successful. In cases that do not respond to behavioral therapy, naloxone has been used successfully (Brown et al. 1987).

Phobias. Dogs have an interesting variety of individual phobias, although potentially phobic stimuli seem to be limited. Dogs do not develop phobias to trees, for example. Many dogs are afraid of thunder, others of gunfire, some of street noises or of riding in a car. Occasionally, the phobia can be traced to a fear and/or pain-producing experience. For example, as a young dog, a cocker spaniel loved to ride in cars, but after the car in which it was riding was involved in a serious accident that injured the dog slightly, the dog was extremely frightened of riding in cars and had to be heavily tranquilized when automobile transportation was necessary.

Storm-related Phobias. Fear of loud thunder is understandable, and most dogs that are afraid of thunder simply slink under a bed. However, some dogs will become frantic and try to escape from the pen or house and, if they succeed, run long distances. Most dogs with storm-related phobias appear to be afraid of the noise of thunder, but in one English setter, the animal appeared to be responding to changes in barometric pressure. It would scratch, chew on its kennel, and generally make escape attempts before any storm, rain or snow. It would calm down if the owner was present. It was a good hunting dog and was not afraid of gunfire. Its early history was unknown, but it may have been abused.

TREATMENT. The dog's fears should be alleviated. Two types of drugs that have been successful are chlorazepate (0.5-1 mg/kg) and buspirone (10–15 mg BID or TID). Dogs that are afraid of thunder should be on medication for the summer storm season because once the dog is frightened drug treatment is rarely effective.

The most commonly used treatment for thunder phobias is progressive desensitization. The dog is made to lie quietly near the owner and a very good quality tape recording of thunder is played to the dog. It is important that the sound will elicit the behavior. The owner should play the sound at high volume a day before desensitization is to begin. If the dog does not react, another stimulus sound will be necessary. At first, the volume is kept very low. If the dog shows no signs of apprehension, it is rewarded with food, and the volume is raised slightly. The daily desensitization sessions should be short, only 10 minutes. Gradually the volume is raised to that occurring during a nearby thunderstorm. Each day the initial volume is just below the loudest volume tolerated on the previous day.

Fear of guns can be treated in a similar manner by placing a starter's pistol inside a nest of cardboard boxes, then rewarding the dog for not flinching when the gun is fired. Gradually, one box at a time is removed until the dog is calm even when the gun is fired without muffling (Hart and Hart 1985).

Progressive desensitization can be used for nonsound-related fears as well. Fear of cars can be treated by first getting the dog to relax by gentle praise near and then in the car. When the dog can sit for a few minutes in a quiet car without trembling, the engine is started, but the car is not moved. When the dog accepts the noise of the engine, the car is moved a few yards. Gradually the rides are lengthened. By breaking the fearful event into as many subevents as possible and patiently exposing the dog to each in turn under relaxed and rewarding circumstances, phobias may be overcome. Owners must be advised not to try to calm the dog because the soothing may reward the dog for acting in a wild or frightened manner.

Running and Car Chasing. The best way to prevent running away and car chasing is to keep dogs on a leash, under voice control, or in a sturdy pen at all times. Giving a dog freedom to roam is giving it freedom to die under the wheels of a car. Once a dog has learned to chase cars or to roam, it can often find ways to escape confinement, so that behavioral means of attenuating those activities as well as restraint should be used. Tortora (1977) has described a unique means of preventing a dog from wandering into the street. The dog was placed on a 20–foot leash so that it could run into the street but not across the street. Children were stationed in trees with plastic bags of water. Whenever the dog darted into the street it was bombed with water. The dog was soon cured of running into the street. Squirting the dog with a water pistol when it chases cars or bicycles has also been suggested (Campbell 1975).

Backyard Problems

Escaping. The most common backyard problem is the dog that escapes. A high enough fence can hold a small dog, but a large dog can sometimes jump or climb nearly any fence, no matter how high and expensive. One suggestion (Campbell 1975) is to erect a small barrier a yard or two in front of the fence so that the dog can not gather momentum for a jump or to dig a ditch for the same purpose. Many dogs respond well to dog doors that enable them to enter and leave the house as they please. Occasionally, the destructive dog and the one that digs in the yard will do neither if it is not barricaded in one location or the other. The best way to deal with a roaming dog or one that escapes from a fenced yard is to keep the dog indoors and walk it on a leash. Roaming can be prevented by boundary training. The dog is taught that it cannot step across a visible or invisible barrier or it will be punished verbally. The dog should be on a leash while it is boundary trained so that it cannot keep running. The most common response of an owner is to punish a dog for running away or for failing to come after it has returned. The dog learns that if it comes when called it will be punished; then it learns not to come.

Dogs usually have one of two motives for roaming: sex or food. Castration reduces the first motivation (Hopkins et al. 1976), and ad libitum feeding will reduce the second. In this case, the risk of obesity is less than the risk that the dog will be hit by a car.

A method that has been suggested for a roaming dog that does not respond to these simple remedies is to fast the dog for 24 hours and then feed it a very palatable food a spoonful at a time every 15 minutes for 2 days and then on a variable schedule of 15 to 45 minutes during the next two days and at increasing intervals on the following days (Hart and Hart 1985). This method relies on the difficulty of extinguishing behavior rewarded on a variable interval schedule (see Chap. 7).

A more difficult problem is the dog that is kept indoors and walked on a leash but sneaks out the door. Teaching the dog that the door will be slammed if it precedes the owner may help.

Digging. Digging holes in the yard is another closely related backyard behavior problem. Dogs that dig are frequently those that are put in the yard as a punishment for bad indoor behavior. If the indoor behavior can be improved, isolation punishment will not be necessary. Dogs also dig holes to keep cool or to catch rodents. Providing a cool shelter should reduce digging for the first reason and eliminating the rodents should reduce digging for the second. Putting chicken wire where the dog digs may also deter it (Dunbar and Bohnenkamp 1985).

Campbell (1975) recommends the same routine for backyard digging, as that recommended for destructiveness because he feels that the etiology of the two behaviors is the same. Koehler (1962) recommends filling the hole with water and submerging the dog's head in it. This method only teaches the dog to dig elsewhere and to fear water. Some dogs dig up flower beds in imitation of their owners. The solution for these dogs is simple: do not let them watch people working in the garden.

A few dogs dig imaginary holes in the house or outdoors. For example, a well-trained Doberman pinscher began to dig holes. He looked as if he were digging mice out of their holes. While engaged in this behavior, he would barely react to his name, would perform none of his obedience routines, and would become increasingly rigid in posture and unresponsive as stimulation (calling, patting, etc.) increased. If the dog was put into a stay position by the owner when it was not digging, it would not hold the stay for more than a few minutes before digging began, whereas, formerly, the dog would maintain a stay for an hour.

For more serious digging problems, dogs should be treated as they are treated for destructiveness in the house. In some cases, provision of a dog door that allows the dog to enter and leave the house at will is sometimes successful (Monks of New Skete 1978).

Jumping upon People. Jumping upon people is a very common minor behavior problem. It is minor as long as the dog is small and it only soils trousers or tears stockings when it jumps up. The problem becomes major when the dog is large and knocks down children, or even adults. Typical solutions are to place a knee in the oncoming dog's chest. This technique may be effective if everyone upon whom the dog jumps can apply this treatment. Otherwise, the dog will only learn not to jump on robust adults and will continue to jump on children and old people. Stepping on the dog's hind feet is also recommended, but that maneuver requires agility and may result in fractured canine toes. Campbell (1975) recommends squatting down when the dog is approaching to jump. The rationale is that the dog will not jump up if the human is at its level. Some dogs still jump up on a squatting person and can do more damage to the head and neck than if the person were standing.

A more suitable reprimand is to hold the dogs' paws when he jumps, walk him backwards on his hind legs, and end by pushing the dog away while saying "off." Only a dozen repetitions are usually needed to teach the dog not to jump on the person who has used that technique.

The best therapy is *down* training in which the dog must lie or sit quietly to obtain any reward—food or a walk. The time required for the dog to lie still is gradually increased, and when the dog will lie for all family members, other people can be enlisted so that the dog learns to sit even when visitors come to the door. This is simply obedience training, with the one command emphasized to rid the dog of the habit of jumping.

Fly Catching. Some dogs develop a behavior that resembles snapping at flies when no flies are present. Snapping at air occurs several times a minute. It probably represents a form of psychomotor epilepsy but needs more thorough study.

Vomiting. Carnivores vomit frequently, which is an advantage to a scavenger animal that may ingest spoiled food. Vomiting food is part of normal canine maternal behavior also. Owners may complain that the dog vomits "for spite" when they are away or to attract attention. Vomiting usually does attract attention so the dog is rewarded for its behavior with the owners' attention. In one case, the dog vomited into the owner's shoe. This behavior is probably akin to urinating on beds in response to the owner's scent, in this case, a maternal response.

Psychosomatic Problems. Fox (1968) and Tanzer and Lyons (1977) present a number of cases of dogs that exhibit sympathy lameness or seem to enjoy receiving medication.

CATS. In general, cats develop behavior problems much less frequently than dogs. Feline destructive behavior appears to fall into three categories: (1) clawing (discussed in Chap. 2); (2) wool sucking (a vice of Siamese cats in particular), which may or may not be related to early weaning, is sometimes transferred to synthetics to the great detriment of sweaters and upholstery, and (3) plant eating, which has been discussed in Chapter 8.

Withdrawal. A change in the cat's environment is not always followed by inappropriate urination. Another response may be withdrawal from the owner. In one case, the cat was taken from a suburban setting, where it was free to go out doors, to a high-rise apartment. It had been an affectionate cat but ignored the owner for weeks after the move. The owner then boarded the cat for two weeks; on its return to the apartment it was again affectionate. The stress of boarding probably was responsible for the cat's adaptation to the less stressful apartment.

HORSES. These large powerful animals that are trained to work closely with humans may develop a wide range of irritating or even dangerous abnormal behaviors commonly referred to as vices. Equine vices may be divided into stable vices and vices while under saddle (or in harness).

Head Shyness. The horse in Figure 9.6 also illustrates head shyness, another common equine vice. Head shyness is usually, but not always, secondary to mismanagement of the horse when it was first handled. Occasionally, progressive desensitization may be used by a gentle and patient owner to overcome the vice, but while the horse is becoming accustomed to handling of its ears and poll it will still be difficult to bridle and attempting to do so will undo the desensitization. Horses will often tolerate handling of their heads and ears best when they are hot, sweaty, and, apparently, itchy. Putting on the bridle with one cheek piece unfastened may be tolerated.

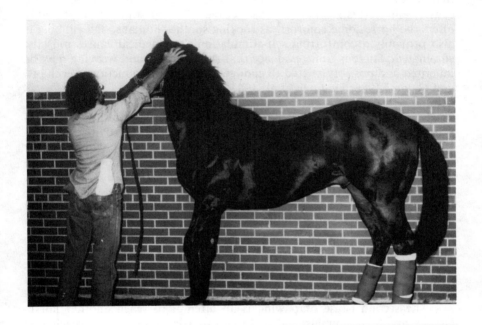

Fig. 9.6. A horse that kicks stall walls, thus aggravating a hock injury. The horse is also head shy.

Stable Vices

Locomotion Vices

STALL WALKING AND WEAVING. Horses that circle their stalls constantly may lose condition or fail to obtain it because they expend more energy walking than they ingest. Their racing performance is usually affected as well. In contrast to oral vices, an increase in roughage in the diet does not reduce the frequency or severity of the problem. Restraining the horse by tying it often converts a stall-walking horse into a weaver that stands in one spot but shifts its weight and its head from side to side. The cause of the behavior appears to be confinement. The treatment is to maintain the animal with other horses on pasture with a run-out shed for shelter. Other treatments are stall toys and more work. The size of the stall does not seem to influence the stall walking behavior because horses given access to an entire barn still circled in one corner. Stress appears to aggravate the problem; horses circle more frequently the evening before a hunt or show.

Weaving may be similar in etiology to stall walking. It is a stereotypy that appears in many zoo animals as well. Confinement and frustration are causes, and horses in pain may also weave. The self-stimulation probably offers the horse some comfort, as rocking soothes humans. Tail rubbing is also probably a comforting self-stimulation, but medical causes must be eliminated. There is some evidence that stall walking and weaving may be inherited and possibly related to endogenous opiate production (Vecchiotti and Galanti 1986; Dodman et al. 1987).

PAWING. Pawing has been described by Odberg (1973) as a response to frustration, a displacement activity that originated from the activity of uncovering food buried under snow. Horses have been noted to paw in a variety of situations: when restrained from moving, when eating grain, when expecting feed, at a recumbent foal that does not stand, and to reach another horse. The most extreme forms of pawing occur in horses that have barrier frustration; that is, they are seeking to escape from their stalls. A typical case is that of a standardbred that had spent most of his life on pasture with a run out shed for shelter. When confined in a box stall for training, he dug a 4-foot deep (1.5 m) hole in his stall. Some horses habituate to stall confinement eventually; others may do extensive damage to themselves and the stalls as they try to escape by jumping over the walls. Pawing is a less dangerous activity, but it damages dirt or clay floors.

The second cause of pawing is an attempt to reach another horse. Horses are herd animals and stallions, in particular, try to reach one another to play, if they are young, or to fight, if they are mature. If dirt floors are replaced with concrete, the horses will stop pawing, but their motivation will not change. The stallions may rear to reach one another over stall walls that do not reach the ceiling or lean out the front of the stalls to make contact if that is possible. A concrete floor can be hazardous when slippery and is more stressful to the horses' limbs and feet because it is a rigid surface.

Pawing in anticipation of food is similar to kicking in anticipation of food and will be discussed under kicking. Pawing at recumbent foals may stimulate the foal but can injure it, especially if the foal is unable to rise, and the pawing continues for some time. The reason horses paw when eating grain is unknown. One hypothesis is that they may be responding to the highly palatable food that can be prehended quickly, two unnatural and therefore frustrating qualities.

STALL KICKING. Aggressive kicking is covered in Chapter 2. Only kicking directed against stall walls will be considered here. Most horses kick stall walls with their hooves; a few knock their hocks against the wall. Both activities produce unwanted concussion on the horses' bones and joints. Kicking can damage the walls as well. A small hole made by kicking is often enlarged by wood chewing. Stall kicking, like weaving, may also be a form of self-stimulation. The horse kicks to hear the sound its hoof makes as it strikes the wood. Sometimes stabling a horse on a wooden floor that makes a similar sound as the horse walks on it will eliminate the kicking.

By far the most common form of pawing and kicking is that which the owner has operantly conditioned. The horse will tend to paw or kick at feeding time because it is frustrated to see or smell food (or the giver of food) but not be able to eat. This is reflected physiologically in the increase in heart rate observed in horses that see feed. The horse is, of course, fed, so the animal's behavior is positively reinforced. It has learned that kicking or pawing are followed by food. The horse paws and food appears. It will begin to paw or kick earlier and earlier. In effect, it is increasing the fixed ratio of number of responses (paws or kicks) for every reward. The longer the horse kicks before feeding, the longer it will take to extinguish the behavior. To extinguish the behavior, the owner should feed the horse only when it does not kick. The process will go much more quickly if many small meals a day are given. The horse must at first only refrain from kicking for 2 seconds before it is given .5 cup of feed. Gradually the criterion is raised so that there must be no kicking for 5, then 10, then 30 seconds before food will be given. Only when the horse will refrain from kicking for a short time for several feedings or trials should a longer time be demanded. The training will go faster if the horse is taught a countercommand such as "stand" for a food reward at the same time.

Trailering Problems. Many horses exhibit undesirable behavior in relation to ground transport. By far the most common undesirable response is failure to load. Both heredity and learning contribute to loading problems. The properties of a trailer that release the horse's innate fears are (1) the dark interior of the trailer; (2) the hollow sound of hooves on the ramp, an indication of poor or insecure footing; (3) the instability of the ramp and vehicle. In addition, horses are generally neophobic; that is, they are afraid of new or strange things.

Experience or learning can also play a part in loading problems. A horse that already has had unpleasant experiences with loading or riding in a trailer will also be understandably difficult to load. Hitting the horse may cause it to leap forward into a trailer, but the animal will associate pain with trailers and may be even more difficult to load on the next occasion. Learning to dislike loading is even more likely to occur if the horse injured itself while resisting loading. For example, the horse may rear back just as it reaches the entrance to the trailer and strike its head on the roof.

The first approach to solving a loading problem is to train the horse to move in response to a touch on its body. This procedure, if correctly done, takes only a few minutes. The handler, preferably a person who has not tried to load the horse, holds the horse with a lead rope and halter. A frightening but nonpainful stimulus, such as a lunge whip with a cloth tied to the end, is used to tap the horse. The horse should move away from the stimulus. The horse can be encouraged to walk forward, to stop, to move the hindquarters toward the handler, and to move the hindquarters away from the handler. After practicing these movements, the horse is walked to the trailer and urged to approach it but not allowed to enter. After a few repetitions, the horse is tapped on the hindquarters to encourage it to enter the trailer. The first entrance may only be partial, only the forefeet. Later the horse can be allowed to enter the trailer with all four feet many times before the trip begins.

The best solution to a loading problem is to desensitize and counter condition the horse by rewarding it for loading itself. Desensitization takes time — days and weeks — and, therefore, is not a solution for the acute problem of how to load a horse, which is covered below. Owners of horses with trailer problems should begin the desensitization program a month or more in advance of the proposed trailer trip. The horse and the trailer should be placed in a paddock. The trailer should be secured so that it does not tip, and the wheels should be blocked. The horse's food, all of it, should be put at the bottom of the trailer ramp. Each day the feed, hay and grain, should be placed a little farther up on the ramp. Next, the feed is placed on the floor of the trailer. Day by day, the feed should be placed farther and farther into the trailer, forcing the horse to load itself in order to eat. The horse should obtain no other food except that in the trailer during the desensitization process. Very few horses will starve to death rather than enter a trailer. Once the horse is loading itself into the trailer, it should be led on to the trailer for all its meals. Some horses will learn to load themselves but will still object to being led onto a trailer. The loading problem may not be completely solved even if the horse loads easily in its home paddock. The same horse may be reluctant to enter the same trailer when it is parked in a strange place. A few days of confining the horse in a paddock at another farm will help the horse generalize the lesson that getting on a trailer is safe and desirable. Another method to encourage a horse to enter a trailer is to put it in an unfamiliar place in such a way that entering the trailer is the only means of escape.

The rear-facing trailers described below also offer a solution for horses that will not enter conventional trailers. The ramp of rear-facing trailers can be made into a horizontal platform. The horse can step onto the platform and can be backed into the trailer. The fact that riding in a rear-facing trailer is different and presumably less stressful than conventional trailers so the horse will not associate riding in it with its previous experiences in conventional trailers may result in easy loading. It should not refuse to back into the trailer the second time it is attempted if it does not dislike or fear the sensations of riding in the trailer.

There are some solutions to an acute trailer problem in which the horse must be loaded that do not cause the horse to associate being loaded with pain. If a horse must be forced onto a trailer, pushing is to be recommended over hitting. Long ropes tied to each side of the trailer held by two people can be crossed behind the horse and used to apply pressure to its rump. The ropes must be soft to avoid rope burns. The ends attached to the trailer should discourage the horse from leaping to either side of the ramp. The ropes should be held so that they can be dropped if the horse becomes entangled. Sometimes two people can join hands behind the horse and use their arms to push against its rump to encourage the horse that is only mildly reluctant to enter the trailer.

Horses are herd animals. Advantage can be taken of equine gregariousness to solve immediate trailer problems. A horse that is reluctant to enter a trailer may be willing to follow another. Naturally this method — using social facilitation of behavior — is unlikely to be successful if the horse is already very excited. The social facilitation characteristic can be used to prevent trailer problems. A foal should be loaded beside its mother several times in its first few months because it will follow its dam onto the trailer as it follows her elsewhere. Of course, this method should not be used if the mare herself does not load. Bad behavior as well as good behavior can be learned by observation.

If the foal's mother is reluctant to load, the foal can still be trained to lead relatively easily after weaning is complete. With the foal correctly haltered and on a lead rope, gently encourage the foal to put one hoof on the ramp for a handful of grain. Gradually encourage it to go farther and farther up the ramp. Talk softly and encouragingly, and immediately reward any progress with a little grain. Full loading may take several sessions. Once the foal has gotten on the trailer the first time, feed it its daily grain ration in the trailer several times before traveling with it. Never use hitting, pushing, or pulling during these training sessions. Repeat the process in as many different vehicles and in as many different sites as possible to help the foal generalize. Before the foal goes on its first long trailer ride, it should take several short trips .5 to 1 mile each, ending with a grazing session. Gradually take the foal on longer trips. The etiology of some horses' trailer problems may be a first trailer experience that consisted of 4 hours of fear and exhaustion while trying to balance on a strange surface. Several short trips help the horse to habituate to the trailer and the sensation of a moving vehicle while the animal is rested. The time involved can save much more time later; a foal properly introduced to trailering should load easily as an adult even when trailering episodes are years apart.

Sedatives such as xylazine can be used in the acute situation. The sedative will be far more effective if the animal has not become excited or frightened before administration. It might also be possible to desensitize a horse to the loading procedure by repeatedly sedating it and loading it several times a day for a week or so, but in some experimental situations, other species cannot remember in the undrugged state what they learned in the drugged state, a phenomenon called state dependent learning. Sedatives have several disadvantages: a sedated horse cannot perform properly and in some states cannot perform legally. Furthermore, a sedated horse is more at risk of losing its balance in a moving trailer. A way to avoid state dependent learning is to gradually decrease the dose of the sedative over several sessions.

Moving Trailer Problems—Scrambling. Some horses enter trailers without hesitating but misbehave when the trailer moves. The horse can be badly injured and/or the trailer can be damaged. There are several causes of struggling. Most are related to the horse's inability to keep its balance in the trailer. The forward movement of the trailer is not a familiar force for a horse to resist, nor one that the horse evolved to counter. For this reason, Creiger (1982) has advocated trailers designed to transport horses facing towards the rear. The horse can brace against the forces of the moving trailer because it is the same action the horse would use to stop itself. Creiger has many examples of horses that could not be trailered in conventional trailers but rode quietly in rear-facing trailers. It is important to note that these trailers are designed to hold horses facing backwards. A conventional trailer is not designed for horses to face backwards and it is dangerous to use such trailers in that manner. Rear-facing trailers are commercially available.

If the horse is struggling because it is losing its balance or is afraid of doing so, simple changes in the trailer may be all that is necessary. Removing the center partition of a two-horse trailer will allow the horse to plant its feet more widely. A layer of sand topped with wood chips on the floor of the trailer will prevent the horse from slipping. Alternative solutions are to transport the horse in a large horse van or a stock trailer.

A horse may be reacting to erratic movements of the trailer, which can be ascertained if the horse scrambles only when a certain person is driving. A rarer cause of misbehavior is electric shock. If the horse only struggles when the brakes are applied, the wiring should be examined.

Desensitization and counterconditioning can be attempted to reduce the frequency and severity of moving trailer problems, but this approach is less successful than with loading problems. The horse should be loaded into the trailer and the towing vehicle's engine started. If, and only if, the horse remains calm, it should be given a grain reward. A person must be in the trailer. The person will be in danger if the horse does struggle. Once the horse appears to accept the sound of the motor calmly, the trailer can be moved a few feet. Again, the horse should be rewarded only if it stands quietly. After many repetitions, the horse can be taken for short drives at slow speeds and rewarded for calm behavior. Finally, high-speed highway travel can be attempted.

Horses may struggle in trailers because they are losing their balance, as mentioned, but some horses may be reacting in anticipation of the journey's end. Horses participating in such high-speed events as barrel racing or games are the most likely to scramble on the way to a competition. These horses could be desensitized by taking them on many trailer rides that end not with competing but with grazing or a leisurely trail ride. Care must be taken to treat the horse in the same manner before a "therapeutic" trailer trip as before a ride to a show. The same tack and the same grooming routine should be used in both cases so the horse cannot discriminate between trips that end in shows and those that end at less exciting destinations.

Horses that load and ride well but do not stand quietly in a stationary trailer are still another problem. They will move, kick, and struggle when the trailer stops at a toll gate or a traffic light. The same procedures that are recommended for horses that struggle during transit, that is, small food rewards for standing quietly during practice trailering sessions, can be used. The reason why horses fret in stationary trailers is unknown but may be the same as those for horses that struggle in the moving trailer. Better footing, driving ability, or a different type of trailer may be necessary to eliminate the problem.

A final problem is the horse that will not leave the trailer. Although horses can sometimes be backed where they will not walk, a horse that reaches back with a hind hoof and encounters nothing but air will be afraid to back off. Trailers with walk-through construction enabling the horse to enter from the back and exit through the front are helpful. To solve the immediate problem, the horse should be allowed to turn its head so it can see where it is backing. It may be necessary to remove the center partition of a two-horse trailer so the horse has room to do so. Horses are more reluctant to back off step-up trailers, those without a ramp. A loading dock of dirt can be made to give the animal a solid place to put its hind feet.

The best approach to the treatment of stable problems and trailer problems is to remove the cause of the problems, that is, to change the horse's motivation. This approach is more apt to be successful than punishment or physical restraint of the horse. Careful early handling and optimal housing can prevent the emergence of these problems.

Saddle Vices

Bucking, Shying, and Grazing. Vices under saddle can vary from bucking and running away to grazing. The former habits are quite dangerous, and treatments as bizarre as injections of succinylcholine have been used to cure horses of bucking, but nothing discourages the child who is learning to ride more than the pony that puts its head down to graze at every opportunity. Muzzling the pony may discourage it from attempts to eat. The habit that is most apt to dislodge the rider is the simple fear response of shying. The good rider knows what objects are most likely to elicit a shying response in a particular horse, is prepared, and tries to calm the animal.

Phobias. Some horses have fears that might be classified as phobias. A horse that was badly stung by bees became uncontrollable whenever it heard any insect buzz. The horse could be desensitized by playing bee noises to it, in the same way that dogs are desensitized to thunder. This technique also can be used to accustom horses to clippers, band music, applause, and the other frightening aspects of parades and horse shows.

Behavior-related Problems. Many horses will move much more quickly toward their stables than away from them. These animals are called "barn rats." Other horses will trot smartly in a group of horses but must be pummeled into moving off by themselves. Such behavior is hardly surprising because horses are herd animals and group, or allelomimetic, behavior is part of their evolutionary heritage. The horse racing back to the barn is rejoining its herd, if other horses are there, or is simply returning to its home range. Consideration of feral horse behavior indicates that many of the vices that appear to be spiteful, lazy, or stupid, are in reality perfectly normal adaptive behaviors for an animal that must find food (graze while bridled), avoid predators (shy), remain with its herd, or rejoin it if separated (barn rat).

Physiologically Based Problems. Other common habits of horses may be explained by their physiology. Horses refuse jumps because they can not see the jump when they are close to it. Some horses are very hard to canter on the left lead but take the right easily because some horses (70%) show definite handedness (Grzimek 1949) even if they have been carefully schooled to canter both clockwise and counterclockwise.

The myriad other vices and training problems of horses are beyond the scope of this book. A behavioral history form for horses is given in Table 9.2.

CATTLE AND OTHER FARM ANIMALS. The behavior problems of farm animals have been reviewed by Fraser (1963) and by Kiley-Worthington (1977).

Cattle

Problems Related to Changes in Management. Dairy cattle present few behavior problems in general probably because dairy farmers have selected for tractable animals as well as high producers. The biggest problems that arise with dairy cattle concern changes in their management. When cattle that had formerly been milked and fed simultaneously in a stanchion barn are placed in free stalls and are not fed concentrates when they are milked, they may enter the milking parlor only reluctantly. Milking time can be prolonged for hours. Apparently, the milking procedure and the reward of relieving pressure on the udder is not enough to induce cattle to be milked.

One solution is to feed grain during milking. Even if the cow requires more concentrates than it can consume while being milked, a portion of its grain can be fed at milking and the rest afterwards. Other management-related behavior problems have involved electrically operated squeeze gates that push the cattle closest to the gate when the cattle ahead balk. Time can be lost and injuries sustained if mechanical devices are not built with the behavior of the normal cow in mind.

Cattle that are always milked or handled from one side may become frightened or aggressive when handled from the other side. A veterinarian or stockperson should note the position of the milking machine outlet in stanchioned cattle and approach the cow from the same side. Milk production may be affected by milking from the "wrong" (unfamiliar) side. Such obvious stresses as isolation and being chased by a dog can lower milk production (Whittlestone et al. 1970), but the herdperson's attitude can also affect production either positively or negatively. If the handler is satisfied with the job, the cows will give more milk and will approach the milking area more quickly (Seabrook 1972).

Kicking. Beef cattle cause more problems for the average veterinarian because they have neither been selected for tractability nor handled often. A beef heifer brought from range or pasture and placed in a box stall is as dangerous as an ill-mannered horse. Such animals should be treated with caution because even when heavily tranquilized they can and will kick. Stock chutes should be used. Cattle do not always "cowkick." They can place a well-aimed blow backward as well as forward. Dairy cattle that kick should be fed only when they stand quietly. The food is the reward for nonkicking behavior.

Calves and adult bulls show a variety of oral "vices" such as bar licking and tongue rolling. Perhaps the most compelling evidence for the protective effect of sterotypies is the Wiepkema et al. (1987) finding that veal calves that indulged in tongue rolling had a lower incidence of abomasal ulcers than calves that had no "vices."

Problems Related to Changes in Environment. Cattle not used to stanchions, such as those living in loose-housing facilities, may find it difficult to lie down or arise when first stanchioned. Hospitalized cattle may be particularly affected because the illness for which they were hospitalized will be compounded by the stress produced by lack of rest.

Pigs. Few pigs are kept as individuals, so most of their behavior problems concern group interactions. Aggression has been dealt with in Chapter 2 and sexual behavior in Chapter 4.

Other problems seen are rubbing the snout on the floor or on the flank of another pig. Bar biting by confined sows is believed to be a consequence of restraint and may release endogenous opiates because treatment of the sow with the opiate blocker naloxone will halt the behavior (Cronin et al. 1986).

Sheep and Goats. The only miscellaneous problem of sheep not discussed elsewhere in this book is wool chewing. This behavior is of unknown etiology.

Goats frequently become nuisances, especially if they are kept as pets and inadequately restrained. The behavior of the goat is not abnormal. It is usually browsing normally, but it may browse on the ornamental flower beds or the crops. Goats are herd animals, and a solitary goat may seek human company if it has been kept as a pet since it was a kid. Normal caprine feeding, investigatory, and allelomimetic behavior may seem very abnormal to the naive goat owner.

References

Abraham, S., R. Baker, D. A. Denton, F. Kraintz, L. Kraintz, and L. Purser. 1973. Components in the regulation of salt balance: Salt appetite studied by operant behaviour. *Aust. J. Exp. Biol. Med. Sci.* 51:65–81.

Abraham, S. F., R. M. Baker, E. H. Blaine, D. A. Denton, and M. J. McKinley. 1975. Water drinking induced in sheep by angiotensin — A physiological or pharmacological effect? *J. Comp. Physiol. Psychol.* 88:503–518.

Adamec, R. E. 1976. The interaction of hunger and preying in the domestic cat (*Felis catus*): An adaptive hierarchy? *Behav. Biol.* 18:263–272.

Adamec, R. E., C. Stark-Adamec, and K. E. Livingston. 1983. The expression of an early developmentally emergent defensive bias in the adult domestic cat (*Felis catus*) in non-predatory situations. *Appl. Anim. Ethol.* 10:89–108.

Adams, D. K. 1929. Experimental studies of adaptive behavior in cats. *Comp. Psychol. Monogr.* 6:1–168.

Adams, T. 1963. Hypothalamic temperature in the cat during feeding and sleep. *Science* 139:609–610.

Adler, L. L., and H. E. Adler. 1977. Ontogeny of observational learning in the dog (*Canis familiaris*). *Dev. Psychobiol.* 10:267–271.

Adrian, E. D. 1943. Afferent areas in the brain of ungulates. *Brain* 66:89–103.

Agrawal, H. C., M. W. Fox, and W. A. Himwich. 1967. Neurochemical and behavioral effects of isolation-rearing in the dog. *Life Sci.* 6:71–78.

Albright, J. L. 1969. Social environment and growth. In *Animal growth and nutrition,* E. S. E. Hafez and I. A. Dyer, eds. Philadelphia: Lea and Febiger. Pp. 106–120.

Albright, J. L., W. P. Gordon, W. C. Black, J. P. Dietrich, W. W. Snyder, and C. E. Meadows. 1966. Behavioral responses of cows to auditory training. *J. Dairy Sci.* 49:104–106.

Aldinger, S. M., V. C. Speer, V. W. Hays, and D. V. Catron. 1959. Effect of saccharin on consumption of starter rations by baby pigs. *J. Anim. Sci.* 18:1350–1355.

Aldis, O. 1975. *Play fighting.* New York: Academic Press.

Alexander, G. 1977. Role of auditory and visual cues in mutual recognition between ewes and lambs in Merino sheep. *Appl. Anim. Ethol.* 3:65–81.

———. 1978. Odour, and the recognition of lambs by Merino ewes. *Appl. Anim. Ethol.* 4:153–158.

Alexander, G., and L. R. Bradley. 1985. Fostering in sheep. IV. Use of restraint. *Appl. Anim. Behav. Sci.* 14:355–364.

Alexander, G., and E. E. Shillito. 1977a. The importance of odour, appearance and voice in maternal recognition of the young in Merino sheep (Ovis aries). Appl. Anim. Ethol. 3:127–135.

———. 1977b. Importance of visual cues from various body regions in maternal recognition of the young in Merino sheep (*Ovis aries*). Appl. Anim. Ethol. 3:137–143.

———. 1978. Maternal responses in Merino ewes to artificially coloured lambs. *Appl. Anim. Ethol.* 4:141–152.

Alexander, G., and E. E. Shillito Walser. 1978. Visual discrimination between ewes by lambs. *Appl. Anim. Ethol.* 4:81–85.

Alexander, G., and D. Stevens. 1981. Recognition of washed lambs by merino ewes. *Appl. Anim. Ethol.* 7:77–86.

_____. 1985. Fostering in sheep. III. Facilitation by the use of odorants. *Appl. Anim. Behav. Sci.* 14:345–354.

Alexander, G., and D. Williams. 1964. Maternal facilitation of sucking drive in newborn lambs. *Science* 146:665–666.

_____. 1966. Teat-seeking activity in lambs during the first hours of life. *Anim. Behav.* 14:166–176.

Alexander, G., J. P. Signoret, and E. S. E. Hafez. 1974. Sexual and maternal behavior. In *Reproduction in farm animals,* E. S. E. Hafez, ed. Philadelphia: Lea and Febiger. Pp. 222–254.

Alexander, G., D. Stevens, R. Kilgour, H. de Langen, B. E. Mottershead, and J. J. Lynch. 1983a. Separation of ewes from twin lambs: Incidence in several breeds. *Appl. Anim. Ethol.* 10:301–317.

Alexander, G., D. Stevens, and L. R. Bradley. 1983b. Washing lambs and confinement as aids to fostering. *Appl. Anim. Ethol.* 10:251–261.

_____. 1985. Fostering in sheep. I. Facilitation by use of textile lamb coats. *Appl. Anim. Behav. Sci.* 14:315–334.

_____. 1988. Maternal behaviour in ewes following caesarian section. *Appl. Anim. Behav. Sci.* 19:273–277.

Allen, B. D., J. F. Cummings, and A. de Lahunta. 1974. The effects of prefrontal lobotomy on aggressive behavior in dogs. *Cornell Vet.* 64:201–216.

Allin, J. T., and E. M. Banks. 1972. Functional aspects of ultrasound production by infant albino rats (*Rattus norvegicus*). *Anim. Behav.* 20:175–185.

Allison, T., and D. V. Cicchetti. 1976. Sleep in mammals: Ecological and constitutional correlates. *Science* 194:732–734.

Altmann, M. 1941. Interrelations of the sex cycle and the behavior of the sow. *J. Comp. Psychol.* 31:481–498.

Ames, D. R., and L. A. Arehart. 1972. Physiological response of lambs to auditory stimuli. *J. Anim. Sci.* 34:994–998.

Anand, B. K., and J. R. Brobeck. 1951. Hypothalamic control of food intake in rats and cats. *Yale J. Biol. Med.* 24:123–140.

Anand, B. K., S. Dua, and K. Shoenberg. 1955. Hypothalamic control of food intake in cats and monkeys. *J. Physiol.* 127:143–152.

Anderson, D. M., and N. S. Urquhart. 1986. Using digital pedometers to monitor travel of cows grazing arid rangeland. *Appl. Anim. Behav. Sci.* 16:11–23.

Anderson, D. M., C. V. Hulet, J. N. Smith, W. L. Shupe, and L. W. Murray. 1987. Heifer disposition and bonding of lambs to heifers. *Appl. Anim. Behav. Sci.* 19:27–30.

Anderson, O. D., and R. Parmenter. 1941. A long-term study of the experimental neurosis in the sheep and dog. *Psychosom. Med. Monogr.* 2(3–4):1–150.

Anderson, R. S. 1974. Obesity in the dog and cat. In *The veterinary annual 1973,* C. S. G. Grunsell and F. W. G. Hill, eds. Bristol: John Wright and Sons. Pp. 182–186.

Andersson, B., and S. M. McCann. 1955. A further study of polydipsia evoked by hypothalamic stimulation in the goat. *Acta Physiol. Scand.* 33:333–346.

Andersson, B., L. Eriksson, and R. Oltner. 1970. Further evidence for angiotensin-sodium interaction in central control of fluid balance. *Life Sci.* 9(1): 1091–1096.

Andersson, B., O. Augustinsson, E. Bademo, J. Junkergard, C. Kvart, G. Nyman, and M. Wiberg. 1987. Systemic and centrally mediated angiotensin II effects in the horse. *Acta Physiol. Scand.* 129:143–149.

Andry, D. K., and M. W. Luttges. 1972. Memory traces: Experimental separation by cyclohex-imide and electroconvulsive shock. *Science* 178:518–520.

Andy, O. J., D. F. Peeler, Jr., and D. P. Foshee. 1967. Avoidance and discrimination learning following hippocampal ablation in the cat. *J. Comp. Physiol. Psychol.* 64:516–519.

Angel, C., O. D. Murphree, and D. C. De Lucia. 1974. The effects of chlordiazepoxide, amphetamine and cocaine on bar-press behavior in normal and genetically nervous dogs. *Res. Nerv. Syst.* 35:220–223.

Anlezark, G. M., T. J. Crow, and A. P. Greenway. 1973. Impaired learning and decreased cortical norepinephrine after bilateral locus coeruleus lesions. *Science* 181:682–684.

Anonymous. 1975. Behaviour of boars. TDC article. *Vet. Rec.* 96:221.

Arave, C. W., and J. L. Albright. 1976. Social rank and physiological traits of dairy cows as influenced by changing group membership. *J. Dairy Sci.* 59:974–981.

Archer, M. 1973. The species preferences of grazing horses. *J. Br. Grassland Soc.* 28:123–128.

Archibald, J. 1974. *Canine surgery,* 2d ed. Santa Barbara: American Veterinary Publications.

Arendt, J., A. M. Symons, C. A. Laud, and S. J. Pryde. 1983. Melatonin can induce early onset of the breeding season in ewes. *J. Endocrinol.* 97:395–400.

Arnold, G. W. 1966a. The special senses in grazing animals. I. Sight and dietary habits in sheep. *Aust. J. Agric. Res.* 17:521–529.

————. 1966b. The special senses in grazing animals. II. Smell, taste, and touch and dietary habits in sheep. *Aust. J. Agric. Res.* 17:531–542.

————. 1984–85. Comparison of the time budgets and circadian patterns of maintenance activities in sheep, cattle and horses grouped together. *Appl. Anim. Behav. Sci.* 13:19–30.

Arnold, G. W., and M. L. Dudzinski. 1978. *Ethology of free-ranging domestic animals.* Amsterdam: Elsevier Scientific Publishing.

Arnold, G. W., and A. Grassia. 1982. Ethogram of agonistic behaviour for Thoroughbred horses. *Appl. Anim. Ethol.* 8:5–25.

Arnold, G. W., and R. A. Maller. 1974. Some aspects of competition between sheep for supplementary feed. *Anim. Prod.* 19:309–319.

Arnold, G. W., and P. D. Morgan. 1975. Behaviour of the ewe and lamb at lambing and its relationship to lamb mortality. *Appl. Anim. Ethol.* 2:25–46.

Arnold, G. W., and P. J. Pahl. 1974. Some aspects of social behaviour in domestic sheep. *Anim. Behav.* 22:592–600.

Arnold, G. W., C. A. P. Boundy, P. D. Morgan, and G. Bartle. 1975. The roles of sight and hearing in the lamb in the location and discrimination between ewes. *Appl. Anim. Ethol.* 1:167–176.

Arnold, G. W., S. R. Wallace, and R. A. Maller. 1979. Some factors involved in natural weaning processes in sheep. *Appl. Anim. Ethol.* 5:43–50.

Arnold, G. W., S. R. Wallace, and W. A. Rea. 1981. Associations between individuals and home-range behaviour in natural flocks of three breeds of domestic sheep. *Appl. Anim. Ethol.* 7:239–257.

Aronson, L. R., and M. L. Cooper. 1966. Seasonal variation in mating behavior in cats after desensitization of glans penis. *Science* 152:226–230.

————. 1974. Olfactory deprivation and mating behavior in sexually experienced male cats. *Behav. Biol.* 11:459–480.

————. 1977. Central versus peripheral genital desensitization and mating behavior in male cats: Tonic and phasic effects. *Ann. N.Y. Acad. Sci.* 290:299–313.

Asa, C. S., D. A. Goldfoot, and O. J. Ginther. 1979. Sociosexual behavior and the ovulatory cycle of ponies (*Equus caballus*) observed in harem groups. *Horm. Behav.* 13:46–65.

————. 1983. Assessment of the sexual behavior of pregnant mares. *Horm. Behav.* 17:405–413.

Aschoff, J. 1965. *Circadian clocks.* Proceedings of the Feldafing Summer School. Amsterdam: North Holland Publishing.

Ashmead, D. H., R. K. Clifton, and E. P. Reese. 1986. Development of auditory localization in dogs: Single source and precedence effect sounds. *Dev. Psychobiol.* 19:91–103.

Atkeson, F. W., A. O. Shaw, and H. W. Cave. 1942. Grazing habits of dairy cattle. *J. Dairy Sci.* 25:779–784.

Auffray, P. 1969. Effets des lésions des noyaux ventro-medians hypothalamiques sur la prise d'aliment chez le porc. *Ann. Biol. Anim. Biochim. Biophys.* 9:513–526.

Auffray, P., and J.-C. Marcilloux. 1983. An analysis of feeding patterns in the adult pig. *Reprod. Nutr. Dev.* 22:517–524.

Back, D. G., B. W. Pickett, J. L. Voss, and G. E. Seidel. 1974. Observations on the sexual behavior of nonlactating mares. *J. Am. Vet. Med. Assoc.* 165:717–720.

Bacon, W. E. 1973. Aversive conditioning in neonatal kittens. *J. Comp. Physiol. Psychol.* 83:306–313.

Bacon, W. E., and W. C. Stanley. 1963. Effect of deprivation level in puppies on performance maintained by a passive person reinforcer. *J. Comp. Physiol. Psychol.* 56:783–785.

Bado, A., M. Rodriguez, M. J. M. Lewin, J. Martinez, and M. Dubrasquet. 1988. Cholecystokinin suppresses food intake in cats: Structure-activity characterization. *Pharmacol. Biochem. Behav.* 31:297–303.

Bado, A., M. J. M. Lewin, and M. Dubrasquet. 1989. Effects of bombesin on food intake and gastric acid secretion in cats. *Am. J. Physiol.* 256:R181–R186.

Baer, K. L., G. D. Potter, T. H. Friend, and B. V. Beaver. 1983. Observation effects on learning in horses. *Appl. Anim. Ethol.* 11:123–129.

Baile, C. A., and J. M. Forbes. 1974. Control of feed intake and regulation of energy balance in ruminants. *Physiol. Rev.* 54:160–214.

Baile, C. A., and W. H. Pfander. 1966. A possible chemo sensitive regulatory mechanism of ovine feed intake. *Am. J. Physiol.* 210:1243–1248.

Baile, C. A., C. W. Simpson, L. F. Krabill, and F. H. Martin. 1972. Adrenergic agonists and antagonists and feeding in sheep and cattle. *Life Sci.* 11(1): 661–668.

Baile, C. A., C. L. McLaughlin, and M. A. Della Fera. 1986. Role of cholecystokinin and opioid peptides in control of food intake. *Physiol. Rev.* 66:171–234.

Baile, C. A., C. L. McLaughlin, F. C. Buonomo, T. J. Lauterio, L. Marson, and M. A. Della Fera. 1987. Opioid peptides and the control of feeding in sheep. *Fed. Proc.* 46:173–177.

Bailey, C. J., and L. W. Porter. 1955. Relevant cues in drive discrimination in cats. *J. Comp. Physiol. Psychol.* 48:180–182.

Bailey, P. J., A. H. Bishop, and C. T. Boord. 1974. Grazing behaviour of steers. *Proc. Aust. Soc. Anim. Prod.* 10:303–306.

Baker, A.E.M., and B. H. Crawford. 1986. Observational learning in horses. *Appl. Anim. Behav. Sci.* 15:7–13.

Baker, A.E.M., and G. E. Seidel. 1985. Why do cows mount other cows? *Appl. Anim. Behav. Sci.* 13:237–241.

Balch, C. C. 1955. Sleep in ruminants. *Nature* 175:940–941.

Baldwin, B. A. 1969. The study of behaviour in pigs. *Br. Vet. J.* 125:281–288.

———. 1977. Ability of goats and calves to distinguish between conspecific urine samples using olfaction. *Appl. Anim. Ethol.* 3:145–150.

———. 1979. Operant studies on shape discrimination in goats. Physiol. Behav. 23:455–459.

———. 1981. Shape discrimination in sheep and calves. *Anim. Behav.* 29:830–834.

Baldwin, B. A., and T. R. Cooper. 1979. The effects of olfactory bulbectomy on feeding behaviour in pigs. *Appl. Anim. Ethol.* 5:153–159.

Baldwin, B. A., and D. L. Ingram. 1968. Factors influencing behavioral thermoregulation in the pig. *Physiol. Behav.* 3:409–415.

Baldwin, B. A., and G. B. Meese. 1977a. The ability of sheep to distinguish between conspecifics by means of olfaction. *Physiol. Behav.* 18:803–808.

———. 1977b. Sensory reinforcement and illumination preference in the domesticated pig. *Anim. Behav.* 25:497–507.

———. 1979. Social behaviour in pigs studied by means of operant conditioning. *Anim. Behav.* 27:947–957.

Baldwin, B. A., and R. F. Parrott. 1979. Studies on intracranial electrical self-stimulation in pigs in relation to ingestive and exploratory behaviour. *Physiol. Behav.* 22:723–730.

———. 1985. Effects of intracerebroventricular injection of naloxone on operant feeding and drinking in pigs. *Pharmacol. Biochem. Behav.* 22:37–40.

Baldwin, B. A., and E. E. Shillito. 1974. The effects of ablation of the olfactory bulbs on parturition and maternal behaviour in Soay sheep. *Anim. Behav.* 22:220–223.

Baldwin, B. A., and D. B. Stephens. 1970. Operant conditioning procedures for producing emotional responses in pigs. *J. Physiol.* 210:127P-128P.

———. 1973. The effects of conditioned behaviour and environmental factors on plasma corticosteroid levels in pigs. *Physiol. Behav.* 10:267–274.

Baldwin, B. A., and S. N. Thornton. 1986. Operant drinking in pigs following intracerebralventricular injections of hypertonic solutions and angiotensin II. *Physiol. Behav.* 36:325–328.

Baldwin, B. A., and J. O. Yates. 1977. The effects of hypothalamic temperature variation and intracarotid cooling on behavioural thermoregulation in sheep. *J. Physiol.* 265:705–720.

Baldwin, B. A., D. J. Conner, and G. B. Meese. 1974. Sensory reinforcement in the pig. *J. Physiol.* 242:27.

Baldwin, B. A., W. L. Grovum, C. A. Baile, and J. R. Brobeck. 1975. Feeding following intraventricular injection of Ca⁺⁺, Mg⁺⁺ or pentobarbital in pigs. *Pharm. Biochem. Behav.* 3:915–918.

Baldwin, B. A., C. L. McLaughlin, and C. A. Baile. 1977. The effect of ablation of the olfactory bulbs on feeding behaviour in sheep. *Appl. Anim. Ethol.* 3:151–161.

Bali, V. R., and J. Hörmeyer. 1986. Eine chirurgische Behandlungs-moglichkeit des Harnspritzens bei kastrierten Katern. *Kleintierpraxis* 31:329–332.

Bandler, R. J., Jr., and J. P. Flynn. 1974. Neural pathways from thalamus associated with regulation of aggressive behavior. *Science* 183:96–99.

Bane, A. 1954. Studies on monozygous cattle twins. XV. Sexual functions of bulls in relation to heredity, rearing intensity and somatic conditions. *Acta Agric. Scand.* 4:95–208.

Banks, E. M. 1964. Some aspects of sexual behavior in domestic sheep, *Ovis aries. Behaviour* 23:249–279.

Bard, P. 1936. Oestrual behavior in surviving decorticate cats. *Am. J. Physiol.* (abstr.) 116:4–5.

Bareham, J. R. 1975. The effect of lack of vision on suckling behaviour of lambs. *Appl. Anim. Ethol.* 1:245–250.

———. 1976. The behaviour of lambs on the first day after birth. *Br. Vet. J.* 132:152–162.

Barnes, R. H., A. U. Moore, and W. G. Pond. 1970. Behavioral abnormalities in young adult pigs caused by malnutrition in early life. *J. Nutr.* 100:149–155.

Barnett, J. H., P. H. Hemsworth, C. G. Winfield, and C. Hansen. 1986. Effects of social environment on welfare status and sexual behaviour of female pigs. I. Effects of group size. *Appl. Anim. Behav. Sci.* 16:249–257.

Baron, A., C. N. Stewart, and J. M. Warren. 1957. Patterns of social interaction in cats (*Felis domestica*). *Behaviour* 11:56–66.

Barondes, S. H., and H. D. Cohen. 1966. Puromycin effect on successive phases of memory storage. *Science* 151:594–595.

Barrett, P., and P. Bateson. 1978. The development of play in cats. *Behaviour* 66:106–120.

Bartoshuk, L. M., M. A. Harned, and L. H. Parks. 1971. Taste of water in the cat: Effects on sucrose preference. *Science* 171:699–701.

Bartus, R. T., R. L. Dean III, B. Beer, and A. S. Lippa. 1982. The cholinergic hypothesis of geriatric memory dysfunction. *Science* 217:408–417.

Bateson, P. 1978. Sexual imprinting and optimal outbreeding. *Nature* 273:659–660.

———. 1979. How do sensitive periods arise and what are they for? *Anim. Behav.* 27:470–486.

Baumgardt, B. R., and A. D. Peterson. 1970. Hyperphagia in sheep induced by infusion of the ventriculo cisternal system with a depressant. *Fed. Proc.* (abstr.) 29:760.

Beach, F. A., Jr. 1937. The neural basis of innate behavior. I. Effects of cortical lesions upon the maternal behavior pattern in the rat. *J. Comp. Psychol.* 24:393–439.

Beach, F. A. 1968. Coital behavior in dogs. III. Effects of early isolation on mating in males. *Behaviour* 30:218–238.

———. 1970a. Coital behavior in dogs. VI. Long-term effects of castration upon mating in the male. *J. Comp. Physiol. Psychol. Monogr.* 70(3):1–32.

———. 1970b. Coital behaviour in dogs. VIII. Social affinity, dominance and sexual preference in the bitch. *Behaviour* 36:131–148.

———. 1974. Effects of gonadal hormones on urinary behavior in dogs. *Physiol. Behav.* 12:1005–1013.

Beach, F. A., and R. W. Gilmore. 1949. Response of male dogs to urine from females in heat. *J. Mammal.* 30:391–392.

Beach, F. A., and J. Jaynes. 1956. Studies on maternal retrieving in rats. II. Effects of practice and previous parturitions. *Am. Natur.* 90:103–109.

Beach, F. A., and R. E. Kuehn. 1970. Coital behavior in dogs. X. Effects of androgenic stimulation during development on feminine mating responses in females and males. *Horm. Behav.* 1:347–367.

Beach, F. A., and B. J. Le Boeuf. 1967. Coital behaviour in dogs. I. Preferential mating in the bitch. *Anim. Behav.* 15:546–558.

Beach, F. A., and A. Merari. 1968. Coital behavior in dogs, IV. Effects of progesterone in the bitch. *Proc. Nat. Acad. Sci.* 61:442–446.

———. 1970. Coital behavior in dogs. V. Effects of estrogen and progesterone on mating and other forms of social behavior in the bitch. *J. Comp. Physiol. Psychol. Monogr.* 70(1) 2:1–22.

Beach, F. A., A. Zitrin, and J. Jaynes. 1955. Neural mediation of mating in male cats. II. Contributions of the frontal cortex. *J. Exp. Zool.* 130:381–401.

_____. 1956. Neural mediation of mating in male cats. I. Effects of unilateral and bilateral removal of the neocortex. *J. Comp. Physiol. Psychol.* 49:321–327.

Beach, F. A., R. E. Kuehn, R. H. Sprague, and J. J. Anisko. 1972. Coital behavior in dogs. XI. Effects of andorgenic stimulation during development on masculine mating responses in females. *Horm. Behav.* 3:143–168.

Beach, F. A., A. I. Johnson, J. J. Anisko, and I. F. Dunbar. 1977. Hormonal control of sexual attraction in pseudohermaphroditic female dogs. *J. Comp. Physiol. Psychol.* 91:711–715.

Beach, F. A., I. F. Dunbar, and M. G. Buehler. 1982. Sexual characteristics of female dogs during successive phases of the ovarian cycle. *Horm. Behav.* 16:414–442.

Beach, F. A., M. G. Buehler, and I. F. Dunbar. 1983. Development of attraction to estrous females in male dogs. *Physiol. Behav.* 31:293–297.

Beamer, W., G. Bermant, and M. T. Clegg. 1969. Copulatory behaviour of the ram, *Ovis aries.* II. Factors affecting copulatory satiation. *Anim. Behav.* 17:706–711.

Beauchemin, K. A., S. Zelin, D. Genner, and J. G. Buchanan-Smith. 1989. An automatic system for quantification of eating and ruminating activities of dairy cattle housed in stalls. *J. Dairy Sci.* 72:2746–2759.

Beaver, B. V. 1980. *Veterinary aspects of feline behavior.* St. Louis: C. V. Mosby.

_____. 1983. Clinical classification of canine aggression. *Appl. Anim. Ethol.* 10:35–43.

Beck, A. M. 1973. *The ecology of stray dogs. A study of free-ranging urban animals.* Baltimore: York Press.

Becker, B. A., J. J. Ford, R. K. Christenson, R. C. Manak, G. L. Hahn, and J. A. DeShazer. 1985. Cortisol response of gilts in tether stalls. *J. Anim. Sci.* 60:264–270.

Becker, R. F., J. E. King, and J. E. Markee. 1962. Studies on olfactory discrimination in dogs. II. Discriminatory behavior in a free environment. *J. Comp. Physiol. Psychol.* 55:773–780.

Beckett, S. D., R. S. Hudson, D. F. Walker, and R. C. Purohit. 1978. Effect of local anesthesia of the penis and dorsal penile neurectomy on the mating ability of bulls. *J. Am. Vet. Med. Assoc.* 173:838–839.

Beilharz, R. G., and D. F. Cox. 1967. Social dominance in swine. Anim. Behav. 15:117–122.

Beilharz, R. G., and P. J. Mylrea. 1963a. Social position and behaviour of dairy heifers in yards. *Anim. Behav.* 11:522–528.

_____. 1963b. Social position and movement orders of dairy heifers. *Anim. Behav.* 11:529–533.

Beilharz, R. G., and K. Zeeb. 1982. Social dominance in dairy cattle. *Appl. Anim. Ethol.* 8:79–97.

Beilharz, R. G., D. F. Butcher, and A. E. Freeman. 1966. Social dominance and milk production in Holsteins. *J. Dairy Sci.* 49:887–892.

Bekoff, M. 1974. Social play and play-soliciting by infant canids. *Am. Zool.* 14:323–340.

_____. 1977. Social communication in canids: Evidence for the evolution of a stereotyped mammalian display. *Science* 197:1097–1099.

Bekoff, M., H. L. Hill, and J. B. Mitton. 1975. Behavioural taxonomy in canids by discriminant function analyses. *Science* 190:1223–1225.

Belkin, M., U. Yinon, L. Rose, and I. Reisert. 1977. Effect of visual environment on refractive error of cats. *Doc. Ophthalmol.* 42:433–437.

Bell, F. R. 1959. Preference thresholds for taste discrimination in goats. *J. Agric. Sci.* 52:125–128.

_____. 1960. The electroencephalogram of goats during somnolence and rumination. *Anim. Behav.* 8:39–42.

_____. 1963. Alkaline taste in goats assessed by the preference test technique. *J. Comp. Physiol. Psychol.* 56:174–178.

Bell, F. R., and H. L. Williams. 1959. Threshold values for taste in monozygotic twin calves. *Nature* 183:345–346.

Bellinger, L. L., and F. E. Williams. 1989. The effect of portal and jugular infused glucose mannitol and saline on food intake in dogs. *Physiology and Behavior* 46:693–698.

Benjamin, R. M., and R. F. Thompson. 1959. Differential effects of cortical lesions in infant and adult cats on roughness discrimination. *Exp. Neurol.* 1:305–321.

Bennett, M., K. A. Houpt, and H. N. Erb. 1988. Effects of declawing on feline behavior. *Companion Anim. Pract.* 2:7–12.

Berg, I. A. 1944. Development of behavior: The micturition pattern in the dog. *J. Exp. Psychol.* 34:343–368.

Berger, J. 1977. Organizational systems and dominance in feral horses in the Grand Canyon. *Behav. Ecol. Sociobiol.* 2:131–146.

———. 1983. Induced abortion and social factors in wild horses. *Nature* 303:59–61.

———. 1986. *Wild horses of the Great Basin.* Chicago: University of Chicago Press.

Berger, J., and C. Cunningham. 1987. Influence of familiarity on frequency of inbreeding in wild horses. *Evolution* 41:229–231.

Berggren-Thomas, B., and W. D. Hohenboken. 1986. The effects of sire-breed, forage availability and weather on the grazing behavior of crossbred ewes. *Appl. Anim. Behav. Sci.* 15:217–228.

Berkson, G. 1968. Maturation defects in kittens. *Am. J. Ment. Defic.* 72:757–777.

Berman, M., and I. Dunbar. 1983. The social behavior of free-ranging urban dogs. *Appl. Anim. Ethol.* 10:5–17.

Bermant, G., M. T. Clegg, and W. Beamer. 1969. Copulatory behaviour of the ram, *Ovis aries.* I. A normative study. *Anim. Behav.* 17:700–705.

Bernston, G. G., M. S. Beattie, and J. M. Walker. 1976. Effects of nicotine and muscarinic compounds on biting attack in the cat. *Pharm. Biochem. Behav.* 5:235–239.

Bertram, B. C. R. 1975. Social factors influencing reproduction in wild lions. *J. Zool.* 177:463–482.

Bielanski, W., and S. Wierzbowski. 1961. Depletion test in stallions. Proc. 4th Int. Cong. Anim. Reprod. pp. 279–282.

Bigelow, J. A., and T. R. Houpt. 1988. Feeding and drinking patterns in young pigs. *Physiol. Behav.* 43:99–109.

Bines, J. A., S. Suzuki, and C. C. Balch. 1969. The quantitative significance of long-term regulation of food intake in the cow. *Br. J. Nutr.* 23:695–704.

Billing, A. E., and M. A. Vince. 1987a. Teat-seeking behaviour in newborn lambs. I: Evidence for the influence of maternal skin temperature. *Appl. Anim. Behav. Sci.* 18:301–313.

———. 1987b. Teat-seeking behaviour in newborn lambs. II. Evidence for the influence of the dam's surface textures and degree of surface yield. *Appl. Anim. Behav. Sci.* 18:315–325.

Bitterman, M. E. 1965. Phyletic differences in learning. *Am. Psychol.* 20:396–410.

Björk, A., N. G. Olsson, E. Christensson, K. Martinsson, and O. Olsson. 1988. Effects of amperozide on biting behavior and performance in restricted-fed pigs following regrouping. *J. Anim. Sci.* 66:669–675.

Black, A. H., G. A. Young, and C. Batenchuk. 1970. Avoidance training of hippocampal theta waves in flaxedilized dogs and its relation to skeletal movement. *J. Comp. Physiol. Psychol.* 70:15–24.

Blair, R., and J. FitzSimons. 1970. A note on the voluntary feed intake and growth of pigs given diets containing an extremely bitter compound. *Anim. Prod.* 12:529–530.

Blakemore, C., and R. C. Van Sluyters. 1975. Innate and environmental factors in the development of the kitten's visual cortex. *J. Physiol.* 248:663–716.

Bland, K. P., and B. M. Jubilan. 1987. Correlation of flehmen by male sheep with female behaviour and oestrus. *Anim. Behav.* 35:735–738.

Blaxter, K. L. 1944. Food preferences and habits in dairy cows. Br. Soc. Anim. Prod., 2d Meeting. Pp. 85–94.

Blaxter, K. L., V. R. Fowler, and J. C. Gill. 1982. A study of the growth of sheep to maturity. *J. Ag. Sci.* 98:405–420.

Bleicher, N. 1962. Behavior of the bitch during parturition. *J. Am. Vet. Med. Assoc.* 140:1076–1082.

Block, M., and I. Zucker. 1976. Circadian rhythms of rat locomotor activity after lesions of midbrain raphe nuclei. *J. Comp. Physiol. A.* 109:235–247.

Blockey, M. A. de B. 1981a. Development of a serving capacity test for beef bulls. *Appl. Anim. Ethol.* 7:307–319.

_____. 1981b. Further studies on the serving capacity test for beef bulls. *Appl. Anim. Ethol.* 7:337–350.

_____. 1981c. Modification of a serving capacity test for beef bulls. *Appl. Anim. Ethol.* 7:321–336.

Blom, A. K., K. Halse, and K. Hove. 1976. Growth hormone, insulin and sugar in the blood plasma of bulls. Interrelated diurnal variations. *Acta Endocrinol.* 82:758–766.

Booth, W. D., and B. A. Baldwin. 1980. Lack of effect on sexual behaviour or the development of testicular function after removal of olfactory bulbs in prepubertal boars. *J. Reprod. Fertil.* 58:173–182.

Borchelt, P. L. 1983. Aggressive behavior of dogs kept as companion animals: Classification and influence of sex, reproductive status and breed. *Appl. Anim. Ethol.* 10:45–61.

Borchelt, P. L., and V. L. Voith. 1981. Elimination behavior problems in cats. *Comp. Contin. Educ. 3:730–737.*

_____. 1985a. Aggressive behavior in dogs and cats. *Comp. Contin. Educ.* 11:949–957.

_____. 1985b. Punishment. *Comp. Contin. Educ.* 9:780–791.

Borchelt, P. L., R. Lockwood, A. M. Beck, and V. L. Voith. 1983. Attacks by packs of dogs involving predation on human beings. *Public Health Rep.* 98:57–66.

Bottoms, G. D., O. F. Roesel, F. D. Rausch, and E. L. Akins. 1972. Circadian variation in plasma cortisol and corticosterone in pigs and mares. *Am. J. Vet. Res.* 33:785–790.

Bouissou, M.-F. 1965. Observations sur la hierarchie sociale chez les bovins domestiques (in French). *Ann. Biol. Anim. Biochim. Biophys.* 5:327–339.

_____. 1971. Effet de l'absence d'informations optiques et de contact physique sur la manifestation des relations hiérarchiques chez les bovins domestiques (in French). *Ann. Biol. Anim. Biochim. Biophys.* 11:191–198.

_____. 1972. Influence of body weight and presence of horns on social rank in domestic cattle. *Anim. Behav.* 20:474–477.

_____. 1978. Effects of injections of testosterone proprionate on dominance relationships in a group of cows. *Hormon. Behav.* 11:388–400.

Bouissou, M.-F., and V. Gaudioso. 1982. Effect of early androgen treatment on subsequent social behavior in heifers. *Hormon. Behav.* 16:132–146.

Bowersox, S. S., T. L. Baker, and W. C. Dement. 1984. Sleep-wakefulness patterns in the aged cat. *Electroencephalogr. Clin. Neurophysiol.* 58:240–252.

Boy, V., and P. Duncan. 1979. Time budgets of Camargue horses. I. Developmental changes in the time budgets of foals. *Behaviour* 71:187–202.

Boyd, L. E. 1980. The natality, foal survivorship, and mare-foal behavior of feral horses in Wyoming's Red Desert. Master's thesis, University of Wyoming, Laramie.

_____. 1988. Time budgets of adult Przewalski horses: Effects of sex, reproductive status and enclosure. *Appl. Anim. Behav. Sci.* 21:19–39.

Boyd, L. E., D. A. Carbonaro, and K. A. Houpt. 1988. The 24–hour time budget of Przewalski horses. *Appl. Anim. Behav. Sci.* 21:5–17.

Braden, I. C. 1965. Effects of hypothalamically implanted estrogen on the maternal sequence of rats. *73d Am. Psychol. Assoc. Proc.* Pp. 187–188.

Brakel, W. J., and R. A. Leis. 1976. Impact of social disorganization on behavior, milk yield, and body weight of dairy cows. *J. Dairy Sci.* 59:716–721.

Braude, R., M. J. Newport, and J. W. G. Porter. 1971. Artificial rearing of pigs. Pt. 3. The effect of heat treatment on the nutritive value of spray-dried whole-milk powder for the baby pig. *Br. J. Nutr.* 25:113–125.

Bray, A. R., and M. Wodzicka-Tomaszewska. 1974. Perinatal behaviour and progesterone and corticosteroid levels in sheep. *Proc. Aust. Soc. Anim. Prod.* 10:318–321.

Breland, K., and M. Breland. 1966. *Animal behavior.* New York: Macmillan.

Brindley, E. L., D. J. Bullock, and F. Maisels. 1989. Effects of rain and fly harassment on feeding behaviour of free-ranging feral goats. *Appl. Anim. Behav. Sci.* 24:31–41.

Bristol, F. 1982. Breeding behaviour of a stallion at pasture with 20 mares in synchronized oestrus. *J. Reprod. Fertil. Suppl.* 32:71–77.

Brobeck, J. 1955. Neural regulation of food intake. *Ann. N.Y. Acad. Sci.* 63:44–55.

Bromiley, R. B. 1948a. Conditioned responses in a dog after removal of neocortex. *J. Comp. Physiol. Psychol.* 41:102–110.

———. 1948b. The development of conditioned responses in cats after unilateral decortication. *J. Comp. Physiol. Psychol.* 41:155–164.

Bronson, F. H., and W. K. Whitten. 1968. Oestrus-accelerating pheromone of mice: Assay, androgen-dependency, and presence in bladder urine. *J. Reprod. Fertil.* 15:131–134.

Bronson, R. T. 1979. Brain weight-body weight scaling in breeds of dogs and cats. *Brain Behav. Evol.* 16:227–236.

Broom, D. M., and J. D. Leaver. 1978. Effects of group-rearing or partial isolation on later social behaviour of calves. *Anim. Behav.* 26:1255–1263.

Brown, R. F., K. A. Houpt, and H. F. Schryver. 1976. Stimulation of food intake in horses by diazepam and promazine. *Pharm. Biochem. Behav.* 5:495–497.

Brown, S. A., S. Crowell-Davis, T. Malcolm, and P. Edwards. 1987. Naloxone-responsive compulsive tail chasing in a dog. *J. Amer. Vet. Med. Assoc.* 190:884–886.

Brownlee, A. 1954. Play in domestic cattle in Britain: An analysis of its nature. *Br. Vet. J.* 110:48–68.

Bryant, M. J. 1975. A note on the effect of rearing experience upon the development of sexual behaviour in ram lambs. *Anim. Prod.* 21:97–99.

Bryant, M. J., and R. Ewbank. 1972. Some effects of stocking rate and group size upon agonistic behaviour in groups of growing pigs. *Br. Vet. J.* 128:64–70.

Bryant, M. J., and T. Tompkins. 1973. Sexual behaviour of sheep. *Vet. Rec.* 93:253.

Buchenauer, V. D., and B. Fritsch. 1980. Zum Farbsehvermögen von Hausziegen (*Capra bircus* L.) (in German). *Z. Tierpsychol.* 53:225–230.

Buddenberg, B. J., C. J. Brown, Z. B. Johnson, and R. S. Honea. 1986. Maternal behavior of beef cows at parturition. *J. Anim. Sci.* 62: 42–47.

Burger, J. F. 1952. Sex physiology of pigs. *Onderstepoort J. Vet. Res. Suppl.* 2:1–218.

Buser, P., and A. Rougeul. 1961. Observations sur le conditionnement instrumental alimentaire chez le chat (in French). In *Brain Mechanisms and Learning.* A. Fessard, R. W. Gerard, J. Konorski, and J. F. Delafresnaye, eds. Springfield, Ill.: Charles C. Thomas. Pp. 527–554.

Busnel, R.-G. 1963. *Acoustic behaviour of animals.* Amsterdam: Elsevier Publishing.

Cairns, R. B., and D. L. Johnson. 1965. The development of interspecies social attachments. *Psychonom. Sci.* 2:337–338.

Campbell, A., Jr. 1978. Deficits in visual learning produced by posterior temporal lesions in cats. *J. Comp. Physiol. Psychol.* 92:45–57.

Campbell, R. G. 1976. A note on the use of a feed flavour to stimulate the feed intake of weaner pigs. *Anim. Prod.* 23:417–419.

Campbell, S. S., and I. Tobler. 1984. Animal sleep: A review of sleep duration across phylogeny. *Neurosci. Behav. Rev.* 8:269–300.

Campbell, W. E. 1975. *Behavior problems in dogs.* Santa Barbara: American Veterinary Publications.

———. 1989. *Better behavior in dogs and cats.* Loveland, Colo.: Alpine Publications.

Campitelli, S., C. Carenzi, and M. Verga. 1982–83. Factors which influence parturition in the mare and development of the foal. *Appl. Anim. Ethol.* 9:7–14.

Canali, E., M. Verga, M. Montagna, and A. Baldi. 1986. Social interactions and induced behavioural reactions in milk-fed female calves. *Appl. Anim. Behav. Sci.* 16:207–215.

Candland, D. K., and D. Milne. 1966. Species differences in approach-behavior as a function of developmental environment. *Anim. Behav.* 14:539–545.

Carlstead, K. 1986. Predictability of feeding: Its effect on agonistic behaviour and growth in grower pigs. *Appl. Anim. Behav. Sci.* 16:25–38.

Caro, T. M. 1980a. Effects of the mother, object play and adult experience on predation in cats. *Behav. Neural. Biol.* 29:29–51.

———. 1980b. Predatory behaviour and social play in kittens. *Behaviour* 76:1–24.

———. 1980c. Predatory behaviour in domestic cat mothers. *Behaviour* 74:128–148.

Carroll, J. K., and H. C. Stanton. 1974. Changes in hepatic tyrosine aminotransferase activitry due to maturation, circadian rhythm or stress in fetal or neonatal swine. *Fed. Proc.* (abstract) 33:370.

Carson, K., and D. G. M. Wood-Gush. 1983a. Equine behaviour: I. A review of the literature on social and dam foal behavior. *Appl. Anim. Ethol.* 10:165–178.

_____. 1983b. The nursing behavior of Thoroughbred foals. *Eq. Vet. J.* 15:257–262.

Carver, D. S., and H. N. Waterhouse. 1962. The variation in the water consumption of cats. Proc. Anim. Care Panel 12:267–270.

Castle, M. E., and R. J. Halley. 1953. The grazing behaviour of dairy cattle at the National Institute for Research in Dairying. *Br. J. Anim. Behav.* 1:139–143.

Castonguay, T. W. 1981. Dietary dilution and intake in the cat. *Physiol. Behav.* 27:547–549.

Cerny, V. A. 1977. Failure of dihydrotestosterone to elicit sexual behaviour in the female cat. *J. Endocrinol.* 75:173–174.

Cervantes, M., R. Ruelas, and C. Beyer. 1983. Serotonergic influences on EEG synchronization induced by milk drinking in the cat. *Pharm. Biochem. Behav.* 18:851–855.

Chakraborty, P. K., W. B. Panko, and W. S. Fletcher. 1980. Serum hormone concentrations and their relationships to sexual behavior at the first and second estrous cycles of the Labrador bitch. *Biol. Reprod.* 22:227–232.

Chase, L. E., P. J. Wangsness, and B. R. Baumgardt. 1976. Feeding behavior of steers fed a complete mixed ration. *J. Dairy Sci.* 59:123–128.

Chemineau, P. 1986. Sexual behavior and gonadal activity during the year in the tropical Creole meat goat. I. Female estrous behavior and ovarian activity. *Reprod. Nutr. Dev.* 26:441–452.

Chepko, B. D. 1971. A preliminary study of the effects of play deprivation on young goats. *Z. Tierpsychol.* 28:517–526.

Chesler, P. 1969. Maternal influence in learning by observation in kittens. *Science* 166:901–903.

Christensen, H. R., G. W. Seifert, and T. B. Post. 1982. The relationship between a serving capacity test and fertility of beef bulls. *Aust. Vet. J.* 58:241–244.

Christie, D. W., and E. T. Bell. 1972. Studies on canine reproductive behaviour during the normal oestrous cycle. *Anim. Behav.* 20:621–631.

Chute, D. L., and D. C. Wright. 1973. Retrograde state dependent learning. *Science* 180:878–880.

Cizek, L. J., R. E. Semple, K. C. Huang, and M. I. Gregersen. 1951. Effect of extracellular electrolyte depletion on water intake in dogs. *Am. J. Physiol.* 164:415–422.

Clark, J. R., R. W. Bell, L. F. Tribble, and A. M. Lennon. 1985. Effects of composition and density of the group on the performance, behaviour and age at puberty in swine. *Appl. Anim. Behav. Sci.* 14:127–135.

Clarke, I. J., and R. J. Scaramuzzi. 1978. Sexual behaviour and LH secretion in spayed androgenized ewes after a single injection of testosterone or oestradiol-17β. *J. Reprod. Fertil.* 52:313–320.

Clegg, M. T., J. A. Santolucito, J. D. Smith, and W. F. Ganong. 1958. The effect of hypothalamic lesions on sexual behavior and estrous cycles in the ewe. *Endocrinol.* 62:790–797.

Clegg, M. T., W. Beamer, and G. Bermant. 1969. Copulatory behaviour of the ram, *Ovis aries*. III. Effects of pre- and postpubertal castration and androgen replacement therapy. *Anim. Behav.* 17:712–717.

Clemens, L. G. 1971. Perinatal hormones and the modification of adult behavior. In *Steroid hormones and brain function,* C. H. Sawyer and R. A. Gorski, eds. Los Angeles: University of California Press. Pp. 203–213.

Clemens, L. G., and L. W. Christensen. 1975. Sexual behaviour. In *The behaviour of domestic animals,* E. S. E. Hafez, ed. 3d ed. Baltimore: Williams and Wilkins. Pp. 108–145.

Clutton-Brock, T. H., P. J. Greenwood, and R. P. Powell. 1976. Ranks and relationships in Highland ponies and Highland cows. *Z. Tierpsychol.* 41:202–216.

Cohen-Tannoudji, J., A. Locatelli, and J. P. Signoret. 1986. Non-pheromonal stimulation by the male of LH release in the anoestrous ewe. *Physiol. Behav.* 36:921–924.

Cohn, R. 1956. A contribution to the study of color vision in cat. *J. Neurophysiol.* 19:416–423.

Cole, D. D., and J. N. Shafer. 1966. A study of social dominance in cats. *Behaviour* 27:39–53.

Cole, D. J. A., J. E. Duckworth, and W. Holmes. 1967. Factors affecting voluntary feed intake in pigs. I. The effect of digestible energy content of the diet on the intake of castrated male pigs housed in holding pens and in metabolism crates. *Anim. Prod.* 9:141–148.

Collard, R. R. 1967. Fear of strangers and play behavior in kittens with varied social experience. *Child Dev.* 38:877–891.

Collias, N. E. 1956. The analysis of socialization in sheep and goats. *Ecology* 37:228–239.

Collis, K. A. 1976. An investigation of factors related to the dominance order of a herd of dairy cows of similar age and breed. *Appl. Anim. Ethol.* 2:167–173.

Collis, K. A., S. J. Kay, A. J. Grant, and A. J. Quick. 1979. The effect on social organization and milk production of minor group alterations in dairy cattle. *Appl. Anim. Ethol.* 5:103–111.

Colpaert, F. C., and M. Callens. 1974. Effects of VMH lesions on CAR learning in cats. *Physiol. Behav.* 12:893–896.

Concannon, P. W., W. Hansel, and W. J. Visek. 1975. The ovarian cycle of the bitch: Plasma estrogen, LH and progesterone. *Biol. Reprod.* 13:112–121.

Cooling, M. J., and M. D. Day. 1975. Drinking behaviour in the cat induced by renin, angiotensin I, II and isoprenaline. *J. Physiol.* 244:325–336.

Coon, K. 1977. *The dog intelligence test.* New York: Avon.

Coppinger, R., J. Lorenz, J. Glendinning, and P. Pinardi. 1983. Attentiveness of guarding dogs for reducing predation on domestic sheep. *J. Range Manage.* 36:275–279.

Corbett, J. L. 1953. Grazing behaviour in New Zealand. *Br. J. Anim. Behav.* 1:67–71.

Cornwell, P., J. M. Warren, and A. J. Nonneman. 1976. Marginal and extramarginal cortical lesions and visual discrimination by cats. *J. Comp. Physiol. Psychol.* 90:986–995.

Corson, S. A., E. O'L. Corson, V. Kirilcuk, J. Kirilcuk, W. Knopp, and L. E. Arnold. 1972. Differential effects of amphetamines on several types of hyperkinetic and normal dogs and on learning disability. *Psychopharmacology* (abstr.) 26 (suppl): 55.

Cory, V. L. 1927. Activities of livestock on the range. Texas Agricultural Experiment Station Bulletin no. 367. Pp. 5–47.

Cowlishaw, S. J., and F. E. Alder. 1960. The grazing preferences of cattle and sheep. *J. Agric. Sci.* (Camb.) 54:257–265.

Crawford, M. A., M. D. Kittleson, and G. D. Fink. 1984. Hypernatremia and adipsia in a dog. *J. Am. Vet. Med. Assoc.* 184:818–821.

Creel, S. R. and J. L. Albright. 1988. The effects of neonatal social isolation on the behavior and endocrine function of Holstein calves. *Appl. Anim. Behav. Sci.* 21:293–306.

Cregier, S. E. 1982. Reducing equine hauling stress: A review. *J. Equine Vet. Sci.* 2:187–198.

Cresswell, E. 1959. A cattle rangemeter. *Anim. Behav.* 7:244.

———. 1960. Ranging behaviour studies with Romney Marsh and Cheviot sheep in New Zealand. *Anim. Behav.* 8:32–28.

Cronin, G. M., P. R. Wiepkema, and J. M. van Ree. 1986. Endorphins implicated in stereotypies of tethered sows. *Experientia* 42:198–199.

Crowell-Davis, S. L. 1983. The behavior of Welsh pony foals and mares. Ph.D. diss., Cornell University, Ithaca, N.Y.

———. 1985. Nursing behaviour and maternal aggression among Welsh ponies (*Equus caballus*). *Appl. Anim. Behav. Sci.* 14:11–25.

———. 1986. Spatial relations between mares and foals of the Welsh pony (*Equus caballus*). *Anim. Behav.* 34:1007–1015.

Crowell-Davis, S. L., and A. B. Caudle. 1989. Coprophagy by foals: Recognition of maternal feces. *Appl. Anim. Behav. Sci.* 24:267–272.

Crowell-Davis, S. L., and K. A. Houpt. 1985a. Coprophagy by foals: Effect of age and possible function. *Equine Vet. J.* 17:17–19.

———. 1985b. The ontogeny of flehmen in horses. *Anim. Behaviour* 33:739–74.

Crowell-Davis, S. L., K. A. Houpt, and J. S. Burnham. 1985a. Snapping by foals of *Equus caballus. Z. Tierpsychol.* 69:42–54.

Crowell-Davis, S. L., K. A. Houpt, and J. Carnevale. l985b. Feeding and drinking behavior of mares and foals with free access to pasture and water. *J. Anim. Sci.* 60:883–889.

Crowell-Davis, S. L., K. A. Houpt, and C. M. Carini. l986. Mutual grooming and nearest-neighbor relationships among foals of *Equus caballus. Appl. Anim. Behav. Sci.* 15:113–123.

Crowell-Davis, S. L., K. A. Houpt, and L. Kane. 1987. Play development in Welsh pony (*Equus caballus*) foal. *Appl. Anim. Behav. Sci.* 18:119–131.

Culler, E., and F. A. Mettler. 1934. Conditioned behavior in a decorticate dog. *J. Comp. Psychol.* 18:291–303.

Culley, M. J. 1938. Grazing habits of range cattle. *J. Forestry* 36:715–717.

Curtis, Q. F. 1937. Experimental neurosis in the pig. *Psychol. Bull.* (Abstr.) 34:723.

Czarkowska, J. 1983. Changes of some postural reflexes during the first postnatal weeks in the dog. *Acta Neurobiol.* 43:27–35.

Dabrowska, B., W. Harmata, Z. Lenkiewicz, Z. Schiffer, and R. J. Wojtusiak. 1981. Colour perception in cows. *Behav. Processes* 6:1–10.

Dabrowska, J. 1971. Dissociation of impairment after lateral and medial prefrontal lesions in dogs. Science 171:1037–1038.

Dallaire, A., and Y. Ruckebusch. 1974a. Sleep and wakefulness in the housed pony under different dietary conditions. *Can. J. Comp. Med.* 38:65–71.

_____. 1974b. Sleep patterns in the pony with observations on partial perceptual deprivation. *Physiol. Behav.* 12:789–796.

Dalton, D. C., M. E. Pearson, and M. Sheard. 1967. The behaviour of dairy bulls kept in groups. *Anim. Prod.* 9:1–5.

Daniels, T. J. 1983a. The social organization of free-ranging urban dogs. I. Non-estrous social behavior. *Appl. Anim. Ethol.* 10:341–363.

_____. 1983b. The social organization of free-ranging urban dogs. II. Estrus groups and the mating system. *Appl. Anim. Ethol.* 10:365–373.

Dantzer, R. 1976. Effect of diazepam on performance of pigs in a progressive ratio schedule. *Physiol. Behav.* 17:161–163.

_____. 1977. Effects of diazepam on conditioned suppression in pigs. *J. Pharmacol.* 8:405–414.

Dantzer, R., and B. A. Baldwin. 1974a. Changes in heart rate during suppression of operant responding in pigs. *Physiol. Behav.* 12:385–391.

_____. 1974b. Effects of chlordiazepoxide on heart rate and behavioural suppression in pigs subjected to operant conditioning procedures. *Psychopharmacology* 37:169–177.

Dantzer, R., P. Mormede, and B. Favre. 1976. Fear-dependent variations in continuous avoidance behaviour of pigs. II. Effects of diazepam on acquisition and performance of Pavlovian fear conditioning and plasma corticosteroid levels. *Psychopharmacology* 49:75–78.

Dards, J. L. 1983. The behaviour of dockyard cats: Interactions of adult males. *Appl. Anim Behav. Sci.* 10:133–153.

Darke, P. G. G. 1978. Obesity in small animals. *Vet. Rec.* 102:545–546.

Davies, D. A. R., P. M. Lerman, and M. M. Crosse. 1974. Food preferences after weaning of artificially reared lambs. *J. Agric. Sci.* (Camb.) 82:469–471.

Davis, C. N., L. E. Davis, and T. E. Powers. 1975. Comparative body compositions of the dog and goat. *Am. J. Vet. Res.* 36:309–311.

Davis, J. L., and R. A. Jensen. 1976. The development of passive and active avoidance learning in the cat. *Dev. Psychobiol.* 9:175–179.

Davis, K. L., J. C. Gurski, and J. P. Scott. 1977. Interaction of separation distress with fear in infant dogs. *Dev. Psychobiol.* 10:203–212.

Dawson, W. M., and R. L. Revens. 1946. Varying susceptibility in pigs to alarm. *J. Comp. Physiol. Psychol.* 39:297–305.

De Boer, J. N. 1977. The age of olfactory cues functioning in chemocommunication among male domestic cats. *Behav. Proc.* 2:209–225.

deLahunta, A. 1977. *Veterinary neuroanatomy and clinical neurology.* Philadelphia: W. B. Saunders.

Delius, K., M. Günderoth-Palmowski, I. Krause, and W. Engelmann. 1984. Effects of lithium salts on the behaviour and the circadian system of *Mesocricetus auratus* W. *J. interdisciplinary Cycle Res.* 15:289–299.

Delude, L. A. 1986. Activity patterns and behavior of sled dogs. *Appl. Anim. Behav. Sci.* 15:161–168.

Denton, D. A. 1967. Salt appetite. In *Handbook of physiology.* Sec. 6: Alimentary canal. Vol. 1, *Control of food and water intake.* C. F. Code and W. Heidel, eds. Washington, D.C.:

American Physiological Society. Pp. 433–459.

———. 1982. *The hunger for salt.* Berlin: Springer-Verlag.

Deswysen, A. G., W. C. Ellis, and K. R. Pond. 1987. Interrelationships among voluntary intake, eating and ruminating behavior and ruminal motility of heifers fed corn silage. *J. Anim. Sci.* 64:835–841.

Deutsch, J. A. 1971. The cholinergic synapse and the site of memory. *Science* 174:788–794.

Devilat, J., W. G. Pond, and P. D. Miller. 1970. Dietary amino acid balance in growing-finishing pigs: Effect on diet preference and performance. *J. Anim. Sci.* 30:536–543.

De Vuyst, A., G. Thines, L. Henriet, and M. Soffie. 1964. Influence des stimulations auditives sur le comportement sexuel du taureau (in French). *Experientia* 20:648–650.

Diakow, C. 1971. Effects of genital desensitization on mating behavior and ovulation in the female cat. *Physiol. Behav.* 7:47–54.

Dickson, D. P., G. R. Barr, and D. A. Wieckert. 1967. Social relationship of dairy cows in a feed lot. *Behaviour* 29:195–203.

Dickson, D. P., G. R. Barr, L. P. Johnson, and D. A. Wieckert. 1970. Social dominance and temperament of Holstein cows. *J. Dairy Sci.* 53:904–907.

Dinger, J. E., and E. E. Noiles. 1986. Effects of controlled exercise on libido in 2 year old stallions. *J. Anim. Sci.* 62:1220–1223.

Dinius, D. A., and C. A. Baile. 1977. Beef cattle response to a feed intake stimulant given alone and in combination with a propionate enhancer and an anabolic agent. *J. Anim. Sci.* 45:147–153.

Divac, I. 1972. Delayed alternation in cats with lesions of the prefrontal cortex and the caudate nucleus. *Physiol. Behav.* 8:519–522.

Dodman, N. H., L. Schuster, M. H. Court, and R. Dixon. 1987. An investigation into the use of narcotic antagonists in the treatment of a sterotypic behaviour pattern (crib-biting) in the horse. *Am. J. Vet. Res.* 48:311–319.

Dodman, N. H., L. Shuster, M. H. Court, and J. Patel. 1988. Use of a narcotic antagonist (nalmefene) to suppress self-mutilative behavior in a stallion. *J. Am. Vet. Med. Assoc.* 192:1585–1587.

Dodgman, N. H., L. Shuster, S. D. White, M. H. Court, D. Parker, and R. Dixon. 1988. Use of narcotic antagonists to modify stereotypic self-licking, self-chewing, and scratching behavior in dogs. *J. Am. Vet. Med. Assoc.* 193:815–819.

Donaldson, L. E., and J. W. James. 1963. A connection between pregnancy and crush order in cows. *Anim. Behav.* 11:286.

Donovan, C. A. 1967. Some clinical observations on sexual attraction and deterrence in dogs and cattle. *Vet. Med./Small Anim. Clin.* 62:1047–1051.

Donovan, G. A., L. Badinga, R. J. Collier, C. J. Wilcox, and R. K. Braun. 1986. Factors influencing passive transfer in dairy calves. *J. Dairy Sci.* 69:754–759.

Doran, C. W. 1943. Activities and grazing habits of sheep on summer ranges. *J. Forest.* 41:253–258.

Doty, R. L., and I. Dunbar. 1974. Attraction of beagles to conspecific urine, vaginal and anal sac secretion odors. *Physiol. Behav.* 12:825–833.

Doty, R. W., L. T. Rutledge, Jr., and R. M. Larsen. 1956. Conditioned reflexes established to electrical stimulation of cat cerebral cortex. *J. Neurophysiol.* 19:401–415.

Doty, R. W., E. C. Beck, and K. A. Kooi. 1959. Effect of brain-stem lesions on conditioned responses of cats. *Exp. Neurol.* 1:360–385.

Dove, H., R. G. Beilharz, and J. L. Black. 1974. Dominance patterns and positional behaviour of sheep in yards. *Anim. Prod.* 19:157–168.

Duckworth, J. E., and D. W. Shirlaw. 1958. A study of factors affecting feed intake and the eating behaviour of cattle. *Anim. Behav.* 6:147–154.

Dufty, J. H. 1971. Determination of the onset of parturition in Hereford cattle. *Aust. Vet. J.* 47:77–82.

———. 1972. Clinical studies on bovine parturition. Maternal causes of dystocia and stillbirth in an experimental herd of Hereford cattle. *Aust. Vet. J.* 48:1–6.

———. 1973. Clinical studies on bovine parturition—Foetal aspects. *Aust. Vet. J.* 49:177–182.

Dunbar, I. F. 1978. Olfactory preferences in dogs: The response of male and female beagles to conspecific urine. *Biol. Behav.* 3:273–286.

Dunbar, I. F., and G. Bohnenkamp. 1985. *Behavior booklets.* Oakland, Calif.: James and Kenneth.

Dunbar, I. F., and M. Buehler. 1980. A masking effect of urine from male dogs. *Appl. Anim. Ethol.* 6:297–301.

Dunbar, I. F., and M. Carmichael. 1981. The response of male dogs to urine from other males. *Behav. Neural. Biol.* 31:465–470.

Duncan, P. 1980. Time-budgets of Camargue horses. II. Time-budgets of adult horses and weaned sub-adults. *Behaviour* 72:26–49.

———. 1982. Foal killing by stallions. *Appl. Anim. Ethol.* 8:567–570.

———. 1985. Time-budgets of Camargue horses. III. Environmental influences. *Behaviour* 92:188–208.

Duncan, P., and P. Cowtan. 1980. An unusual choice of habitat helps Camargue horses to avoid blood-sucking horse-flies. *Biol. Behav.* 5:55–60.

Duncan, P., P. H. Harvey, and S. M. Wells. 1984. On lactation and associated behaviour in a natural herd of horses. *Anim. Behav.* 32:255–263.

Dunne, H. W., and A. D. Leman. 1975. *Diseases of swine,* 4th ed. Ames, Iowa: Iowa State University Press.

Durrer, J. L., and J. P. Hannon. 1962. Seasonal variations in caloric intake of dogs living in an arctic environment. *Am. J. Physiol.* 202:375–378.

Dworkin, S. 1939. Conditioning neuroses in dog and cat. *Psychosom. Med.* 1:388–396.

Dykman, R. A., O. D. Murphree, and J. E. Peters. 1969. Like begets like: Behavioral test, classical autonomic and motor conditioning and operant conditioning in two strains of pointer dogs. *Ann. N.Y. Acad. Sci.* 159:976–1007.

Eaton, R. L. 1970. Group interactions, spacing and territoriality in cheetahs. *Z. Tierpsychol.* 27:481–491.

Eccles, R. 1982. Autonomic innervation of the vomeronasal organ of the cat. *Physiol. Behav.* 28:1011–1015.

Eckstein, P., and S. Zuckerman. 1956. The oestrous cycle in the mammalia. In *Marshall's physiology of reproduction,* 3d ed. A. S. Parkes, ed. London: Longmans, Green and Co. Pp. 226–396.

Edwards, S. A. 1982. Factors affecting time to first suckling in dairy calves. *Anim. Prod.* 34:339–346.

Edwards, S. A., and D. M. Broom. 1982. Behavioural interactions of dairy cows with their newborn calves and the effects of parity. *Anim. Behav.* 30:525–535.

Ehret, C. F., V. R. Potter, and K. W. Dobra. 1975. Chronotypic action of theophylline and of phenobarbital as circadian Zeitgebers in the rat. *Science* 188:1212–1215.

Eldridge, F., and Y. Suzuki. 1976. A mare mule—Dam or foster mother? *J. Hered.* 67:353–360.

Ellendorff, F., N. Parvizi, D. K. Pomerantz, A. Hartjen, A. Konig, D. Smidt, and F. Elsaesser. 1975. Plasma luteinizing hormone and testosterone in the adult male pig: 24 hour fluctuations and the effect of copulation. *J. Endocrinol.* 67:403–410.

Elliot, O., and J. A. King. 1960. Effect of early food deprivation upon later consummatory behavior in puppies. *Psychol. Rep.* 6:391–400.

Elliot, O., and J. P. Scott. 1961. The development of emotional distress reactions to separation, in puppies. *J. Genet. Psychol.* 99:3–22.

Elliott, J. A., M. H. Stetson, and M. Menaker. 1972. Regulation of testis function in golden hamsters: A circadian clock measures photoperiodic time. *Science* 178:771–773.

Ely, F., and W. E. Petersen. 1941. Factors involved in the ejection of milk. *J. Dairy Sci.* 24:211–223.

England, G. J. 1954. Observations on the grazing behaviour of different breeds of sheep at Pantyrhuad Farm, Carmarthenshire. *Br. J. Anim. Behav.* 2:56–60.

Entingh, D. 1971. Perseverative responding and hyperphagia following entorhinal lesions in cats. *J. Comp. Physiol. Psychol.* 75:50–58.

Esselmont, R. J., R. G. Glencross, M. J. Bryant, and G. S. Pope. 1980. A quantitative study of pre-ovulatory behaviour in cattle (British Friesian heifers). *Appl. Anim. Ethol.* 6:1–17.

Estes, R. D. 1972. The role of the vomeronasal organ in mammalian reproduction. *Mammalia* 36:315–341.

Evans, J. W., C. M. Winget, C. De Roshia, and D. C. Holley. 1976. Ovulation and equine body temperature and heart rate circadian rhythms. *J. Interdisciplin. Cycle Res.* 7:25–37.

Everitt, G. C., and D. S. M. Phillips. 1971. Calf rearing by multiple suckling and the effects on lactation performance of the cow. *Proc. N. Z. Soc. Anim. Prod.* 31:22–40.

Everson, C. A., B. M. Bergmann, and A. Rechtschaffen. 1989. Sleep deprivation in the rat. III. Total sleep deprivation. *Sleep* 12:13–21.

Ewbank, R. 1963. Predicting the time of parturition in the normal cow: A study of the precalving drop in body temperature in relation to the external signs of imminent calving. *Vet. Rec.* 75:367–370.

_____. 1964. Observations on the suckling habits of twin lambs. *Anim. Behav.* 12:34–37.

_____. 1967a. Behavior of twin cattle. *J. Dairy Sci.* 50:1510–1512.

_____. 1967b. Nursing and suckling behaviour amongst Clun Forest ewes and lambs. *Anim. Behav.* 15:251–258.

_____. 1973. Abnormal behaviour and pig nutrition. An unsuccessful attempt to induce tail biting by feeding a high energy, low fibre vegetable protein ration. *Br. Vet. J.* 129:366–369.

Ewbank, R., and M. J. Bryant. 1972. Aggressive behaviour amongst groups of domesticated pigs kept at various stocking rates. *Anim. Behav.* 20:21–28.

Ewbank, R., and A. C. Mason. 1967. A note on the sucking behaviour of twin lambs reared as singles. *Anim. prod.* 9:417–420.

Ewbank, R., and G. B. Meese. 1971. Aggressive behaviour in groups of domesticated pigs on removal and return of individuals. *Anim. Prod.* 13:685–693.

Ewbank, R., G. B. Meese, and J. E. Cox. 1974. Individual recognition and the dominance hierarchy in the domesticated pig. The role of sight. *Anim. Behav.* 22:473–480.

Ewer, R. F. 1959. Suckling behaviour in kittens. *Behaviour* 15:146–162.

_____. 1973. *The Carnivores.* Ithaca, N. Y.: Cornell University Press.

Fagen, R. M., and T. K. George. 1977. Play behavior and exercise in young ponies (*Equus caballus L.*). *Behav. Ecol. Sociobiol.* 2:267–269.

Farner, D. S. 1961. Comparative physiology: Photoperiodicity. *Ann. Rev. Physiol.* 23:71–96.

Fay, P. K., V. T. McElligott, and K. M. Havstad. 1989. Containment of free-ranging goats using pulsed-radio-wave-activated shock collars. *Appl. Anim. Behav. Sci.* 23:165–171.

Feaver, J., M. Mendl, and P. Bateson. 1986. A method for rating the individual distinctiveness of domestic cats. *Anim. Behav.* 34:1016–1025.

Feddes, J. J. R., B. A. Young, and J. A. DeShazer. 1989. Influence of temperature and light on feeding behaviour in pigs. *Appl. Anim. Behav. Sci.* 23:215–222.

Feist, J. D., and D. R. McCullough. 1975. Reproduction in feral horses. *J. Reprod. Fertil. Suppl.* 23:13–18.

_____. 1976. Behavior patterns and communication in feral horses. *Z. Tierpsychol.* 41:337–371.

Feldberg, W. 1959. A physiological approach to the problem of general anaesthesia and of loss of consciousness. *Br. Med. J.* 2:771–782.

Feldmann, B. M. 1974. The problem of urban dogs. *Science* 185:903.

Feldmann, B. M., and T. H. Carding. 1973. Free-roaming urban pets. *Health Serv. Rep.* 88:956–962.

Ferrell, F. 1984. Preference for sugars and nonnutritive sweetners in young beagles. *Neurosci. Biobehav. Rev.* 8:199–203.

Fillenz, M. 1972. Hypothesis for a neuronal mechanism involved in memory. *Nature* 238:41–43.

Finger, K. H., and H. Brummer. 1969. Beobachtungen über das Saugverhalten mutterlos aufgezogener Kalber (in German). *Duet. Tierarztl. Wochenschr.* 76:665–667.

Firth, E. C. 1980. Bilateral ventral accessory neurectomy in windsucking horses. *Vet. Rec.* 106:30–32.

Fisher, R. B., and M.L.G. Gardner. 1976. A diurnal rhythm in the absorption of glucose and water by isolated rat small intestine. *J. Physiol.* 254:821–825.

Fiske, J. C., and G. D. Potter. 1979. Discrimination reversal learning in yearling horses. *J. Anim. Sci.* 49:583–588.

Fitzgerald, K. M., and I. Zucker. 1976. Circadian organization of the estrous cycle of the golden hamster. *Proc. Nat. Acad. Sci.* 73:2923–2927.

Fitzsimons, J. T., and E. Szczepanska-Sadowska. 1974. Drinking and antidiuresis elicited by isoprenaline in the dog. *J. Physiol.* 239:251–267.

Fleischer, S., and B. M. Slotnick. 1978. Disruption of maternal behavior in rats with lesions of the septal area. *Physiol. Behav.* 21:189–200.

Fleming, A., F. Vaccarino, L. Tambosso, and P. Chee. 1979. Vomeronasal and olfactory system modulation of maternal behavior in the rat. *Science* 203:372–374.

Fletcher, I. C., and D. R. Lindsay. 1968. Sensory involvement in the mating behaviour of domestic sheep. *Anim. Behav.* 16:410–414.

Flexner, J. B., and L. B. Flexner. 1971. Pituitary peptides and the suppression of memory by puromycin. *Proc. Natl. Acad. Sci.* 68:2519–2521.

Folman, Y., and R. Volcani. 1966. Copulatory behaviour of the prepubertally castrated bull. *Anim. Behav.* 14:572–573.

Fonberg, E. 1976. The relation between alimentary and emotional amygdalar regulation. In *Hunger: Basic mechanisms and clinical implications.* D. Novin, W. Wyrwicka, and G. A. Bray, eds. New York: Raven Press. Pp. 61–75.

Fontenot, J. P., and R. E. Blaser, 1965. Symposium on factors influencing the voluntary intake of herbage by ruminants: Selection and intake by grazing animals. *J. Anim. Sci.* 24:1202–1208.

Foot, J. Z., and A. J. F. Russel. 1978. Pattern of intake of three roughage diets by nonpregnant, nonlactating Scottish blackface ewes over a long period and the effects of previous nutritional history on current intake. *Anim. Prod.* 26:203–215.

Forbes, J. M. 1986. *The voluntary food intake of farm animals.* London: Butterworths.

Ford, J. J., and R. K. Christenson. 1981. Glucocorticoid inhibition of estrus in ovariectomized pigs: Relationship to progesterone action. *Horm. Behav.* 15:427–435.

Ford, J. J., and H. S. Teague. 1978. Effect of floor space restriction on age at puberty in gilts and on performance of barrows and gilts. *J. Anim. Sci.* 47:828–832.

Forssell, G. 1926. The new surgical treatment against crib-biting. *Vet. J.* 82:538–548.

Foss, I., and G. Flottorp. 1974. A comparative study of the development of hearing and vision in various species commonly used in experiments. *Acta Oto-Laryngol.* 77:202–214.

Foster, J. A., M. Morrison, S. J. Dean, M. Hill, and H. Frenk. 1981. Naloxone suppresses food/water consumption in the deprived cat. *Pharm. Biochem. Behav.* 14:419–421.

Foutz, A. S., M. M. Mitler, and W. C. Dement. 1980. Narcolepsy. *Vet. Clin. N. Amer. (Small Anim. Pract.)* 10:65–80.

Fowler, D. G., and L. D. Jenkins. 1976. The effects of dominance and infertility of rams on reproductive performance. *Appl. Anim. Ethol.* 2:327–337.

Fox, M. W. 1968. *Abnormal behavior in animals.* Philadelphia: W. B. Saunders.

——. 1970. Reflex development and behavioral organization. In *Developmental neurobiology,* W. A. Himwich, ed. Springfield, Ill.: Charles C. Thomas. Pp. 553–580.

——. 1971. *Integrative development of brain and behavior in the dog.* Chicago: University of Chicago Press.

——. 1972. *Understanding your Dog.* New York: Coward, McCann and Geoghegan., 240 p.

——. 1975. The behaviour of cats. In *The Behaviour of Domestic Animals,* 3d ed. E. S. E. Hafez, ed. Baltimore: Williams and Wilkins. Pp. 410–436.

Fox, M. W., and M. Bekoff. 1975. The behaviour of dogs. In *The behaviour of domestic animals,* 3d ed. E. S. E. Hafez, ed. Baltimore: Williams and Wilkins. Pp. 370–409.

Fox, M. W., and J. W. Spencer. 1967. Development of the delayed response in the dog. *Anim. Behav.* 15:162–168.

Fox, M. W., and G. Stanton. 1967. A developmental study of sleep and wakefulness in the dog. *J. Small Anim. Pract.* 8:605–611.

Fox, M. W., and D. Stelzner. 1966a. Approach/withdrawal variables in the development of social behaviour in the dog. *Anim. Behav.* 14:362–366.

——. 1966b. Behavioural effects of differential early experience in the dog. *Anim. Behav.* 14:273–281.

Fox, M. W., A. M. Beck, and E. Blackman. 1975. Behavior and ecology of a small group of urban dogs (*Canis familiaris*). *Appl. Anim. Ethol.* 1:119–137.

Francis-Smith, K., and D. G. M. Wood-Gush. 1977. Coprophagia as seen in Thoroughbred foals. *Eq. Vet. J.* 9:155–157.

Fraser, A. F. 1963. Behavior disorders in domestic animals. *Cornell Vet.* 53:213–223.

_____. 1964. Observations on the pre-coital behaviour of the male goat. *Anim. Behav.* 12:31–33.

_____. 1968. *Reproductive behaviour in ungulates.* New York: Academic Press.

_____. 1974. *Farm animal behaviour.* Baltimore: Williams and Wilkins.

Fraser, D. 1973. The nursing and suckling behaviour of pigs. I. The importance of stimulation of the anterior teats. *Br. Vet. J.* 129:324–336.

_____. 1974a. The behaviour of growing pigs during experimental social encounters. *J. Agric. Sci.* (Camb.) 82:147–163.

_____. 1974b. The vocalizations and other behaviour of growing pigs in an "open field" test. *Appl. Anim. Ethol.* 1:3–16.

_____. 1975a. The effect of straw on the behaviour of sows in tether stalls. *Anim. Prod.* 21:59–68.

_____. 1975b. The nursing and suckling behaviour of pigs. III. Behaviour when milk ejection is elicited by manual stimulation of the udder. *Br. Vet. J.* 131:416–426.

_____. 1975c. The nursing and suckling behaviour of pigs. IV. The effect of interrupting the sucking stimulus. *Br. Vet. J.* 131:549–559.

_____. 1975d. The "teat order" of suckling pigs. II. Fighting during suckling and the effects of clipping the eye teeth. *J. Agric. Sci.* (Camb.) 84:393–399.

_____. 1977. Some behavioural aspects of milk ejection failure by sows. *Br. Vet. J.* 133:126–133.

_____. 1987. Attraction to blood as a factor in tail-biting by pigs. *Appl. Anim. Behav. Sci.* 17:61–68.

Fratta, W., G. Mereu, P. Chessa, E. Paglietti, and G. Gessa. 1976. Benzodiazepine-induced voraciousness in cats and inhibition of amphetamine-anorexia. *Life Sci.* 18:1157–1166.

Freedman, D. G. 1958. Constitutional and environmental interactions in rearing of four breeds of dogs. *Science* 127:585–586.

Freedman, D. G., J. A. King, and E. Elliot. 1961. Critical period in the social development of dogs. *Science* 133:1016–1017.

Freeman, N. C. G., and J. S. Rosenblatt. 1978a. The interrelationship between thermal and olfactory stimulation in the development of home orientation in newborn kittens. *Dev. Psychobiol.* 11:437–457.

_____. 1978b. Specificity of litter odors in the control of home orientation among kittens. *Dev. Psychobiol.* 11:459–468.

Fretz, P. B. 1977. Behavioral virilization in a brood mare. *Appl. Anim. Ethol.* 3:277–280.

Friend, D. W. 1973. Self-selection of feeds and water by unbred gilts. *J. Anim. Sci.* 37:1137–1141.

Friend, T. H., and C. E. Polan. 1974. Social rank, feeding behavior, and free stall utilization by dairy cattle. *J. Dairy Sci.* 57:1214–1220.

Friend, T. H., L. O'Connor, D. Knabe, and G. Dellmeier. 1989. Preliminary trials of a sound-activated device to reduce crushing of piglets by sows. *Appl. Anim. Behav. Sci.* 24:23–29.

Fuller, C. A., F. M. Sulzman, and M. C. Moore-Ede. 1978. Thermoregulation is impaired in an environment without circadian time cues. *Science* 199:794–796.

Fuller, J. L. 1956. Photoperiodic control of estrus in the basenji. *J. Hered.* 47:179–180.

_____. 1967. Experiential deprivation and later behavior. *Science* 158:1645–1652.

Fuller, J. L., C. A. Easler, and E. M. Banks. 1950. Formation of conditioned avoidance responses in young puppies. *Am. J. Physiol.* 160:462–466.

Fuller, J. L., H. E. Rosvold, and K. H. Pribram. 1957. The effect on affective and cognitive behavior in the dog of lesions of the pyriform-amygdala-hippocampal complex. *J. Comp. Physiol. Psychol.* 50:89–96.

Gadbury, J. C. 1975. Some preliminary field observations on the order of entry of cows into herringbone parlours. *Appl. Anim. Ethol.* 1:275–281.

Gaebelein, C. J., R. A. Galosy, L. Botticelli, J. L. Howard, and P. A. Obrist. 1977. Blood pressure and cardiac changes during signalled and unsignalled avoidance in dogs. *Physiol. Behav.* 19:69–74.

Gandelman, R., M. X. Zarrow, V. H. Denenberg, and M. Myers. 1971. Olfactory bulb removal eliminates maternal behavior in the mouse. *Science* 171:210–211.

Ganjam, V. K., and R. M. Kenney. 1975. Androgens and oestrogens in normal and cryptorchid stallion. *J. Reprod. Fertil. Suppl.* 23:67–73.

Garbarg, M., C. Julien, and J.-C. Schwartz. 1974. Circadian rhythm of histamine in the pineal gland. *Life Sci.* 14:539–543.

Garcia, J., W. G. Hankins, and K. W. Rusiniak. 1974. Behavioral regulation of the milieu interne in man and rat. *Science* 185:824–831.

Garcia, M. C., S. M. McDonnell, R. M. Kenney, and H. G. Osborne. 1986. Bull sexual behavior tests: Stimulus cow affects performance. *Appl. Anim. Behav. Sci.* 16:1–10.

Gardner, L. P. 1937a. The responses of cows in a discrimination problem. *J. Comp. Psychol.* 23:35–57.

Gardner, L. P. 1937b. The responses of cows to the same signal in different positions. *J. Comp. Psychol.* 23:333–350.

_____. 1937c. The responses of horses in a discrimination problem. *J. Comp. Psychol.* 23:13–34.

_____. 1937d. Responses of horses to the same signal in different positions. *J. Comp. Psychol.* 23:305–332.

_____. 1945. Responses of sheep in a discrimination problem with variations of the position of the signal. *J. Comp. Physiol. Psychol.* 38:343–351.

Garner, F. H. 1963. The palatability of herbage plants. *J. Br. Grassland Soc.* 18:79–89.

Gary, L. A., G. W. Sherritt, and E. B. Hale. 1970. Behavior of Charolais cattle on pasture. *J. Anim. Sci.* 30:203–206.

Geist, V. 1971. *Mountain sheep. A study in behavior and evolution.* Chicago: University of Chicago Press.

George, J. M., and I. A. Barger. 1974. Observations of bovine parturition. *Proc. Aust. Soc. Anim. Prod.* 10:314–317.

Ghosh, B., D. K. Choudhuri, and B. Pal. 1984. Some aspects of the sexual behaviour of stray dogs, *Canis familiaris. Appl. Anim. Behav. Sci. 13:113–127.*

Gibbs, J., R. C. Young, and G. P. Smith. 1973. Cholecystokinin decreases food intake in rats. *J. Comp. Physiol. Psychol.* 84:488–495.

Giebel, H.–D. 1958. Visuelles Lernvermögen bei Einhufern (in German). *Zool. Jahrb.* 67:487–520.

Gilbert B. J., Jr., and C. W. Arave. 1986. Ability of cattle to distinguish among different wavelengths of light. *J. Dairy Sci.* 69:825–832.

Gill, J. C., and W. Thomson. 1956. Observations on the behaviour of suckling pigs. *Br. J. Anim. Behav.* 4:46–51.

Gill, J. C., K. Skwarlo, and A. Flisinska-Bojanowska. 1974. Diurnal and seasonal changes in carbohydrate metabolism in the blood of Thoroughbred horses. *J. Interdisciplin. Cycle Res.* 5:355–361.

Girden, E., F. A. Mettler, G. Finch, and E. Culler. 1936. Conditioned responses in a decorticate dog to acoustic, thermal and tactile stimulation. *J. Comp. Psychol.* 21:367–385.

Glassman, R. B. 1970. Cutaneous discrimination and motor control following somatosensory cortical ablations. *Physiol. Behav.* 5:1009–1019.

Gleitman, H. 1974. Getting animals to understand the experimenter's instructions. *Anim. Learn. Behav.* 2:1–5.

Glencross, R. G., R. J. Esslemont, M. J. Bryant, and G. S. Pope. 1981. Relationships between the incidence of pre-ovulatory behaviour and the concentrations of oestradiol-17β and progesterone in bovine plasma. *Appl. Anim. Ethol.* 7:141–148.

Goatcher, W. D., and D. C. Church. 1970a. Taste responses in ruminants. I. Reactions of sheep to sugars, saccharin, ethanol and salts. *J. Anim. Sci.* 30:777–783.

_____. 1970b. Taste responses in ruminants. II. Reactions of sheep to acids, quinine, urea and sodium hydroxide. *J. Anim. Sci.* 30:784–790.

_____. 1970c. Taste responses in ruminants. III. Reactions of pygmy goats, normal goats, sheep and cattle to sucrose and sodium chloride. *J. Anim. Sci.* 31:364–372.

_____. 1970d. Taste responses in ruminants. IV. Reactions of pygmy goats, normal goats, sheep and cattle to acetic acid and quinine hydrochloride. *J. Anim. Sci.* 31:373–382.

Goddard, M. E., and R. G. Beilharz. 1982–83. Genetics of traits which determine the suitability of dogs as guide-dogs for the blind. *Appl. Anim. Ethol.* 9:299–315.

Goelet, P., and E. R. Kandel. 1986. Tracking the flow of learned information from the membrane receptor to the genome. *Trends Neurosci.* 9:492–499.

Goldberg, J. M., and W. D. Neff. 1961. Frequency discrimination after bilateral ablation of cortical auditory areas. *J. Neurophysiol.* 24:119–128.

Gonyou, H. W., and J. M. Stookey. 1985. Behavior of parturient ewes in group-lambing pens with and without cubicles. *Appl. Anim. Behav. Sci.* 14:163–171.

Gonyou, H. W., and W. R. Stricklin. 1984. Diurnal behavior of feedlot bulls during winter and spring in northern latitudes. *J. Anim. Sci.* 58:1075–1083.

Goodwin, M., K. M. Gooding, and F. Regnier. 1979. Sex pheromone in the dog. *Science* 203:559–561.

Gould, S. J. 1978. Women's brains. *Natur. Hist.* 87(8):44–50.

Grace, J., and M. Russek. 1969. The influence of previous experience on the taste behavior of dogs toward sucrose and saccharin. *Physiol. Behav.* 4:553–558.

Grandage, J. 1972. The erect dog penis: A paradox of flexible rigidity. *Vet. Rec.* 91:141–147.

Graves, H. B. 1984. Behavior and ecology of wild and feral swine (*Sus scrofa*). *J. Anim. Sci.* 58:482–492.

Green, J. D., C. D. Clemente, and J. de Groot. 1957. Rhinencephalic lesions and behavior in cats. *J. Comp. Neurol.* 108:505–536.

Green, J. S., R. A. Woodruff, and T. T. Tueller. 1984. Livestock-guarding dogs for predator control: Costs, benefits and practicality. *Wildl. Soc. Bull.* 12:44–50.

Greene, W. A., L. Mogil, and R. H. Foote. 1978. Behavioral characteristics of freemartins administered estradiol, estrone, testosterone, and dihydrotestosterone. *Horm. Behav.* 10:71–84.

Greet, T. R. C. 1982. Windsucking treated by myectomy and neurectomy. *Equine Vet. J.* 14:299–301.

Grossman, M. I., G. M. Cummins, and A. C. Ivy. 1947. The effect of insulin on food intake after vagotomy and sympathectomy. *Am. J. Physiol.* 149:100–102.

Grubb, P. 1974a. The rut and behaviour of Soay rams. In *Island survivors: The ecology of the Soay sheep of St. Kilda.* P. A. Jewell, C. Milner, and J. Morton Boyd, eds. London: Athlone Press of the University of London. Pp. 195–223.

_____. 1974b. Social organization of Soay sheep and the behaviour of ewes and lambs. In *Island survivors: The ecology of the Soay sheep of St. Kilda.* P. A. Jewell, C. Milner, and J. Morton Boyd, eds. London: Athlone Press of the University of London. Pp. 131–159.

Grzimek, B. 1949. Rangordnungsversuche mit Pferden (in German). *Z. Tierpsychol.* 6:455–464.

Grzimek, B. 1952. Versuche uber das Farbsehen von Pflanzenessern. I. Das farbige Sehen (und die Sehscharfe) von Pferden (in German). *Z. Tierpsychol.* 9:23–39.

Gubernick, D. J. 1980. Maternal "imprinting" or maternal "labelling" in goats? *Anim. Behav.* 28:124–129.

Gubernick, D. J., K. C. Jones, and P. H. Klopfer. 1979. Maternal imprinting in goats. *Anim. Behav.* 27:314–315.

Gurr, M. I., J. Kirtland, M. Phillip, and M. P. Robinson. 1977. The consequences of early overnutrition for fat cell size and number: The pig as an experimental model for human obesity. *Int. J. Obes.* 1:151–170.

Gustavson, C. R., J. Garcia, W. G. Hankins, and K. W. Rusiniak. 1974. Coyote predation control by aversive conditioning. *Science* 184:581–583.

Haag, E. L., R. Rudman, and K. A. Houpt. 1980. Avoidance, maze learning and social dominance in ponies. *J. Anim. Sci.* 50:329–335.

Hafez, E. S. E. 1952. Studies on the breeding season and reproduction of the ewe. V. Mating behaviour and pregnancy diagnosis. *J. Agric. Sci.* 42:255–265.

_____. 1974. *Reproduction in farm animals.* Philadelphia: Lea and Febiger.

_____. 1975. *The behaviour of domestic animals.* 3d ed. Baltimore: Williams and Wilkins.

Hafez, E. S. E., and M.-F. Bouissou. 1975. The behaviour of cattle. In *The behaviour of domestic animals,* 3d ed. E. S. E. Hafez, ed. Baltimore: Williams and Wilkins. Pp. 203–245.

Hafez, E. S. E., and J. A. Lineweaver. 1968. Suckling behaviour in natural and artificially fed neonate calves. *Z. Tierpsychol.* 25:187–198.

Hafez, E. S. E., and J. P. Signoret. 1969. The behaviour of swine. In *The behaviour of domestic animals,* 2d ed. E. S. E. Hafez, ed. Baltimore: Williams and Wilkins. Pp. 349–390.

Hafez, E. S. E., M. W. Schein, and R. Ewbank. 1969. The behaviour of cattle. In *The behaviour of domestic animals,* 2d ed. E. S. E. Hafez, ed. Baltimore: Williams and Wilkins. Pp. 235–295.

Hale, E. B. 1966. Visual stimuli and reproductive behavior in bulls. *J. Anim. Sci.* (suppl.) 25:36–44.

Hale L. A., and S. E. Huggins. 1980. The electroencephalogram of the normal 'grade' pony in sleep and wakefulness. *Comp. Biochem. Physiol.* 66A:251–257.

Hall, S.J.G. 1986. Chillingham cattle: Dominance and affinities and access to supplementary food. *Ethology* 71:201–215.

———. 1989. Chillingham cattle: Social and maintenance behaviour in an ungulate that breeds all year round. *Anim. Behav.* 38:215–225.

Hamilton, G. V. 1911. A study of trial and error reactions in mammals. *J. Anim. Behav.* 1:33–66.

Hamilton, L. W. 1969. Active avoidance impairment following septal lesions in cats. *J. Comp. Physiol. Psychol.* 69:420–431.

Hamm, D. 1977. A new surgical procedure to control crib-biting. *Proc. 23d Ann. Meet. American Assoc. Equine Pract.* Pp. 301–302.

Hammell, D. L., D. D. Kratzer, and W. J. Bramble. 1975. Avoidance and maze learning in pigs. *J. Anim. Sci.* 40:573–579.

Hancock, J. 1950. Grazing habits of dairy cows in New Zealand. *Emp. J. Exp. Agric.* 18:249–263.

———. 1954. Studies of grazing behaviour in relation to grassland management. I. Variations in grazing habits of dairy cattle. *J. Agric. Sci.* 44:420–433.

Hansel, W., and K. McEntee. 1977. Female reproductive processes. In *Duke's physiology of domestic animals,* 9th ed. M. Swenson, ed. Ithaca, N.Y.: Cornell University Press. Pp. 772–800.

Hansen, K. E., and S. E. Curtis. 1980. Prepartal activity of sows in stall or pen. *J. Anim. Sci.* 51:456–460.

Hardison, W. A., H. L. Fisher, G. C. Graf, and N. R. Thompson. 1956. Some observations on the behavior of grazing lactating cows. *J. Dairy Sci.* 39:1735–1741.

Harker, J. E. 1964. Diurnal rhythms and homeostatic mechanisms. In *Homeostasis and Feedback Mechanisms.* Symp. Soc. Exp. Biol., no. 18:283–300.

Harker, K. W., J. I. Taylor, and D. H. L. Rollinson. 1954. Studies on the habits of Zebu cattle. I. Preliminary observations on grazing habits. *J. Agric. Sci.* 44:193–198.

———. 1956. Studies on the habits of Zebu cattle. V. Night paddocking and its effect on the animal. *J. Agric. Sci.* 47:44–49.

Harlow, H. F., and P. Settlage. 1939. The effect of curarization of the fore part of the body upon the retention of conditioned responses in cats. *J. Comp. Psychol.* 27:45–48.

Harlow, H. F., M. K. Harlow, and E. W. Hansen. 1963. The maternal affectional system of rhesus monkeys. In *Maternal behavior in mammals.* H. L. Rheingold, ed. New York: John Wiley and Sons. Pp. 254–281.

Harrington, F. H. 1986. Timber wolf howling playback studies: Discrimination of pup from adult howls. *Anim. Behav.* 34:1575–1577.

Harrington, F. H., and L. D. Mech. 1978. Howling at two Minnesota wolf pack summer homesites. *Can. J. Zool.* 56:2024–2028.

Harris, D., P. J. Imperato, and B. Oken. 1974. Dog bites—an unrecognized epidemic. *Bull. N. Y. Acad. Med.* 50:981–1000.

Harrison, J., and J. Buchwald. 1983. Eyeblink conditioning deficits in the old cat. *Neurobiol. Aging* 4:45–51.

Hart, B. L. 1968. Role of prior experience in the effects of castration on sexual behavior of male dogs. *J. Comp. Physiol. Psychol.* 66:719–725.

_____. 1970. Mating behavior in the female dog and the effects of estrogen on sexual reflexes. *Horm. Behav.* 1:93–104.

_____. 1974a. Environmental and hormonal influences on urine marking behavior in the adult male dog. *Behav. Biol.* 11:167–176.

_____. 1974b. Medial preoptic-anterior hypothalamic area and sociosexual behavior of male dogs: A comparative neuropsychological analysis. *J. Comp. Physiol. Psychol.* 86:328–349.

_____. 1978. *Feline behavior. A practitioner monograph.* Santa Barbara: Veterinary Practice Publishing.

_____. 1980. Objectionable urine spraying and urine marking in cats: Evaluation of progestin treatment in gonadectomized males and females. *J. Am. Vet. Med. Assoc.* 177:529–533.

_____. 1981. Olfactory tractotomy for control of objectionable urine spraying and urine marking in cats. *J. Am. Vet. Med. Assoc.* 179:231–234.

_____. 1986. Medial preoptic-anterior hypothalamic lesions and sociosexual behavior of male goats. *Physiol. Behav.* 36:301–305.

Hart, B. L., and R. E. Barrett. 1973. Effects of castration on fighting, roaming, and urine spraying in adult male cats. *J. Am. Vet. Med. Assoc.* 163:290–292.

Hart, B. L., and L. Cooper. 1984. Factors relating to urine spraying and fighting in prepubetally gonadectomized male and female cats. *J. Am. Vet. Med. Assoc.* 184:1255–1258.

Hart, B. L., and L. A. Hart. 1985. *Canine and feline behavioral therapy.* Philadelphia: Lea and Febiger.

_____. 1988. *The perfect puppy: How to choose your dog by its behavior.* New York: W. H. Freeman and Company.

Hart, B. L., and C. M. Haugen. 1971. Scent marking and sexual behavior maintained in anosmic male dogs. *Commun. Behav. Biol.* 6:131–135.

Hart, B. L., and T. O. A. C. Jones. 1975. Effects of castration on sexual behavior of tropical male goats. *Horm. Behav.* 6:247–258.

Hart, B. L., and M. G. Leedy. 1983. Female sexual response in male cats facilitated by olfactory bulbectomy and medial preoptic/anterior hypothalamic lesion. *Behav. Neurosci.* 97:608–614.

_____. 1985. Analysis of the catnip reaction: Mediation by olfactory system, not vomeronasal organ. *Behav. Neural Biol.* 44:38–46.

Hart, B. L., and V. L. Voith. 1978. Changes in urine spraying, feeding and sleep behavior of cats following medial preoptic— anterior hypothalamic lesions. *Brain Res.* 145:406–409.

Hart, B. L., C. M. Haugen, and D. M. Peterson. 1973. Effects of medial preoptic-anterior hypothalamic lesions on mating behavior of male cats. *Brain Res.* 54:177–191.

Hartman, D. A., and W. G. Pond. 1960. Design and use of a milking machine for sows. *J. Anim. Sci.* 19:780–785.

Haskins, R. 1977. Effect of kitten vocalizations on maternal behavior. *J. Comp. Physiol. Psychol.* 91:830–838.

_____. A casual analysis of kitten vocalization: An observational and experimental study. *Anim. Behav.* 27:726–736.

Hatch, R. C. 1972. Effect of drugs on catnip (*Nepeta cataria*)-induced pleasure behavior in cats. *Am. J. Vet. Res.* 33:143–155.

Haugse, C. N., W. E. Dinusson, D. O. Erickson, J. N. Johnson, and M. L. Buchanan. 1965. A day in the life of a pig. *N. Dak. Farm Res.* 23(12):18–23.

Hawkes, J., M. Hedges, P. Daniluk, H. F. Hintz, and H. F. Schryver. 1985. Feed preferences of ponies. *Equine Vet. J.* 17:20–22.

Hawking, F. 1971a. Circadian rhythms in monkeys, dogs and other animals. *J. Interdisciplinary Cycle Res.* 2:153–156.

_____. 1971b. Circadian rhythms of parasites. *J. Interdisciplinary Cycle Res.* 2:157–160.

Hayes, K.E.N., and O. J. Ginther. 1989. Relationships between estrous behavior in pregnant mares and the presence of a female conceptus. *J. Equine Vet. Sci.* 9:316–318.

Hayman, R. H. 1964. Exercise of mating preference by a Merino ram. *Nature* 203:160–162.

Heady, H. F. 1964. Palatability of herbage and animal preference. *J. Range Manag.* 17:76–82.

Hebel, R. 1976. Distribution of retinal ganglion cells in five mammalian species (pig, sheep, ox, horse, dog). *Anat. Embryol.* 150:45–51.

Hedlund, L., and J. Rolls. 1977. Behavior of lactating cows during total confinement. *J. Dairy Sci.* 60:1807–1812.

Hedlund, L., M. M. Lischko, M. D. Rollag, and G. D. Niswender. 1977. Melatonin: Daily cycle in plasma and cerebrospinal fluid of calves. *Science* 195:686–687.

Hegsted, D. M., S. N. Gershoff, and E. Lentini. 1956. The development of palatability tests for cats. *Am. J. Vet. Res.* 17:733–737.

Hein, A., and R. Held. 1967. Dissociation of the visual placing response into elicited and guided components. *Science* 158:390–392.

Hein, M. A. 1935. Grazing time of beef steers on permanent pastures. *J. Am. Soc. Agron.* 27:675–679.

Heird, J. C., A. M. Lennon, and R. W. Bell. 1981. Effects of early experience on the learning ability of yearling horses. *J. Anim. Sci.* 53:1204–1209.

Heird, J. C., D. D. Whitaker, R. W. Bell, C. B. Ramsey, and C. E. Lokey. 1986. The effects of handling at different ages on the subsequent learning ability of 2–year-old horses. *Appl. Anim. Behav. Sci.* 15:15–25.

Heitzman, R. J. 1978. The use of hormones to regulate the utilization of nutrients in farm animals: Current farm practices. *Proc. Nutr. Soc.* 37:289–293.

Hemsworth, P. H., C. G. Winfield, and P. D. Mullaney. 1976. A study of the development of the teat order in piglets. *Appl. Anim. Ethol.* 2:225–233.

Hemsworth, P. H., R. G. Beilharz, and D. B. Galloway. 1977a. Influence of social conditions during rearing on the sexual behaviour of the domestic boar. *Anim. Prod.* 24:245–251.

Hemsworth, P. H., C. G. Winfield, R. G. Beilharz, and D. B. Galloway. 1977b. Influence of social conditions post-puberty on the sexual behaviour of the domestic male pig. *Anim. Prod.* 25:305–309.

Hemsworth, P. H., J. K. Findlay, and R. G. Beilharz. 1978. The importance of physical contact with other pigs during rearing on the sexual behaviour of the male domestic pig. *Anim. Prod.* 27:201–207.

Hemsworth, P. H., G. M. Cronin, C. Hansen, and C. G. Winfield. 1984. The effects of two oestrus detection procedures and intense boar stimulation near the time of oestrus on mating efficiency of the female pig. *Appl. Anim. Behav. Sci.* 12:339–347.

Hemsworth, P. H., J. L. Barnett, C. Hansen, and C. G. Winfield. 1986a. Effects of social environment on welfare status and sexual behaviour of female pigs. II. Effects of space allowance. *Appl. Anim. Behav. Sci.* 16:259–267.

Hemsworth, P. H., C. G. Winfield, J. L. Barnett, B. Schirmer, and C. Hansen. 1986b. A comparison of the effects of two estrus detection procedures and two housing systems on the oestrus dectection rate of female pigs. *Appl. Anim. Behav. Sci.* 16:345–351.

Hemsworth, P. H., J. L. Barnett, and C. Hansen. 1987. The influence of inconsistent handling by humans on the behaviour, growth and corticosteroids of young pigs. *Appl. Anim. Behav. Sci.* 17:245–252.

Hemsworth, P. H., C. Hansen, and C. G. Winfield. 1989. The influence of mating conditions on the sexual behaviour of male and female pigs. *Appl. Anim. Behav. Sci.* 23:207–214.

Hendricks, J. C., A. R. Morrison, G. L. Farnbach, S. A. Steinberg, and G. Mann. 1981. A disorder of rapid eye movement sleep in a cat. *J. Am. Vet. Med. Assoc.* 178:55–57.

Hepper, P. G. 1986. Sibling recognition in the domestic dog. *Anim. Behav.* 34: 288–289.

Herbel, C. H., and A. B. Nelson. 1966a. Activities of Hereford and Santa Gertrudis cattle on a southern New Mexico range. *J. Range Manage.* 19:173–176.

———. 1966b. Species preference of Hereford and Santa Gertrudis cattle on a southern New Mexico range. *J. Range Manage* 19:177–181.

Herd, R. M. 1988. A technique for cross-mothering beef calves which does not affect growth. *Appl. Anim. Behav. Sci.* 19:239–244.

Hernandez-Peon, R., and H. Brust-Carmona. 1961. Functional role of subcortical structures in habituation and conditioning. In *Brain mechanisms and learning.* A. Fessard, R. W. Gerard, J. Konorski, and J. F. Delafresnaye, eds. Springfield, Ill.: Charles C. Thomas. Pp. 393–412.

Herrenkohl, L. R., and P. A. Rosenberg. 1972. Hypothalamic deafferentiation during early and late pregnancy suppresses milk ejection but not nursing behavior in the primiparous rat. *Am. Zool.* (abstr.) 12:651.

Herring, S. W., and R. P. Scapino. 1973. Physiology of feeding in miniature pigs. *J. Morphol.* 141:427–460.

Hersher, L., J. B. Richmond, and A. U. Moore. 1963. Maternal behavior in sheep and goats. In *Maternal behavior in mammals.* H. L. Rheingold, ed. New York: John Wiley and Sons. Pp. 203–232.

Hess, E. H. 1972. "Imprinting" in a natural laboratory. *Sci. Am.* 227(2): 24–32.

Hetzer, H. O., and W. R. Harvey. 1967. Selection for high and low fatness in swine. *J. Anim. Sci.* 26:1244–1251.

Hill, J. O., E. J. Pavlik, G. L. Smith III, G. M. Burghardt, and P. B. Coulson. 1976. Species-characteristic responses to catnip by undomesticated felids. *J. Chem.* Ecol. 2:239–253.

Hinch, G. N., J. J. Lynch, and C. J. Thwaites. 1982–83. Patterns and frequency of social interactions in young grazing bulls and steers. *Appl. Anim. Ethol.* 9:15–30.

Hirsch, E., C. Dubose, and H. L. Jacobs. 1978. Dietary control of food intake in cats. *Physiol. Behav.* 20:287–295.

Hite, M., H. M. Hanson, N. R. Bohidar, P. A. Conti, and P. A. Mattis. 1977. Effect of cage size on patterns of activity and health of beagle dogs. *Lab. Anim. Sci.* 27:60–64.

Hoagland, T. A., and M. A. Diekman. 1982. Influence of supplemental lighting during increasing day length on libido and reproductive hormones in prepubertal boars. *J. Anim. Sci.* 55:1483–1489.

Hoffman, M. P., and H. L. Self. 1973. Behavioral traits of feedlot steers in Iowa. *J. Anim. Sci.* 37:1438–1445.

Hoffman, R. 1985. On the development of social behavior in immature males of a feral horse population (*Equus przewalski f. caballus*). *Zeitschrift Saugetierkunde* 50:302–314.

Holder, J. M. 1960. Observations on the grazing behaviour of lactating dairy cattle in a subtropical environment. *J. Agric. Sci.* 55:261–267.

Holmes, J. H., and L. J. Cizek. 1951. Observations on sodium chloride depletion in the dog. *Am. J. Physiol.* 164:407–414.

Holmes, L. N., G. K. Song, and E. O. Price. 1987. Head partitions facilitate feeding by subordinate horses in the presence of dominant pen-mates. *Appl. Anim. Behav. Sci.* 19:179–182.

Hopkins, S. G., T. A. Schubert, and B. L. Hart. 1976. Castration of adult male dogs: Effects on roaming, aggression, urine marking, and mounting. *J. Am. Vet. Med. Assoc.* 168:1108–1110.

Hoppenbrouwers, T., and M. B. Sterman. 1975. Development of sleep state patterns in the kitten. *Exp. Neurol.* 49: 822–838.

Horn, G., S. P. R. Rose, and P. P. G. Bateson. 1973. Experience and plasticity in the central nervous system. *Science* 181:506–514.

Houpt, K. A., 1977a. Horse behavior: Its relevancy to the equine practitioner. *J. Equine Med. Surg.* 1:87–94.

_____. 1977b. The physiology of hunger and palatability in animals. In *Proceedings of the Cornell Conference for Feed Manufacturers.* Ithaca, N.Y.: Cornell University. Pp. 113–119.

_____. 1983. Disruption of the human-companion animal bond: Aggressive behavior in dogs. In *New perspectives on our lives with companion animals.* A. H. Katcher and A. M. Beck, eds. Pp. 197–204.

Houpt, K. A., and H. Hintz. 1978. Palatability and canine food preferences. *Canine Pract.* 5:29–35.

_____. 1982–83. Some effects of maternal deprivation on maintenance behavior, spatial relationships and responses to environmental novelty in foals. *Appl. Anim. Ethol.* 9:221–230.

Houpt, K. A., and T. R. Houpt. 1976. Comparative aspects of the ontogeny of taste. *Chem. Senses Flavor* 2:219–228.

_____. 1977. The neonatal pig: A biological model for the development of taste preferences and controls of ingestive behavior. In *Taste and development. The genesis of sweet prefer-*

ence. J. M. Weiffenbach, ed. Bethesda, Md: U.S. Department of Health, Education, and Welfare, Public Health Service, National Institute of Health. Pp. 86–98.

_____. 1988. Social and illumination preferences of mares. *J. Anim. Sci.* 66:2159–2164.

Houpt, K. A., and R. R. Keiper. 1982. The position of the stallion in the equine dominance hierarchy of feral and domestic ponies. *J. Anim. Sci.* 54:945–950.

Houpt, K. A., and G. Wollney, 1989. Frequency of masturbation and time budgets of dairy bulls used for semen production. *Appl. Animal. Behav. Sci.* 24:217–225.

Houpt, K. A., and T. R. Wolski. 1979. Equine maternal behavior and aberrations. *Equine Pract.* 1:7–20.

_____. 1980. Stability of equine hierarchies and prevention of dominance related aggression. *Equine Vet. J.* 12:18–24.

Houpt, K. A., T. R. Houpt, and W. G. Pond. 1977. Food intake controls in the suckling pig: Glucoprivation and gastrointestinal factors. *Am. J. Physiol.* 232:E510–E514.

Houpt, K. A., K. Law, and V. Martinisi. 1978a. Dominance hierarchies in domestic horses. *Appl. Anim. Ethol.* 4:273–283.

Houpt, K. A., H. F. Hintz, and P. Shepherd. 1978b. The role of olfaction in canine food preferences. *Chem. Senses Flavour* 3:281–290.

Houpt, K. A., T. R. Houpt, and W. G. Pond. 1979a. The pig as a model for the study of obesity and of control of food intake: A review. *Yale J. Biol. Med.* 52:307–329.

Houpt, K. A., B. Coren, H. F. Hintz, and J. E. Hilderbrant. 1979b. Effect of sex and reproductive status on sucrose preference, food intake, and body weight of dogs. *J. Am. Vet. Med. Assoc.* 174:1083–1085.

Houpt, K. A., M. S. Parsons, and H. F. Hintz. 1982. Learning ability of orphan foals, of normal foals and of their mothers. *J. Anim. Sci.* 55:1027–1032.

Houpt, K. A., M. F. O'Connell, T. A. Houpt, and D. A. Carbonaro. 1986. Night-time behavior of stabled and pastured periparturient ponies. *Appl. Anim. Behav. Sci.* 15:103–111.

Houpt, K. A., W. Rivera, and L. Glickstein. 1989. The flehmen response of bulls and cows. *Theriogenology* 32:343–350.

Houpt, K. A., D. M. Zahorik, and J. A. Swartzman-Andert. 1990. Taste aversion learning in horses. *J. Anim. Sci.* 68:2340–2344.

Houpt, T. R. 1974. Stimulation of food intake in ruminants by 2-deoxy-D-glucose and insulin. *Am. J. Physiol.* 227:161–167.

_____. 1984. Controls of feeding in pigs. *J. Anim. Sci.* 59:1345–1353.

Houpt, T. R., and H. H. Hance. 1969. Effect of 2–deoxy-D glucose on food intake by the goat, rabbit and dog. *Fed. Proc.* (abstr.) 28:648.

Houpt, T. R., S. M. Anika, and K. A. Houpt. 1979c. Preabsorptive intestinal satiety controls of food intake in pigs. *Am. J. Physiol.* 236:R328–R337.

Houpt, T. R., B. A. Baldwin, and K. A. Houpt. 1983a. Effects of duodenal osmotic loads on spontaneous meals in pigs. *Physiol. Behav.* 30:787–795.

Houpt, T. R., K. A. Houpt, and A. A. Swan. 1983b. Duodenal osmoconcentration and food intake in pigs after ingestion of hypertonic nutrients. *Am. J. Physiol.* 245:R181–F189.

Houpt, T. R., L. C. Weixler, and D. W. Troy. 1986. Water drinking induced by gastric secretagogues in pigs. *Am. J. Physiol.* 251:R157–R164.

Hudson, S. J. 1977. Multiple fostering of calves onto nurse cows at birth. *Appl. Anim. Ethol.* 3:57–63.

Hudson, S. J., and M. M. Mullord. 1977. Investigations of maternal bonding in dairy cattle. *Appl. Anim. Ethol.* 3:271–276.

Hughes, G. P., and D. Reid. 1951. Studies on the behaviour of cattle and sheep in relation to the utilization of grass. *J. Agric. Sci.* 41:350–366.

Hughes, P. E., P. H. Hemsworth, and C. Hansen. 1985. The effects of supplementary olfactory and auditory stimuli on the stimulus value and mating success of the young boar. *Appl. Anim. Behav. Sci.* 14:245–252.

Hulet, C. V. 1966. Behavioral, social and psychological factors affecting mating time and breeding efficiency in sheep. *J. Anim. Sci.* (suppl.) 25:5–20.

Hulet, C. V., R. L. Blackwell, S. K. Ercanbrack, D. A. Price, and L. O. Wilson. 1962. Mating behavior of the ewe. *J. Anim. Sci.* 21:870–874.

Hulet, C. V., R. L. Blackwell, and S. K. Ercanbrack. 1964. Observations on sexually inhibited rams. *J. Anim. Sci.* 23:1095–1097.

Hulet, C. V., G. Alexander, and E. S. E. Hafez. 1975. The behaviour of sheep. In *The behaviour of domestic animals,* 3d ed. E.S.E. Hafez, ed. Baltimore: Williams and Wilkins. Pp. 246–294.

Hulet, C. V., D. M. Anderson, J. N. Smith, W. L. Shupe, C. A. Taylor, Jr., and L. W. Murray. 1989. Bonding of goats to sheep and cattle for protection from predators. *Appl. Anim. Behav. Sci.* 22:261–267.

Hunter, R. F., and G. E. Davies. 1963. The effect of method of rearing on the social behaviour of Scottish blackface hoggets. *Anim. Prod.* 5:183–194.

Hunter, W. S. 1917. The delayed reaction in a child. *Psychol. Rev.* 24:74–87.

Hurnik, J. F., G. J. King, and H. A. Robertson. 1975. Estrous and related behaviour in postpartum Holstein cows. *Appl. Anim. Ethol.* 2:55–68.

Hutson, G. D. 1980. The effect of previous experience on sheep movement through yards. *Appl. Anim. Ethol.* 6:233–240.

———. 1985. The influence of barley food rewards on sheep movement through a handling system. *Appl. Anim. Behav. Sci.* 14:263–273.

Ibuka, N., and H. Kawamura. 1975. Loss of circadian rhythm in sleep-wakefulness cycle in the rat by suprachiasmatic nucleus lesions. *Brain Res.* 96:76–81.

Igel, G. J., and A. D. Calvin. 1960. The development of affectional responses in infant dogs. *J. Comp. Physiol. Psychol.* 53:302–305.

Illius, A. W., N. B. Haynes, and G. E. Lamming. 1976. Effects of ewe proximity on peripheral plasma testosterone levels and behaviour in the ram. *J. Reprod. Fert.* 48:25–32.

Im, H. S., R. H. Barnes, D. A. Levitsky, and W. G. Pond. 1973. Postnatal malnutrition and regional cholinesterase activities in brain of pigs. *Brain Res.* 63:461–465.

Ingram, D. L., and M. J. Dauncy. 1985. Circadian rhythms in the pig. *Comp. Biochem. Physiol.* 82(A):1–5.

Ingram, D. L., and K. F. Legge. 1974. Effects of environmental temperature on food intake in growing pigs. *Comp. Biochem. Physiol.* 48(A):573–581.

Ingram, D. L., and D. B. Stephens. 1979. The relative importance of thermal, osmotic and hypovolaemic factors in the control of drinking in the pig. *J. Physiol.* 293:501–512.

Ingram, D. L., M. J. Dauncy and K. F. Legge. 1985. Synchronization of motor activity in young pigs to a non-circadian rhythm without affecting food intake and growth. *Comp. Biochem. Physiol.* 80(A):363–368.

Inselman-Temkin, B. R., and J. P. Flynn. 1973. Sex-dependent effects of gonadal and gonadotropic hormones on centrally-elicited attack in cats. *Brain Res.* 60:393–410.

Irwin, M. R., D. R. Melendy, M. S. Amoss, and D. P. Hutcheson. 1979. Roles of predisposing factors and gonadal hormones in the buller syndrome of feedlot steers. *J. Am. Vet. Med. Assoc.* 174:367–370.

Izard, M. K., and J. G. Vandenbergh. 1982. Priming pheromones from oestrous cows increase synchronization of oestrus in dairy heifers after PGF-2α injection. *J. Reprod. Fert.* 66:189–196.

Jacklet, J. W. 1977. Neuronal circadian rhythm: Phase shifting by a protein synthesis inhibitor. *Science* 198:69–71.

Jackson, B., and A. Reed. 1969. Catnip and the alteration of consciousness. *J. Am. Med. Assoc.* 207:1349–1350.

Jackson, H. M., and D. W. Robinson. 1971. Evidence for hypothalamic α and β adrenergic receptors involved in the control of food intake of the pig. *Br. Vet. J.* 127: li-liii.

Jackson, S. A., R. A. Rich, and S. L. Ralston. 1984. Feeding behavior and feed efficiency in groups of horses as a function of feeding frequency and use of alfalfa hay cubes. *J. Anim. Sci.* 59 (Suppl. 1):152–153.

Jalowiec, J. E., J. Panksepp, H. Shabshelowitz, A. J. Zolovick, W. Stern, and P. J. Morgane. 1973. Suppression of feeding in cats following 2-deoxy-D-glucose. *Physiol. Behav.* 10:805–807.

James, W. T. 1951. Social organization among dogs of different temperaments, terriers and beagles, reared together. *J. Comp. Physiol. Psychol.* 44:71–77.

James, W. T., and T. F. Gilbert. 1955. The effect of social facilitation on food intake of puppies fed separately and together for the first 90 days of life. *Br. J. Anim. Behav.* 3:131–133.

Janowitz, H. D., and M. I. Grossman. 1949a. Effect of variations in nutritive density on intake of food of dogs and rats. *Am. J. Physiol.* 158:184–193.

_____. 1949b. Some factors affecting the food intake of normal dogs and dogs with esophagostomy and gastric fistula. *Am. J. Physiol.* 159:143–148.

Janowitz, H. D., M. E. Hanson, and M. I. Grossman. 1949. Effect of intravenously administered glucose on food intake in the dog. *Am. J. Physiol.* 156:87–91.

Jasper, H. H., and J. Tessier. 1971. Acetylcholine liberation from cerebral cortex during paradoxical (REM) sleep. *Science* 172:601–602.

Jensen, P. 1980. An ethogram of social interacion patterns in group-housed dry sows. *Appl. Anim. Ethol.* 6:341–350.

_____. 1984. Effects of confinement on social interaction patterns in dry sows. *Appl. Anim. Behav. Sci.* 12:93–101.

_____. 1986. Observations on the maternal behavior of free-ranging domestic pigs. *Appl. Anim. Behav. Sci.* 16:131–142.

Jensen, P., K. Floren, and B. Hobroh. 1987. Peri-parturient changes in behaviour in free-ranging domestic pigs. *Appl. Anim. Behav. Sci.* 17:69–76.

Jeppesen, L. E. l982a. Teat-order in groups of piglets reared on an artificial sow. I. Formation of teat-order and influence of milk yield on teat preference. *Appl. Anim. Ethol.* 8:335–345.

_____. l982b. Teat-order in groups of piglets reared on an artificial sow. II. Maintenance of teat-order with some evidence for the use of odour cues. *Appl. Anim. Ethol.* 8:347–355.

Jewell, P. A., S. J. G. Hall, and M. M. Rosenberg. 1986. Multiple mating and siring success during natrual oestrus in the ewe. *J. Reprod. Fert.* 77:81–89.

Joby, R., J. E. Jemmett, and A. S. Miller. 1984. The control of undesirable behaviour in male dogs using megestrol acetate. *J. Small Anim. Pract.* 25:567–572.

John, E. R. 1972. Switchboard versus statistical theories of learning and memory. *Science* 177:850–864.

John, E. R., P. Chesler, F. Bartlett, and I. Victor. 1968. Observation learning in cats. *Science* 159:1589–1491.

Johnson, B. F., and C. Chura. 1974. Diurnal variation in the effect of tolbutamide. *Am. J. Med. Sci.* 268:93–96.

Johnson, R. H., and R. F. Thompson. 1969. Role of association cortex in auditory-visual conditional learning in the cat. *J. Comp. Physiol. Psychol.* 69:485–491.

Johnsson, A., W. Engelmann, B. Pflug, and W. Klemke. 1980. Influence of litium ions on human circadian rhythms. *Z. Naturforsch.* 35:503–507.

Johnstone-Wallace, D. B., and K. Kennedy. 1944. Grazing management practices and their relationship to the behaviour and grazing habits of cattle. *J. Agric. Sci.* 34:190–197.

Jones, C. G., K. D. Maddever, D. L. Court, and M. Phillips. 1966. The time taken by cows to eat concentrates. *Anim. Prod.* 8:489–497.

Jordan, W. A., E. E. Lister, J. M. Wauthy, and J. C. Comeau. 1973. Voluntary roughage intake by nonpregnant and pregnant or lactating beef cows. *Can. J. Anim. Sci.* 53:733–738.

Jouvet, M. 1967. Neurophysiology of the states of sleep. *Physiol. Rev.* 47:117–177.

Kalmus, H. 1955. The discrimination by the nose of the dog of individual human odours and in particular of the odours of twins. *Br. J. Anim. Behav.* 3:25–31.

Kanarek, R. B. 1975. Availability and caloric density of the diet as determinants of meal patterns in cats. *Physiol. Behav.* 15:611–618.

Kanno, Y. 1977. Experimental studies on body temperature rhythm in dogs. I. Application of cosinor method to body temperature rhythm in dogs. *Jap. J. Vet. Sci.* 39:69–76.

Karas, G. G., R. I. Willham, and D. F. Cox. 1962. Avoidance learning in swine. *Psychol. Rep.* 11:51–54.

Kare, M. R., W. C. Pond, and J. Campbell. 1965. Observations on the taste reactions in pigs. *Anim. Behav.* 13:265–269.

Karlander, S., J. Mansson, and G. Tufvesson. 1965. Buccostomy as a method of treatment for aerophagia (windsucking) in the horse. *Nordisk Vet.-Med.* 17:455–458.

Karn, H. W., and H. R. Malamud. 1939. The behavior of dogs on the double alternation problem in the temporal maze. *J. Comp. Psychol.* 27:461–466.

Karn, J. F., and D. C. Clanton. 1974. Electronically controlled individual cattle feeding. *J. Anim. Sci.* (abstract) 39:136.

Katz, R. J., and E. Thomas. 1976. Effects of para-chlorophenylalanine upon brain stimulated affective attack in the cat. *Pharmacol. Biochem. Behav.* 5:392–94.

Keiper, R. R. 1976. Social organization of feral ponies. *Proc. Penn. Acad. Sci.* 50:69–70.

_____. 1981. Time budgets of EIA positive ponies. *Equine Pract.* 3:6–10.

_____. 1985. *The Assateague ponies.* Centreville, Md.: Tidewater Press.

Keiper, R. R., and J. Berger. 1982–83. Refuge-seeking and pest avoidance by feral horses in desert and island environments. *Appl. Anim. Ethol.* 9:111–120.

Keiper, R. R., and K. A. Houpt. 1984. Reproduction in feral horses: An eight-year study. *Amer. J. Vet. Res.* 45:991–995.

Keiper, R. R., and M. A. Keenan. 1980. Nocturnal activity patterns of feral horses. *J. Mammal.* 61:116–118.

Keiper, R. R., and H. H. Sambraus. 1986. The stability of equine dominance hierarchies and the effects of kinship, proximity and foaling status on hierarchy rank. *Appl. Anim. Behav. Sci.* 16:121–130.

Kendrick, K. M., and B. A. Baldwin. 1986a. The activity of neurones in the lateral hypothalamus and zona incerta of the sheep responding to the sight or approach of food is modified by learning and satiety and reflects food preferences. *Brain Res.* 375:320–328.

_____ 1986b. Characterization of neuronal responses in the zona incerta of the subthalamic region of the sheep during the ingestion of food and liquid. *Neurosci. Lett.* 63:237–242.

Kendrick, K. M., E. B. Keverne, B. A. Baldwin, and D. F. Sharman. 1986. Cerebrospinal fluid levels of acetylcholinesterase, monoamines and oxytocin during labour, parturition, vaginal stimulation, lamb separartion and suckling in sheep. *Neuroendocrinol.* 44:149–156.

Kendrick, K. M., E. B. Keverne, and B. A. Baldwin. 1987. Intracerebroventricular oxytocin stimulates maternal behaviour in sheep. *Neuroendocrinology.* 46:56–61.

Kennedy, J. M., and B. A. Baldwin. 1972. Taste preferences in pigs for nutritive and non-nutritive sweet solutions. *Anim. Behav.* 20:706–718.

Kenny, F. J., and P. V. Tarrant. 1987. The behaviour of young Friesian bulls during social re-grouping at an abattoir. Influence of an overhead electrified wire grid. *Appl. Anim. Behav. Sci.* 18:233–246.

Kent, J. P. 1984. A note on multiple fostering of calves on to nurse cows at a few days post-partum. *Appl. Anim. Behav. Sci.* 12:183–186.

Kerruish, B. M. 1955. The effect of sexual stimulation prior to service on the behaviour and conception rate of bulls. *Br. J. Anim. Behav.* 3:125–130.

Keverne, E. B., F. Levy, P. Poindron, and D. R. Lindsay. 1982. Vaginal stimulation: An important determinant of maternal bonding in sheep. *Science* 219:81–83.

Key, C., and R. M. MacIver. 1977. Factors affecting sexual preferences in sheep. *Appl. Anim. Ethol.* (abstr.) 3:291.

Khalaf, F. 1969. Hyperphagia and aphagia in swine with induced hypothalamic lesions. *Res. Vet. Sci.* 10:514–517.

Kiddy, C. A., D. S. Mitchell, D. J. Bolt, and H. W. Hawk. 1978. Detection of estrus-related odors in cows by trained dogs. *Biol. Reprod.* 19:389–395.

Kiley, M. 1972. The vocalizations of ungulates, their causation and function. *Z. Tierpsychol.* 31:171–222.

_____. 1976. Fostering and adoption in beef cattle. *Br. Cattle Breeders* Club Dig. 31:42–55.

Kiley-Worthington, M. 1976. The tail movements of ungulates, canids and felids with particular reference to their causation and function as displays. *Behaviour* 56:69–115.

_____. 1977. *Behavioural problems of farm animals.* Stocksfield, England: Oriel Press.

Kiley-Worthington, M., and P. Savage. 1978. Learning in dairy cattle using a device for economical management of behaviour. *Appl. Anim. Ethol.* 4:119–124.

Kilgour, R. 1972. Some observations on the suckling activity of calves on nurse cows. *Proc. N. Z. Soc. Anim. Prod.* 32:132–136.

_____. 1975. The open-field test as an assessment of the temperament of dairy cows. *Anim. Behav.* 23:615–624.

_____. 1981. Use of the Hebb-Williams closed-field test to study the learning ability of Jersey cows. *Anim. Behav.* 29:850–860.

Kilgour, R., and D. N. Campin. 1973. The behaviour of entire bulls of different ages at pasture. *Proc. N. Z. Soc. Anim. Prod.* 33:125–138.

Kilgour, R., and T. H. Scott. 1959. Leadership in a herd of dairy cows. *Proc. N. Z. Soc. Anim. Prod.* 19:36–43.

Kilgour, R., and C. G. Winfield. 1977. Pen-mating of pedigree sheep. *N. Z. J. Agric.* 134:25–27.

Kilgour, R., C. G. Winfield, K. J. Bremner, M. M. Mullord, H. De Langen, and S. J. Hudson. 1976. Behaviour of early-weaned calves in indoor individual cubicles and group pens. *N. Z. Vet. J.* 23:119–123.

Kilgour, R., B. H. Skarsholt, J. F. Smith, K. J. Bremner, and M.C.L. Morrison. 1977. Observations on the behaviour and factors influencing the sexually active group in cattle. *Proc. N. Z. Soc. Anim. Prod.* 37:128–135.

Kimble, D. P., L. Rogers, and C. W. Hendrickson. 1967. Hippocampal lesions disrupt maternal, not sexual, behavior in the albino rat. *J. Comp. Physiol. Psychol.* 63:401–407.

King, J. E., R. F. Becker, and J. E. Markee. 1964. Studies on olfactory discrimination in dogs. 3. Ability to detect human odour trace. *Anim. Behav.* 12:311–315.

Kirkpatrick, J. F., R. Vail, S. Devous, S. Schwend, C. B. Baker, and L. Wiesner. 1976. Diurnal variation of plasma testosterone in wild stallions. *Biol. Reprod.* 15:98–101.

Kirkwood, R. N., J. M. Forbes, and P. E. Hughes. 1981. Influence of boar contact on attainment of puberty in gilts after removal of the olfactory bulbs. *J. Reprod. Fertil.* 61:193–196.

Kitchell, R. L. 1972. Dogs know what they like. *Friskies Res. Dig.* 8(3):1–4.

Kitsikis, A., and A. G. Roberge. 1981. Changes in brain biogenic amines in cats performing a symmetrically reinforced go-nogo visual discrimination task. *Behav. Neural Biol.* 32:133–147.

Kleinman, D., and E. R. John. 1975. Contradiction of auditory and visual information by brain stimulation. *Science* 187:271–273.

Klemm, W. R., C. J. Sherry, L. M. Schake, and R. F. Sis. 1983. Homosexual behavior in feedlot steers: An aggression hypothesis. *Appl. Anim. Ethol.* 11:187–195.

Kling, A., and D. Coustan. 1964. Electrical stimulation of the amygdala and hypothalamus in the kitten. *Exp. Neurol.* 10:81–89.

Kling, A., J. K. Kovach, and T. J. Tucker. 1969. The behaviour of cats. In *The behaviour of domestic animals,* 2d ed. E. S. E. Hafez, ed. Baltimore: Williams and Wilkins. Pp. 482–512.

Klingel, H. 1974. A comparison of the social behaviour of the Equidae. In *The behaviour of ungulates and its relation to management.* V. Geist, and F. Walther, eds. Morges, Switzerland: International Union for Conservation of Nature and Natural Resources. Pp. 124–132.

Klopfer, F. D. 1961. Early experience and discrimination learning in swine. *Am. Zool.* (abstr.) 1:366.

_____. 1966. Visual learning in swine. In *Swine in biomedical research.* L. K. Bustad, R. O. McClellan, and M. P. Burns, eds. Richland, Wash.: Bettelle Memorial Institute Pacific Northwest Laboratory. Pp. 559–574.

Klopfer, P. H., and J. Gamble. 1966. Maternal "imprinting" in goats: The role of chemical senses. *Z. Tierpsychol.* 23:588–592.

Knecht, C. D., J. E. Oliver, R. Redding, R. Selcer, and G. Johnson. 1973. Narcolepsy in a dog and a cat. *J. Am. Vet. Med. Assoc.* 162:1052–1053.

Knight, T. W., and P. R. Lynch. 1980. Source of ram pheromones that stimulate ovulation in the ewe. *Anim. Reprod. Sci.* 3:133–136.

Koehler, W. 1962. *The Koehler method of dog training.* New York: Howell Book House.

Koepke, J. E., and K. H. Pribram. 1971. Effect of milk on the maintenance of sucking behavior in kittens from birth to six months. *J. Comp. Physiol. Psychol.* 75:363–377.

Kolb, B., and A. J. Nonneman. 1975. The development of social responsiveness in kittens. *Anim. Behav.* 23:368–374.

Kondo, S., N. Kawakami, H. Kohama, and S. Nishino. 1983/84. Changes in activity spatial pattern and social behavior in calves after grouping. *Appl. Anim. Ethol.* 11:217–228.

Konorski, J. 1961. The physiological approach to the problem of recent memory. In *Brain mechanisms and learning*. A. Fessard, R. W. Gerard, J. Konorski, and J. F. Delafresnaye, eds. Springfield, Ill.: Charles C. Thomas. Pp. 115–132.

Konrad, K. W., and M. Bagshaw. 1970. Effect of novel stimuli on cats reared in a restricted environment. *J. Comp. Physiol. Psychol.* 70:157–164.

Korda, P., and J. Brewinska. 1977. Effect of stimuli emitted by sucklings on tactile contact of the bitches with sucklings and on number of licking acts. *Acta Neurobiol. Exp.* 37:99–115.

Kovach, J. K., and A. Kling. 1967. Mechanisms of neonate sucking behaviour in the kitten. *Anim. Behav.* 15:91–101.

Kovalcik, K. and M. Kovalcik. 1986. Learning ability and memory testing in cattle of different ages. *Appl. Anim. Behav. Sci.* 15:27–29.

Krabill, L. F., P. J. Wangsness, and C. A. Baile. 1978. Effects of elfazepam on digestibility and feeding behavior in sheep. *J. Anim. Sci.* 46:1356–1359.

Kratzer, D. D. 1969. Effects of age on avoidance learning in pigs. *J. Anim. Sci.* 28:175–179.

———. 1971. Learning in farm animals. *J. Anim. Sci.* 32:1268–1273.

Kratzer, D. D., W. M. Netherland, R. E. Pulse, and J. P. Baker. 1977. Maze learning in quarter horses. *J. Anim. Sci.* 45:896–902.

Krebs, J. R., and N. B. Davies. 1978. *Behavioural ecology. An evolutionary approach*. Sunderland, Mass.: Sinauer Associates.

Kristal, M. B., A. C. Thompson, S. B. Heller, and B. R. Komisaruk. 1986. Placenta ingestion enhances analgesia produced by vaginal/cervical stimulation in rats. *Physiol. Behav.* 36:1017–1020.

Kropp, J. R., J. W. Holloway, D. F. Stephens, L. Knori, R. D. Morrison, and R. Totusek. 1973. Range behavior of Hereford, Hereford x Holstein and Holstein non-lactating heifers. *J. Anim. Sci.* 36:797–802.

Kuhn, C. M., S. R. Butler, and S. M. Schanberg. 1978. Selective depression of serum growth hormone during maternal deprivation in rat pups. *Science* 201:1043–1036.

Kuhn, G., and W. Hardegg. 1988. Effects of indoor and outdoor maintenance of dogs upon food intake, body weight, and different blood parameters. *Z. Versuchstierkd.* 31:205–214.

Kuipers, M., and T. S. Whatson. 1979. Sleep in piglets: An observational study. *Appl. Anim. Ethol.* 5:145–151.

Kuo, Z. Y. 1930. The genesis of the cat's responses to the rat. *J. Comp. Psychol.* 11:1–35.

Kurz, J. C., and R. L. Marchinton. 1972. Radiotelemetry studies of feral hogs in South Carolina. *J. Wildl. Manage.* 36:1240–1248.

Kushida, C. A., B. M. Bergmann, and A. Rechtschaffen. 1989. Sleep deprivation in the rat: IV. Paradoxical sleep deprivation. *Sleep* 12:22–30.

Lamb, M. E. 1975. Physiological mechanisms in the control of maternal behavior in rats: A review. *Psychol. Bull.* 82:104–119.

Lammers, G. J., and A. De Lange. 1986. Pre- and post-farrowing behaviour in primiparous domesticated pigs. *Appl. Anim. Behav.* 15:31–43.

Lampkin, G. H., J. Quarterman, and M. Kidner. 1958. Observations on the grazing habits of grade and Zebu steers in a high altitude temperate climate. *J. Agric. Sci.* 50:211–218.

Larsen, H. J. 1963. Feeding habits of grazing and green feeding cows. *J. Anim. Sci.* (abstr.) 22:1134.

Lashley, K. S. 1926. Studies of cerebral function in learning. VII. The relation between cerebral mass, learning, and retention. *J. Comp. Neurol.* 41:1–58.

Laundre, J. 1977. The daytime behavior of domestic cats in a free-roaming population. *Anim. Behav.* 25:990–998.

Laut, J. E., K. A. Houpt, H. F. Hintz, and T. R. Houpt. 1985. The effects of caloric dilution on meal patterns and food intake of ponies. *Physiol. Behav*, 35:549–554.

Le Boeuf, B. J. 1967. Interindividual associations in dogs. *Behaviour* 29:268–295.

————. 1970. Copulatory and aggressive behavior in the prepuberally castrated dog. *Horm. Behav.* 1:127–136.

Leedy, M. G., and B. L. Hart. 1985. Female and male sexual responses in female cats with ventromedial hypothalamic lesions. *Behav. Neurosci.* 99:936–941.

Lees, J. L., and M. Weatherhead. 1970. A note on mating preferences of Clun Forest ewes. *Anim. Prod.* 12:173–175.

Lehner, P. N., C. McCluggage, D. R. Mitchell, and D. H. Neil. 1983. Selected parameters of the Fort Collins, Colorado, dog population, 1979–1980. *Appl. Anim. Ethol.* 10:19–25.

Le Neindre, P., P. Poindron, and C. Delouis. 1979. Hormonal induction of maternal behavior in non-pregnant ewes. *Physiol. Behav.* 22:731–734.

Lenhardt, M. L. 1977. Vocal contour cues in maternal recognition of goat kids. *Appl. Anim. Ethol.* 3:211–219.

Leon, M. 1974. Maternal pheromone. *Physiol. Behav.* 13:441–453.

Levine, A. S., C. E. Sievert, J. E. Morley, B. A. Gosnell, and S. E. Silvis. 1984. Peptidergic regulation of feeding in the dog (*Canis familiaris*). *Peptides* 5:675–679.

Levy, F., and P. Poindron. 1987. The importance of amniotic fluids for the establishment of maternal behaviour in experienced and inexperienced ewes. *Anim. Behav.* 35:1188–1192.

Levy, F., P. Poindron, and P. Le Neindre. 1983. Attraction and repulsion by amniotic fluids and their olfactory control in the ewe around parturition. *Physiol. Behav.* 31:687–692.

Lewis, N. J., and J. F. Hurnick. 1985. The development of nursing behaviour in swine. *Appl. Anim. Behav.* 14:225–232.

Lewis, R. C., and J. D. Johnson. 1954. Observations of dairy cow activities in loose-housing. *J. Dairy Sci.* 37:269–275.

Leyhausen, P. 1973. Addictive behavior in free ranging animals. In Bayer-symposium IV, *Psychic Dependence.* L. Goldberg, and F. Hoffmeister, eds. Berlin: Springer-Verlag. Pp. 58–64.

————. 1975. *Verhaltensstudien an Katzen.* Berlin: Paul Parey. Available as Leyhausen, P. 1979. *Cat behavior.* B. A. Tonkin, trans. New York: Garland STPM Press.

————. 1979. *Cat behavior.* New York: Garland STPM Press.

Liberg, O. 1983. Courtship behaviour and sexual selection in the domestic cat. *Appl. Anim. Ethol.* 10:117–132.

Lichtenstein, P. E. 1950a. Studies of anxiety: I. The production of a feeding inhibition in dogs. *J. Comp. Physiol. Psychol.* 43:16–29.

————. 1950b. Studies of anxiety. II. The effects of lobotomy on a feeding inhibition in dogs. *J. Comp. Physiol. Psychol.* 43:419–427.

Lickliter, R. E. 1984a. Hiding behavior in domestic goat kids. *Appl. Anim. Behav.* 12:245–251.

————. 1984b. Mother-infant spatial relationships in domestic goats. *Appl. Anim. Behav.* 13:93–100.

————. 1985. Behavior associated with parturition in the domesticated goat. *Appl. Anim. Behav.* 13:335–345.

————. 1987. Activity patterns and companion preferences of domestic goat kids. *Appl. Anim. Behav.* 19:137–145.

Lickliter, R. E., and J. R. Heron. 1984. Recognition of mother by newborn goats. *Appl. Anim. Behav.* 12:187–192.

Liddell, H. S. 1926a. The effect of thyroidectomy on some unconditioned responses of the sheep and goat. *Am. J. Physiol.* 75:579–590.

————. 1926b. A laboratory for the study of conditioned motor reflexes. *Am. J. Psychol.* 37:418–419.

————. 1954. Conditioning and emotions. *Sci. Am.* 190(1): 48–57.

Liddell, H. S., and O. D. Anderson. 1931. A comparative study of the conditioned motor reflex in the rabbit, sheep, goat, and pig. *Am. J. Physiol.* (abstr.) 97:539–540.

Liddell, H. S., W. T. James, and O. D. Anderson. 1934. The comparative physiology of the conditioned motor reflex based on experiments with the pig, dog, sheep, goat and rabbit. *Comp. Psychol. Monogr.* 11:1–89.

Lidfors, L., and P. Jensen. 1988. Behaviour of free-ranging beef cows and calves. *Appl. Anim. Behav.* 20:237–247.

Lien, J., and F. D. Klopfer. 1978. Some relations between stereotyped suckling in piglets and exploratory behaviour and discrimination reversal learning in adult swine. *Appl. Anim. Ethol.* 4:223–233.

Lindahl, I. L. 1964. Time of parturition in ewes. *Anim. Behav.* 12:231–234.

Lindsay, D. R. 1965. The importance of olfactory stimuli in the mating behaviour of the ram. *Anim. Behav.* 13:75–78.

_____. 1966a. Mating behaviour of ewes and its effect on mating efficiency. *Anim. Behav.* 14:419–424.

_____. 1966b. Modification of behavioural oestrus in the ewe by social and hormonal factors. *Anim. Behav.* 14:73–83.

Lindsay, D. R., and I. C. Fletcher. 1968. Sensory involvement in the recognition of lambs by their dams. *Anim. Behav.* 16:415–417.

_____. 1972. Ram-seeking activity associated with oestrous behaviour in ewes. *Anim. Behav.* 20:452–456.

Lindsay, D. R., and T. J. Robinson. 1961a. Studies on the efficiency of mating in the sheep. I. The effect of paddock size and number of rams. *J. Agric. Sci.* 57:137–140.

_____. 1961b. Studies on the efficiency of mating in the sheep. II. The effect of freedom of rams, padddock size, and age of ewes. *J. Agric. Sci.* 57:141–145.

Line, S. W., B. L. Hart, and L. Sanders. 1985. Effect of prepubertal versus postpubertal castration on sexual and aggressive behavior in male horses. *J. Am. Vet. Med. Assoc.* 186:249–251.

Liptrap, R. M., and J. I. Raeside. 1978. A relationship between plasma concentrations of testosterone and corticosteroids during sexual and aggressive behaviour in the boar. *J. Endocrinol.* 76:75–85.

Littlejohn, A., and R. Munro. 1972. Equine recumbency. *Vet. Rec.* 90:83–85.

Lockwood, R. 1987. Pit bull terriers. *Arthrozoos* 1:193–194.

Lofgreen, G. P., J. H. Meyer, and J. L. Hull. 1957. Behavior patterns of sheep and cattle being fed pasture or soilage. *J. Anim. Sci.* 16:773–780.

Lohse, C. L. 1974. Preferences of dogs for various meats. *J. Am. Anim. Hosp. Assoc.* 10:187–192.

Lorenz, K. Z. 1952. *King Solomon's ring. New light on animal ways.* New York: Thomas Y. Crowell.

_____. 1957. Companionship in bird life. In *Instinctive behavior. The development of a modern concept.* C. H. Schiller and K. S. Lashley, eds. New York: International Universities Press. Pp. 83–128.

Lowman, B. G., M. S. Hankey, N. A. Scott, and D. W. Deas. 1981. Influence of time of feeding on time of parturition in beef cows. *Vet. Rec.* 109:557–559.

Lubar, J. F. 1964. Effect of medial cortical lesions on the avoidance behavior of the cat. *J. Comp. Physiol. Psychol.* 58:38–46.

Lucas, E. A., E. W. Powell, and O. D. Murphree. 1977. Baseline sleep-wake patterns in the pointer dog. *Physiol. Behav.* 19:285–291.

Lunstra, D. D., G. W. Boyd, and L. R. Corah. 1989. Effects of natural mating stimuli on serum luteinizing hormone, testosterone and estradiol-17β in yearling beef bulls. *J. Anim. Sci.* 67:3277–3288.

Lustgarten, C., G. D. Bottoms, and J. R. Shaskas. 1973. Experimental adrenalectomy of pigs. *Am. J. Vet. Res.* 34:279–282.

Lynch, J. J., and G. Alexander. 1976. The effect of gramineous windbreaks on behaviour and lamb mortality among shorn and unshorn Merino sheep during lambing. *Appl. Anim. Ethol.* 2:305–325.

Lynch, J. J., and J. F. McCarthy. 1967. The effect of petting on a classically conditioned emotional response. *Behav. Res. Ther.* 5:55–62.

Macdonald, D. 1981. The behaviour and ecology of farm cats. In *The ecology and control of feral cats.* Potters Bar, Hertfordshire: Universities Federation for Animal Welfare. Pp. 23–29.

Macfarlane, J. S. 1974. The effect of two post-weaning management systems on the social and sexual behaviour of Zebu bulls. *Appl. Anim. Ethol.* 1:31–34.

Mackenzie, S. A., E. A. B. Oltenacu, and E. Leighton. 1985. Heritable estimate for temperament scores in German shepherd dogs and its genetic correlation with hip dysplasia. *Behav. Genet.* 15:475–482.

Mackenzie, S. A., E. A. B. Oltenacu, and K. A. Houpt. 1986. Canine behavioral genetics — A review. *Appl. Anim. Behav.* 15:365–393.

Mader, D. R., and E. O. Price. 1980. Discrimination learning in horses: effects of breed, age and social dominance. *J. Anim. Sci.* 50:962–965.

_____. 1984. The effects of sexual stimulation on the sexual performance of hereford bulls. *J. Anim. Sci.* 59:294–300.

Maddison, S., and B. A. Baldwin. 1983. Diencephalic neuronal activity during acquisition and ingestion of food in sheep. *Brain Res.* 278:195–206.

Maier, N. R. F., and T. C. Schneirla. 1964. *Principles of animal psychology.* New York: Dover Publications.

Maier, S. F., and M. E. P. Seligman. 1976. Learned helplessness: Theory and evidence. *J. Exp. Psychol. (General)* 105:3–46.

Marcella, K. L. 1983. A note on canine aggression towards veterinarians. *Appl. Anim. Ethol.* 10:155–157.

Marcuse, F. L., and A. U. Moore. 1944. Tantrum behavior in the pig. *J. Comp. Psychol.* 37:235–241.

_____. 1946. Motor criteria of discrimination. *J. Comp. Psychol.* 39:25–27.

Marinier, S. L., A. J. Alexander, and G. H. Waring. 1988. Flehmen behaviour in the domestic horse: Discrimination of conspecific odours. *Appl. Anim. Behav.* 19:227–237.

Marten, G. C., and J. D. Donker. 1964. Selective grazing induced by animal excreta. I. Evidence of occurrence and superficial remedy. *J. Dairy Sci.* 47:773–776.

Martin, F. H., J. R. Seoane, and C. A. Baile. 1973a. Feeding in satiated sheep elicited by intraventricular injections of CSF from fasted sheep. *Life Sci.* 13:177–184.

Martin, J. T. 1975. Movement of feral pigs in North Canterbury, New Zealand. *J. Mammal.* 56:914–915.

Martin, P. 1984. The time and energy costs of play behaviour in the cat. *Z. Tierpsychol.* 64:298–312.

_____. 1986. An experimental study of weaning in the domestic cat. *Behaviour* 99:221–249.

Martin, P., and P. Bateson. 1985. The ontogeny of locomotor play behaviour in the domestic cat. *Anim. Behav.* 33:502–510.

Martin, R. J., J. L. Gobble, T. H. Hartsock, H. B. Graves, and J. H. Ziegler. 1973b. Characterization of an obese syndrome in the pig. *Proc. Soc. Exp. Biol. Med.* 143:198–203.

Martins, T. 1949. Disgorging of food to the puppies by the lactating dog. *Physiol. Zool.* 22:169–172.

Marx, J. L. 1975. Learning and behavior (I): Effects of pituitary hormones. *Science* 190:367–370.

Mason, E. 1970. Obesity in pet dogs. *Vet. Rec.* 86:612–616.

Matsumoto, J., K. Sogabe, and Y. Hori-Santiago. 1972. Sleep in parabiosis. *Experientia* 28:1043–1044.

Matthies, H. 1974. The biochemical basis of learning and memory. *Life Sci.* 15:2017–2031.

Mattner, P. E., A. W. H. Braden, and K. E. Turnbull. 1967. Studies in flock mating of sheep. 1. Mating behaviour. *Aust. J. Exp. Agric. Anim. Husb.* 7:103–109.

Mayes, E., and P. Duncan. 1986. Temporal patterns of feeding in free-ranging horses. *Behaviour* 96:105–129.

McBane, S. 1987. *Behaviour problems of horses.* North Promfret, Vt.: David and Charles.

McBride, G. 1963. The "teat order" and communication in young pigs. *Anim. Behav.* 11:53–56.

McBride, G., J. W. James, and N. Hodgens. 1964. Social behaviour of domestic animals. IV. Growing pigs. *Anim. Prod.* 6:129–139.

McBride, G., J. W. James, and G. S. F. Wyeth. 1965. Social behaviour of domestic animals. VII. Variation in weaning weight in pigs. *Anim. Prod.* 7:67–74.

McCall, C. A. 1989. The effect of body condition of horses on discrimination learning abilities. *Appl. Anim. Behav. Sci.* 22:327–334.

McCall, C. A., G. D. Potter, T. H. Friend, and R. S. Ingram. 1981. Learning abilities in yearling horses using the Hebb-Williams closed field maze. *J. Anim. Sci.* 53:928–933.

McCall, C. A., G. D. Potter, and J. L. Kreider. 1985. Locomotor, vocal and other behavioural responses to varying methods of weaning foals. *Appl. Anim. Behav.* 14:27–35.

McCleary, R. A. 1961. Response specificity in the behavioral effects of limbic system lesions in the cat. *J. Comp. Physiol. Psychol.* 54:605–613.

McConnell, P. B. 1990. Acoustic structure and receiver response in domestic dogs, *Canis familiaris. Anim. Behav.* 39:897–904.

McConnell, P. B., and J. R. Baylis. 1985. Interspecific communication in cooperative herding: Acoustic and visual signals from human shepherds and herding dogs. *Z. Tierpsychol.* 67:302–328.

McDonald, C. L., R. G. Beilharz, and J. C. McCutchan. 1981. Training cattle to control by electric fences. *Appl. Anim. Ethol.* 7:113–121.

McDonnell, S. M. l986. Reproductive behavior of the stallion. *Veterinary Clinics of North America* 2(3): 535–556.

McDonnell, S. M., M. C. Garcia, and R. M. Kenney. 1987. Imipramine-induced erection, masturbation, and ejaculation in male horses. *Pharmacol. Biochem. Behav.* 27:187–191.

McDonnell, S. M., R. M. Kenney, P. E. Meckley, and M. C. Garcia. 1985. Conditioned suppression of sexual behavior in stallions and reversal with diazepam. *Physiol. Behav.* 34:951–956.

McDonnell, S. M., R. M. Kenney, P. E. Meckley, and M. C. Garcia. l986. Novel environment suppression of stallion sexual behavior and effects of diazepam. *Physiol. Behav.* 37:503–505.

McDougall, K. D., and W. McDougall. 1931. Insight and foresight in various animals — monkey, racoon, rat, and wasp. *J. Comp. Psychol.* 11:237–273.

McEachron, D. L., D. F. Kripke, R. Hawkins, E. Haus, D. Pavlinac, and L. Deftos. 1982. Lithium delays biochemical circadian rhythms in rats. *Neuropsychobiology* 8:12–29.

McGeer, E. G., and P. L. McGeer. 1966. Circadian rhythm in pineal tyrosine hydroxylase. *Science* 153:73–74.

McGinty, D. J., and M. B. Sterman. 1968. Sleep suppression after basal forebrain lesions in the cat. *Science* 160:1253–1255.

McGinty, D. J., M. Stevenson, T. Hoppenbrouwers, R. M. Harper, M. B. Sterman, and J. Hodgman. 1977. Polygraphic studies of kitten development: Sleep state patterns. *Dev. Psychobiol.* 10:455–469.

McGlone, J. J. 1985a. Olfactory cues and pig agonistic behavior: Evidence for a submissive pheromone. *Physiol. Behav.* 34:195–198.

_____. 1985b. A quantitative ethogram of aggressive and submissive behaviors in recently regrouped pigs. *J. Anim. Sci.* 61:559–565.

_____. 1986. Influence of resources on pig aggression and dominance. *Behav. Process* 12:135–144.

McGlone, J. J., and S. E. Curtis. 1985. Behavior and performance of weanling pigs in pens equipped with hide areas. *J. Anim. Sci.* 60:20–24.

McGlone, J. J., and J. L. Morrow. 1987. Individual differences among mature boars in T-maze preference for estrous or nonestrous sows. *Appl. Anim. Behav.* 17:77–82.

McGlone, J. J., K. W. Kelley, and C. T. Gaskins. 1980. Lithium and porcine aggression. *J. Anim. Sci.* 51:447–455.

McGrath, J. T. 1960. *Neurologic examination of the dog.* Philadelphia: Lea and Febiger.

McGuire, R. A., W. M. Rand, and R. J. Wurtman. 1973. Entrainment of the body temperature rhythm in rats: Effect of color and intensity of environmental light. *Science* 181:956–957.

McKinley, P. E. 1982. Cluster analysis of the domestic cat's vocal repertoire. Ph.D. diss., University of Maryland, College Park.

McLaughlin, C. L., L. F. Krabill, G. C. Scott, and C. A. Baile. 1976. Chemical stimulants of feeding animals. *Fed. Proc.* (abstr.) 35:579.

McLaughlin, C. L., C. A. Baile, L. L. Buckholtz, and S. K. Freeman. 1983. Preferred flavors and performance of weanling pigs. *J. Anim. Sci.* 56:1287–1293.

McPhee, C. P., G. McBride, and J. W. James. 1964. Social behaviour of domestic animals. III. Steers in small yards. *Anim. Prod.* 6:9–15.

Mech, L. D. 1975. Hunting behavior in two similar species of social canids. In *The wild Canids. Their systematics, behavioral ecology and evolution*. M. W. Fox, ed. New York: Van Nostrand Reinhold. Pp. 363–368.

Meese, G. B., and B. A. Baldwin. 1975a. The effects of ablation of the olfactory bulbs on aggressive behaviour in pigs. *Appl. Anim. Ethol.* 1:251–262.

———. 1975b. Effects of olfactory bulb ablation on maternal behaviour in sows. *Appl. Anim. Ethol.* 1:379–386.

Meese, G. B., and R. Ewbank. 1973a. The establishment and nature of the dominance hierarchy in the domesticated pig. *Anim. Behav.* 21:326–334.

———. 1973b. Exploratory behaviour and leadership in the domesticated pig. *Br. Vet. J.* 129:251–262.

Meese, G. B., D. J. Conner, and B. A. Baldwin. 1975. Ability of the pig to distinguish between conspecific urine samples using olfaction. *Physiol. Behav.* 15:121–125.

Meier, G. W. 1961. Infantile handling and development in Siamese kittens. *J. Comp. Physiol. Psychol.* 54:284–286.

Meier, G. W., and J. L. Stuart. 1959. Effects of handling on the physical and behavioral development of Siamese kittens. *Psychol. Rep.* 5:497–501.

Melrose, D. R., H. C. B. Reed, and R. L. S. Patterson. 1971. Androgen steroids associated with boar odour as an aid to the detection of oestrus in pig artificial insemination. *Br. Vet. J.* 127:497–502.

Melzack, R. 1962. Effects of early perceptual restriction on simple visual discrimination. *Science* 137:978–979.

Melzack, R., and T. H. Scott. 1957. The effects of early experience on the response to pain. *J. Comp. Physiol. Psychol.* 50:155–161.

Mendl, M., and E. S. Paul. 1989. Observation of nursing and sucking behaviour as an indicator of milk transfer and parental investment. *Anim. Behav.* 37:513–515.

Merrick, A. W., and D. W. Scharp. 1971. Electroencephalography of resting behavior in cattle, with observations on the question of sleep. *Am. J. Vet. Res.* 32:1893–1897.

Mersmann, H. J., M. D. MacNeil, S. C. Seideman, and W. G. Pond. 1987. Compensatory growth in finishing pigs after feed restriction. *J. Anim. Sci.* 64:752–764.

Mertens, D. R. 1987. Predicting intake and digestibility using mathematical models of ruminal function. *J. Anim. Sci.* 64:1548–1558.

Metz, J. H. M. 1985. The reaction of cows to a short-term deprivation of lying. *Appl. Anim. Behav.* 13:301–307.

Metz, J. H. M., and P. Mekking. 1984. Crowding phenomena in dairy cows as related to available idling space in a cubicle housing system. *Appl. Anim. Behav.* 12:63–78.

Meyer, J. H., G. P. Lofgreen, and J. L. Hull. 1957. Selective grazing by sheep and cattle. *J. Anim. Sci.* 16:766–772.

Michael, R. P. 1961. "Hypersexuality" in male cats without brain damage. *Science* 134:553–554.

———. 1973. The effects of hormones on sexual behavior in the female cat and rhesus monkey. In *Handbook of physiology*. Sec. 7, *Endocrinology*. Vol. 2, *Female reproductive system*. Pt. 1. R. O. Greep, and E. B. Astwood, eds. Washington, D.C.: American Physiological Society. Pp. 187–221.

Miles, R. C. 1958. Learning in kittens with manipulatory, exploratory, and food incentives. *J. Comp. Physiol. Psychol.* 51:39–42.

Miller, E. R., S. Vathana, F. F. Green, J. R. Black, D. R. Romsos, and D. E. Ullrey. 1974. Dietary caloric density and caloric intake in the pig. *J. Anim. Sci.* (abstr.) 39:980.

Miller, N. E. 1969. Learning of visceral and glandular responses. *Science* 163:434–445.

Miller, R. R. 1981. Male aggression, dominance, and breeding behavior in Red Desert feral horses. *Z. Tierpsychol.* 57:340–351.

Miller, R. R., and R. H. Denniston, II. 1979. Interband dominance in feral horses. *Z. Tierpsychol.* 51:41–47.

Mistlberger, R. E., T. A. Houpt, and M. C. Moore-Ede. 1990. Food-anticipatory rhythms under 24–hour schedules of limited access to single macronutrients. *J. Biol. Rhythms* 5:35–46.

Mitler, M. M., O. Soave, and W. C. Dement. 1976. Narcolepsy in seven dogs. *J. Am. Vet. Assoc.* 168:1036–1038.

Moelk, M. 1944. Vocalizing in the house-cat: A phonetic and functional study. *Am. J. Psychol.* 57:184–205.

Molliver, M. E. 1963. Operant control of vocal behavior in the cat. *J. Exp. Anal. Behav.* 6:197–202.

Moltz, H. 1960. Imprinting: Empirical basis and theoretical significance. *Psychol. Bull.* 57:291–314.

Monks of New Skete. 1978. *How to be your dog's best friend.* Boston: Little Brown.

Monteiro, L. S. 1972. The control of appetite in lactating cows. *Anim. Prod.* 14:263–281.

Montgomery, G. G. 1957. Some aspects of the sociality of the domestic horse. *Trans. Kansas Acad. Sci.* 60:419–424.

Moore, A. U., and F. L. Marcuse. 1945. Salivary, cardiac and motor indices of conditioning in two sows. *J. Comp. Psychol.* 38:1–16.

Moore, C. L., W. G. Whittlestone, M. Mullord, P. N. Priest, R. Kilgour, and J. L. Albright. 1975. Behavior responses of dairy cows trained to activate a feeding device. *J. Dairy Sci.* 58:1531–1535.

Moore, D. T., and R. A. McCleary. 1976. Fornix damage enhances successive, but not simultaneous, object-discrimination learning in cats. *J. Comp. Physiol. Psychol.* 90:109–118.

Moore, R. M., R. B. Zehmer, J. I. Moulthrop, and R. L. Parker. 1977. Surveillance of animal-bite cases in the United States, 1971–1972. *Arch. Environ. Health* 32:267–270.

Moore-Ede, M. C., F. M. Sulzman, and C. A. Fuller. 1982. *The clocks that time us.* Cambridge, Mass.: Harvard University Press.

Moorefield, J. G., and H. H. Hopkins. 1951. Grazing habits of cattle in a mixed-prairie pasture. *J. Range Manage.* 4:151–157.

Morag, M. 1967. Influence of diet on the behaviour pattern of sheep. *Nature* 213:110.

Morgan, M., and K. A. Houpt. 1989. Feline behavior problems: The influence of declawing. *Anthrozoos* 3:50–53.

Morgan, P. D., and G. W. Arnold. 1974. Behavioural relationships between Merino ewes and lambs during the four weeks after birth. *Anim. Prod.* 19:169–176.

Morgan, P. D., C. A. P. Boundy, G. W. Arnold, and D. R. Lindsay. 1975. The roles played by the senses of the ewe in the location and recognition of lambs. *Appl. Anim. Ethol.* 1:139–150.

Mormede, P., and R. Dantzer. 1977a. Effects of dexamethasone on fear conditioning in pigs. *Behav. Biol.* 21:225–235.

―――. 1977b. Experimental studies on avoidance behaviour in pigs. *Appl. Anim. Ethol.* 3:173–185.

Morrison, A. R. 1983. A window on the sleeping brain. *Sci. Amer.* 248:86–94.

Morrison, S. R., H. F. Hintz, and R. L. Givens. 1968. A note on effect of exercise on behaviour and performance of confined swine. *Anim. Prod.* 10:341–344.

Morrow, D. A. 1976. Fat cow syndrome. *J. Dairy Sci.* 59:1625–1629.

Morrow, D. A., D. Hillman, A. W. Dade, and H. Kitchen. 1979. Clinical investigation of a dairy herd with the fat cow syndrome. *J. Am. Vet. Med. Assoc.* 174:161–167.

Morrow-Tesch, J., and J. J. McGlone. 1990. Sensory systems and nipple attachment behavior in neonatal pigs. *Physiol. Behav.* 47:1–4.

Moulton, D. G., E. H. Ashton, and J. T. Eayrs. 1960. Studies in olfactory acuity. 4. Relative detectability of n-aliphatic acids by the dog. *Anim. Behav.* 8:117–128.

Mount, L. E. 1979. *Adaptation to thermal environment: Man and his productive animals.* Baltimore, Md.: University Park Press.

Moyer, K. E. 1968. Kinds of aggression and their physiological basis. *Comm. Behav. Biol.* 2(A):65–87.

Mugford, R. A. 1977. External influences on the feeding of carnivores. In *The chemical senses and nutrition,* M. R. Kare and O. Maller, eds. New York: Academic Press. Pp. 25–50.

Munkenbeck, N. 1983. A test for color vison and a spectral sensitivity curve in the sheep (*Ovis aries*). Master's thesis, Cornell University, Ithaca, N.Y.

Munro, J. 1956. Observations on the suckling behaviour of young lambs. *Br. J. Anim. Behav.* 4:34–36.

Murphey, R. M., F. A. M. Duarte, W. Coelho Novaes, and M. C. Torres Penedo. 1981. Age group differences in bovine investigatory behavior. *Dev. Psychobiol.* 14:117–125.

Murphree, O. D., J. E. Peters, and R. A. Dykman. 1969. Behavioral comparisons of nervous, stable, and crossbred pointers at ages 2, 3, 6, 9, and 12 months. *Cond. Reflex.* 4:20–23.

Myers, G. C. 1916. The importance of primacy in the learning of a pig. *J. Anim. Behav.* 6:64–69.

Myers, R. D. 1964. Emotional and autonomic responses following hypothalamic chemical stimulation. *Can. J. Psychol.* 18:6–14.

Myers, R. D., and D. C. Mesker. 1960. Operant responding in a horse under several schedules of reinforcement. *J. Exp. Anal. Behav.* 3:161–164.

Mylrea, P. J., and R. G. Beilharz. 1964. The manifestation and detection of oestrous in heifers. *Anim. Behav.* 12:25–30.

Nagamachi, Y. 1972. Effect of satiety center damage on food intake, blood glucose and gastric secretion in dogs. *Am. J. Dig. Dis.* n.s. 17:139–148.

Nagy, J., and L. Decsi. 1976. Effect of some intracerebrally administered drugs on a conditioned reflex in the cat. *Physiol. Behav.* 16:21–25.

Natoli, E. 1985. Spacing pattern in a colony of urban stray cats *(Felis catus L)* in the historic centre of Rome. *Appl. Anim. Behav.* 14:289–304.

Neathery, M. W. 1971. Acceptance of orphan lambs by tranquilized ewes *(Ovis aries)*. *Anim. Behav.* 19:75–79.

Neff, W. D., and I. T. Diamond. 1958. The neural basis of auditory discrimination. In *Biological and biochemical bases of behavior.* H. F. Harlow, and C. N. Woolsey, eds. Madison: University of Wisconsin Press. Pp. 101–126.

Neitz, J., T. Geist, and G. H. Jacobs. 1989. Color vision in the dog. *Visual Neurosci.* 3:119–125.

Newberry, R. C., and D. G. M. Wood-Gush. 1985. The suckling behavior of domestic pigs in a semi-natural environment. *Behaviour* 95:11–25.

⸻. 1986. Social relationships of piglets in a seminatural environment. *Anim. Behav.* 34:1311–1318.

Nikitopoulou, G., and J. L. Crammer. 1976. Change in diurnal temperature rhythm in manic-depressive illness. *Br. Med. J.* 1:1311–1314.

Noble, M., and C. K. Adams. 1963. Conditioning in pigs as a function of the interval between CS and US. *J. Comp. Physiol. Psychol.* 56:215–219.

Noda, K., and K. Chikamori. 1976. Effect of ammonia via prepyriform cortex on regulation of food intake in the rat. *Am. J. Physiol.* 231:1263–1266.

Nonneman, A. J., and J. M. Warren. 1977. Two-cue learning by brain-damaged cats. *Physiol. Psychol.* 5:397–402.

Norman, R. J., J. S. Buchwald, and J. R. Villablanca. 1977. Classical conditioning with auditory discrimination of the eye blink in decerebrate cats. *Science* 196:551–553.

Noyes, L. 1976. A behavioural comparison of gnotobiotic with normal neonate pigs, indicating stress in the former. *Appl. Anim. Ethol.* 2:113–121.

Numan, M., and H. G. Smith. 1984. Maternal behavior in rats: Evidence for the involvement of preoptic projections to the ventral tegmental area. *Behav. Neurosci.* 98:712–727.

O'Brien, P. H. 1984a. Feral goat home range: Influence of social class and environmental variables. *Appl. Anim. Behav.* 12:373–385.

⸻. 1984b. Leavers and stayers: Maternal postpartum strategies in feral goats. *Appl. Anim. Behav.* 12:233–243.

⸻. 1988. Feral goat social organization: A review and comparative analysis. *Appl. Anim. Behav.* 21:209–221.

O'Donnell, T. G., and G. A. Walton. 1969. Some observations on the behaviour and hill-pasture utilization of Irish cattle. *J. Br. Grassland Soc.* 24:128–133.

Oberg, R.G.E., and I. Divac. 1975. Dissociative effects of selective lesions in the caudate nucleus of cats and rats. *Acta Neurobiol. Exp.* 35:647–659.

Oberosler, R., C. Carenzi, and M. Verga. 1982. Dominance hierarchies of cows on Alpine pastures as related to phenotype. *Appl. Anim. Ethol.* 8:67–77.

Occupational Safety and Health Administration, U.S. Department of Labor. 1972. Occupational safety and health standards. *Fed. Reg.* 37(202), Pt. 2: 22102–22356.

Odberg, F. O. 1973. An interpretation of pawing by the horse (*Equus caballus* Linnaeus), displacement activity and original functions. *Saugetierkund. Mitteil.* 21:1–12.

Odberg, F. O., and K. Francis-Smith. 1977. Studies on the formation of ungrazed eliminative areas in fields used by horses. *Appl. Anim. Ethol.* 3:27–34.

Odde, K. G., G. H. Kiracofe, and R. R. Schalles. 1985. Suckling behavior in range beef calves. *J. Anim. Sci.* 61:307–309.

Offord, K. P., L. D. Satter, and D. A. Wieckert. 1969. Study of behavioral conditioning and feed intake in dairy heifers. *J. Dairy Sci.* (abstr.) 52:918.

Olm, D. D., and K. A. Houpt. 1988. Feline house-soiling problems. *Appl. Anim. Behav.* 20:335–345.

Olmstead, C. E., J. R. Villablanca, R. J. Marcus, and D. L. Avery. 1976. Effects of caudate nuclei or frontal cortex ablations in cats. IV. Bar pressing, maze learning, and performance. *Exp. Neurol.* 53:670–693.

Owen, J. B., and W. J. Ridgman. 1967. The effect of dietary energy content on the voluntary intake of pigs. *Anim. Prod.* 9:107–113.

Owen, R. R., F. J. McKeating, and D. W. Jagger. 1980. Neurectomy in windsucking horses. *Vet. Rec.* 106:134–135.

Owens, J. L., T. N. Edey, B. M. Bindon, and L. R. Piper. 1984–85. Parturient behaviour and calf survival in a herd selected for twinning. *Appl. Anim. Behav.* 13:321–333.

Packwood, J., and B. Gordon. 1975. Stereopsis in normal domestic cat, Siamese cat, and cat raised with alternating monocular occlusion. *J. Neurophysiol.* 38:1485–1499.

Palazzolo, D. L., and S. K. Quadri. 1987. The effects of aging on the circadian rhythm of serum cortisol in the dog. *Exp. Gerontol.* 22:379–387.

Palen, G. F., and G. V. Goddard. 1966. Catnip and oestrous behaviour in the cat. *Anim. Behav.* 14:372–377.

Panaman, R. 1981. Behavior and ecology of free-ranging female farm cats (*Felis catus L.*). *Z. Tierpsychologie* 56:59–73.

Pappas, T. N., R. L. Melendez, K. M. Strah, and H. T. Debas. 1985. Cholecystokinin is not a peripheral satiety signal in the dog. *Am. J. Physiol.* 249:G733–G738.

Pappenheimer, J. R., G. Koski, V. Fencl, M. L. Karnovsky, and J. Krueger. 1975. Extraction of sleep-promoting factor S from cerebrospinal fluid and from brains of sleep-deprived animals. *J. Neurophysiol.* 38:1299–1311.

Parrott, R. F., and B. A. Baldwin. 1984. Olfactory stimuli and intermale aggression in androgen-treated castrated sheep. *Aggress. Behav.* 10:115–122.

Parrott, R. F., and W. D. Booth. 1984. Behavioural and morphological effects of 5 α-dihydrotestosterone and oestradiol-17β in the prepubertally castrated boar. *J. Reprod. Fert.* 71:453–461.

Parsons, S. D., and G. L. Hunter. 1967. Effect of the ram on duration of oestrus in the ewe. *J. Reprod. Fert.* 14:61–70.

Pavlov, I. P. 1927. *Conditioned reflexes. An investigation of the physiological activity of the cerebral cortex.* London: Oxford University Press.

Pekas, J. C. 1983. A method for direct gastric feeding and the effect of voluntary ingestion in young swine. *Appetite* 4:23–30.

―――. 1985. Animal growth during liberation from appetite suppression. *Growth* 49:19–27.

Penfield, W. 1958. *The excitable cortex in conscious man.* Liverpool, England: Liverpool University Press.

Penny, R.H.C., F.W.G. Hill, J. E. Field, and J. T. Plush. 1972. Tailbiting in pigs: A possible sex incidence. *Vet. Rec.* 91:482–483.

Pepelko, W. E., and M. T. Clegg. 1965a. Influence of season of the year upon patterns of sexual behavior in male sheep. *J. Anim. Sci.* 24:633–637.

―――. 1965b. Studies of mating behaviour and some factors influencing the sexual response in the male sheep *Ovis aries. Anim. Behav.* 13:249–258.

Perez, O., N. Jimenez de Perez, P. Poindron, P. Le Neindre, and J. P. Ravault. 1985. Influence of management conditions after calving on mother-young relationships and PRL response to mammaray stimulation in the cow. *Reprod. Nutr. Dev.* 25:605–618.

Persson, N. 1962. Self-stimulation in the goat. *Acta Physiol. Scand.* 55:276–285.

Petersen, H. V., K. Vestergaard, and P. Jensen. 1989. Integration of piglets into social groups of free-ranging domestic pigs. *Appl. Anim. Behav.* 23:223–236.

Petit, M. 1972. Emploi du temps des troupeaux de vaches-mères et de leurs veaux sur les pâturages d'altitude de l'Aubrac (in French). *Ann. Zootech.* 21:5–27.

Pettijohn, T. F., T. W. Wong, P. D. Ebert, and J. P. Scott. 1977. Alleviation of separation distress in 3 breeds of young dogs. *Dev. Psychobiol.* 10:373–381.

Pettyjohn, J. D., J. P. Everett, Jr., and R. D. Mochrie. 1963. Responses of dairy calves to milk replacer fed at various concentrations. *J. Dairy Sci.* 46:710–714.

Pfaffenberger, C. J., and J. P. Scott. 1959. The relationship between delayed socialization and trainability in guide dogs. *J. Genet. Psychol.* 95:145–155.

Pickens, R. W., and R. L. Kelley. 1967. Visual discrimination learning by the visually naive split-brain cat. *Psychonom. Sci.* 7:305–306.

Pickett, B. W., L. C. Faulkner, and J. L. Voss. 1975. Effect of season on some characteristics of stallion semen. *J. Reprod. Fert.* (suppl.) 23:25–28.

Pickett, B. W., J. L. Voss, and E. L. Squires. 1977. Impotence and abnormal sexual behavior in the stallion. *Theriogenology* 8:329–347.

Pollard, J. S., M. D. Baldock, and R. F. Lewis. 1971. Learning rate and use of visual information in five animal species. *Aust. J. Psychol.* 23:29–34.

Pond, W. G., and J. H. Maner. 1974. *Swine production in temperate and tropical environments.* San Francisco: W. H. Freeman and Co.

Prescott, C. W. 1973. Reproduction patterns in the domestic cat. *Aust. Vet. J.* 49:126–129.

Price, E. O., and V. M. Smith. 1984. The relationhip of male-male mounting to mate choice and sexual preference in male dairy goats. *Appl. Anim. Behav.* 13:71–82.

Price, E. O., J. Thos, and G. B. Anderson. 1981. Maternal responses of confined beef cattle to single versus twin calves. *J. Anim. Sci.* 53:934–939.

Price, E. O., M. Dunbar, and M. R. Dally. 1984a. Behavior of ewes and lambs subjected to restraint fostering. *J. Anim. Sci.* 58:1084–1089.

Price, E. O., G. C. Dunn, J. A. Talbot, and M. R. Dally. 1984b. Fostering lambs by odor transfer: The substitution experiment. *J. Anim. Sci.* 59:301–307.

Price, E. O., V. M. Smith, and L. S. Katz. 1984c. Sexual stimulation of male dairy goats. *Appl. Anim. Behav.* 13:83–92.

Price, E. O., C. L. Martinez, and B. L. Coe. 1984–85. The effects of twinning on mother-offspring behavior in range beef cattle. *Appl. Anim. Behav.* 13:309–320.

Price, E. O., L. S. Katz, G. P. Moberg, and S. J. R. Wallach. 1986. Inability to predict sexual and aggressive behavior by plasma concentration of testosterone and luteinizing hormone in Hereford bulls. *J. Anim. Sci.* 62:613–617.

Price, E. O., L. S. Katz, S. J. R. Wallach, and J. J. Zenchak. 1988. The relationship of male-male mounting to the sexual preferences of young rams. *Appl. Anim. Behav.* 21:347–355.

Prince, J. H. 1977. The eye and vision. In *Duke's physiology of domestic animals,* 9th ed. M. J. Swenson, ed. Ithaca, N.Y.: Cornell University Press. Pp 696–712.

Provenza, F. D., and J. C. Malechek. 1986. A comparison of food selection and foraging behavior in juvenile and adult goats. *Appl. Anim. Behav.* 16:49–61.

Purcell, D., C. W. Arave, and J. L. Walters. 1988. Relationship of three measures of behavior to milk production. *Appl. Anim. Behav.* 21:307–313.

Putnam, P. A., and R. E. Davis. 1963. Ration effects on drylot steer feeding patterns. *J. Anim. Sci.* 22:437–443.

Ralston, S. L. 1984. Controls of feeding in horses. *J. Anim. Sci.* 59:1354–1361.

Ralston, S. L., and C. A. Baile. 1983. Effects of intragastric loads of xylose, sodium chloride and corn oil on feeding behavior of ponies. *J. Anim. Sci.* 56:302–308.

Ralston, S. L., G. Van den Broek, and C. A. Baile. 1979. Feed intake patterns and associated blood glucose, free fatty acid and insulin changes in ponies. *J. Anim. Sci.* 49:838–845.

Ralston, S. L., D. E. Freeman, and C. A. Baile. 1983. Volatile fatty acids and the role of the large intestine in the control of feed intake in ponies. *J. Anim. Sci.* 57:815–825.

Ramsay, D. J., B. J. Rolls, and R. J. Wood. 1975. The relationship between elevated water intake and oedema associated with congestive cardiac failure in the dog. *J. Physiol.* 244:303–312.

_____. Wood. 1977. Thirst following water deprivation in dogs. *Am. J. Physiol.* 232:R93–R100.

Randall, G.C.B. 1972. Observations on parturition in the sow. I. Factors associated with the delivery of the piglets and their subsequent behaviour. *Vet. Rec.* 90:178–182.

Randall, R. P., W. A. Schurg, and D. C. Church. 1978. Response of horses to sweet, salty, sour and bitter solutions. *J. Anim. Sci.* 47:51–55.

Randall, W., and V. Lakso. 1968. Body weight and food intake rhythms and their relationship to the behavior of cats with brain stem lesions. *Psychonom. Sci.* 11:33–34.

Randall, W., R. Swenson, V. Parsons, J. Elbin, and M. Trulson. 1975. The influence of seasonal changes in light on hormones in normal cats and in cats with lesions of the superior colliculi and pretectum. *J. Interdisciplin. Cycle Res.* 6:253–266.

Randt, C. T., D. Quartermain, M. Goldstein, and B. Anagnoste. 1971. Norepinephrine biosynthesis inhibition: Effects on memory in mice. *Science* 172:498–499.

Rashotte, M. E., J. C. Smith, T. Austin, T. W. Castonguay, and L. Jonsson. 1984. Twenty-four-hour free feeding patterns of dogs eating dry food. *Neurosci. Biobehav. Rev.* 8:205–210.

Rasmussen, O. G., E. M. Banks, T. H. Berry, and D. E. Becker. 1962. Social dominance in gilts. *J. Anim. Sci.* 21:520–522.

Ray, D. E., and C. B. Roubicek. 1971. Behavior of feedlot cattle during two seasons. *J. Anim. Sci.* 33:72–76.

Redding, R. W. 1975. Prefrontal lobotomy. In *Current techniques in small animal surgery.* M. J. Bojrab, ed. Philadelphia: Lea and Febiger. Pp. 3–9.

Reed, H. C. B., D. R. Melrose, and R. L. S. Patterson. 1974. Androgen steroids as an aid to the detection of oestrus in pig artificial insemination. *Br. Vet. J.* 130:61–67.

Reinhardt, V., F. M. Mutiso, and A. Reinhardt. 1978a. Resting habits of Zebu cattle in a nocturnal enclosure. *Appl. Anim. Ethol.* 4:261–271.

_____. 1978b. Social behaviour and social relationships between female and male prepubertal bovine calves (*Bos indicus*). *Appl. Anim. Ethol.* 4:43–54.

Reis, D. J. 1974. The chemical coding of aggression in brain. *Adv. Behav. Biol.* 10:125–150.

Remmers, J. E., and H. Gautier. 1972. Neural and mechanical mechanisms of feline purring. *Respir. Physiol.* 16:351–361.

Rensch, B. 1956. Increase of learning capability with increase of brain-size. *Am. Natur.* 90:81–95.

Reppert, S. M., H. G. Artman, S. Swaminathan, and D. A. Fisher. 1981. Vasopressin exhibits a rhythmic daily pattern in cerebrospinal fluid, but not in blood. *Science* 213:1256–1257.

Rheingold, H. L. 1963. Maternal behavior in the dog. In *Maternal behavior in mammals.* H. L. Rheingold, ed. New York: John Wiley and Sons. Pp. 169–202.

Rheingold, H. L., and C. O. Eckerman. 1971. Familiar and nonsocial stimuli and the kitten's response to a strange environment. *Dev. Psychobiol.* 4:71–89.

Riches, J. H., and R. H. Watson. 1954. The influence of the introduction of rams on the incidence of oestrus in Merino ewes. *Aust. J. Agric. Res.* 5:141–147.

Ringo, J., M. L. Wolbarsht, H. G. Wagner, R. Crocker, and F. Amthor. 1977. Trichromatic vision in the cat. *Science* 198:753–755.

Riol, J. A., J. M. Sanchez, V. G. Eguren, and V. R. Gaudioso. 1989. Colour perception in fighting cattle. *Appl. Anim. Behav.* 23:199–206.

Roberts, S. J. 1971. *Veterinary obstetrics and genital diseases (theriogenology),* 2d ed. Ithaca, N.Y.: Stephen J. Roberts.

Roberts, W. W. 1958a. Both rewarding and punishing effects from stimulation of posterior hypothalamus of cat with same electrode at same intensity. *J. Comp. Physiol. Psychol.* 51:400–407.

_____. 1958b. Rapid escape learning without avoidance learning motivated by hypothalamic stimulation in cats. *J. Comp. Physiol. Psychol.* 51:391–399.

Robinson, D. W. 1975. Food intake regulation in pigs. IV. The influence of dietary threonine imbalance on food intake, dietary choice and plasma acid patterns. *Br. Vet. J.* 131:595–600.

Rogers, V. P., G. T. Hartke, and R. L. Kitchell. 1967. Behavioral technique to analyze a dog's ability to discriminate flavors in commercial food products. In *Olfaction and taste II, Japan.* Y. Hayashi, ed. Oxford: Pergamon Press. Pp. 353–359.

Rohde, K. A., and H. W. Gonyou. 1987. Strategies of teat-seeking behavior in neonatal pigs. *Appl. Anim. Behav.* 19:57–72.

Romsos, D. R., and D. Ferguson. 1983. Regulation of protein intake in adult dogs. *J. Am. Vet. Med. Assoc.* 182:41–43.

Rose, G. H., and J. P. Collins. 1975. Light-dark discrimination and reversal learning in early postnatal kittens. *Dev. Psychobiol.* 8:511–518.

Rose, J. E. 1968. Discussion following paper, Cortical representation, by E. F. Evans. In *Hearing Mechanisms in Vertebrates.* A. V. S. De Reuck, and J. Knight, eds. Ciba Foundation Symposium. London: Churchill. Pp. 287–295.

Rosenblatt, J. S. 1965a. The basis of synchrony in the behavioral interaction between the mother and her offspring in the laboratory rat. In *Determinants of infant behaviour III.* B. M. Foss, ed. New York: John Wiley and Sons. Pp. 3–45.

———. 1965b. Effects of experience on sexual behavior in male cats. In *Sex and behavior.* F. A. Beach, ed. New York: John Wiley and sons. Pp. 416–439.

———. 1971. Suckling and home orientation in the kitten: A comparative developmental study. In *The biopsychology of development.* E. Tobach, L. R. Aronson, and E. Shaw, eds. New York: Academic Press. Pp. 345–410.

Rosenblatt, J. S., and L. R. Aronson. 1958a. The decline of sexual behavior in male cats after castration with special reference to the role of prior sexual experience. *Behaviour* 12:285–338.

———. 1958b. The influence of experience on the behavioural effects of androgen in prepuberally castrated male cats. *Anim. Behav.* 6:171–182.

Rosenblatt, J. S., and T. C. Schneirla. 1962. The behaviour of cats. In *The behaviour of domestic animals.* E. S. E. Hafez, ed. Baltimore: Williams and Wilklins. Pp. 453–488.

Rosenkilde, C. E., and I. Divac. 1976. Time-discrimination performance in cats with lesions in prefrontal cortex and caudate nucleus. *J. Comp. Physiol. Psychol.* 90:343–352.

Rosillon-Warner, A., and R. Paquay. 1984–85. Development and consequences of teat-order in piglets. *Appl. Anim. Behav.* 13:47–58.

Ross, S. 1951. Sucking behavior in neonate dogs. *J. Abnorm. Soc. Psychol.* 46:142–149.

Ross, S., and J. Berg. 1956. Stability of food dominance relationships in a flock of goats. *J. Mammal.* 37:129–131.

Ross, S., and J. P. Scott. 1949. Relationship between dominance and control of movement in goats. *J. Comp. Physiol. Psychol.* 42:75–80.

Ross, S., J. P. Scott, M. Cherner, and V. H. Denenberg. 1960. Effects of restraint and isolation on yelping in puppies. *Anim. Behav.* 8:1–5.

Rossdale, P. D. 1967. Clinical studies on the newborn thoroughbred foal. I: Perinatal behaviour. *Br. Vet. J.* 123:470–481.

Roth, L. L., and R. D. Lisk. 1968. Effects of hypothalamic implants of progesterone on parturition, lactation, and maternal behavior in the rat. *76th Am. Psychol. Assoc. Proc.* 3:267–268.

Roy, J.H.B., K.W.G. Shillam, and J. Palmer. 1955. The outdoor rearing of calves on grass with special reference to growth rate and grazing behaviour. *J. Dairy Res.* 22:252–269.

Rozin, P. 1967. Thiamine specific hunger. In *Handbook of Physiology.* Sec. 6, *Alimentary canal.* Vol. 1, *Control of food and water intake.* C. F. Code and W. Heidel, eds. Washington, D.C.: American Physiological Society. Pp. 411–431.

Rozin, P., and J. W. Kalat. 1971. Specific hungers and poison avoidance as adaptive specializations of learning. *Psychol. Rev.* 78:459–486.

Rozkowska, E., and E. Fonberg. 1973. Salivary reactions after ventromedial hypothalamic lesions in dogs. *Acta Neurobiol. Exp.* 33:553–562.

Rubel, E. W. 1971. A comparison of somatotopic organization in sensory neocortex of newborn kittens and adult cats. *J. Comp. Neurol.* 143:447–480.

Rubenstein, D. I. 1981. Behavioural ecology of island feral horses. *Equine Vet. J.* 13:27–34.

Ruckebusch, Y. 1972. The relevance of drowsiness in the circadian cycle of farm animals. *Anim. Behav.* 20:637–643.

_____. 1974. Sleep deprivation in cattle. *Brain Res.* 78:4495–499.

_____. 1975. The hypnogram as an index of adaptation of farm animals to changes in their environment. *Appl. Anim. Ethol.* 2:3–18.

Ruckebusch, Y., and M. Gaujoux. 1976. Sleep patterns of the laboratory cat. *Electroencephalogr. Clin. Neurophysiol.* 41:483–490.

Ruckebusch, Y., R. W. Dougherty, and H. M. Cook. 1974. Jaw movements and rumen motility as criteria for measurement of deep sleep in cattle. *Am. J. Vet. Res.* 35:1309–1312.

Ruckebusch, Y., M. Gaujoux, and B. Eghbali. 1977. Sleep cycles and kinesis in the foetal lamb. *Electroencephalogr. Clin. Neurophysiol.* 42:226–237.

Rudge, M. R. 1970. Mother and kid behaviour in feral goats (*Capra hircus L.*). *Z. Tierpsychol.* 27:687–692.

Rushen, J. 1984. Stereotyped behaviour, adjunctive drinking and the feeding periods of tethered sows. *Anim. Behav.* 32:1059–1067.

Russek, M. 1970. Gluco-ammonia receptors in liver. *Fed. Proc.* (abstr.) 29:658.

Russek, M., and P. J. Morgane. 1963. Anorexic effect of intraperitoneal glucose in the hypothalamic hyperphagic cat. *Nature* 199:1004–1005.

Ryder, M. L. 1976. Seasonal changes in the coat of the cat. *Res. Vet. Sci.* 21:280–283.

Sacks, J. J., R. W. Sattin, and S. E. Bonzo. 1989. Dog bite-related fatalities from 1979 to 1988. *J. Amer. Med. Assoc.* 262:1489–1492.

Salter, R. E., and R. J. Hudson. 1979. Feeding ecology of feral horses in western Alberta. *J. Range Manage.* 32:221–225.

Salzinger, K., and M. B. Waller. 1962. The operant control of vocalization in the dog. *J. Exp. Anal. Behav.* 5:383–389.

Sambraus, H. H., and D. Sambraus. 1975. Prägung von Nutztieren auf Menschen (in German). *Z. Tierpsychol.* 38:1–17.

Sandler, B. E., G. A. Van Gelder, W. B. Buck, and G. G. Karas. 1968. Effect of dieldrin exposure on detour behavior in sheep. *Psychol. Rep.* 23:451–455.

Sandler, B. E., G. A. Van Gelder, D. D. Elsberg, G. G. Karas, and W. B. Buck. 1969. Dieldrin exposure and vigilance behavior in sheep. *Psychonom. Sci.* 15:261–262.

Sandler, B. E., G. A. Van Gelder, G. G. Karas, and W. B. Buck. 1971. An operant feeding device for sheep. *J. Exp. Anal. Behav.* 15:95–96.

Sasaki, H., and N. Yoshii. 1984. Conditioned responses in the visual cortex of dogs. II. During sleep. *Electroenceph. Clin. Neurophysiol.* 58:448–456.

Sato, S. 1982. Leadership during actual grazing in a small herd of cattle. *Appl. Anim. Ethol.* 8:53–65.

Schafer, M. 1975. *The language of the horse.* New York: Arco Publishing Co.

Schake, L. M., and J. K. Riggs. 1969. Activities of lactating beef cows in confinement. *J. Anim. Sci.* 28:568–572.

Schaller, G. B. 1972. *The Serengeti lion.* Chicago: University of Chicago Press.

Schein, M. W., and M. H. Fohrman. 1955. Social dominance relationships in a herd of dairy cattle. *Br. J. Anim. Behav.* 3:45–55.

Schloeth, R. 1961. Das Sozialleben des Camargue-Rindes. Qualitative und quantitative Untersuchungen über die sozialen Beziehungen—insbesondere die soziale Rangordnung—des halbwilden franzosischen Kampfrindes (in German). *Z. Tierpsychol.* 18:574–627.

Schmidt, M. J., and H. Markowitz. 1977. Behavioral engineering as an aid in the maintenance of healthy zoo animals. *J. Am. Vet. Med. Assoc.* 171:966–969.

Schmisseur, W. E., J. L. Albright, W. M. Dillon, E. W. Kehrberg, and W.H.M. Morris. 1966. Animal behavior responses to loose and free stall housing. *J. Dairy Sci.* 49:102–104.

Schmitt, M. 1973. Circadian rhythmicity in responses of cells in the lateral hypothalamus. *Am. J. Physiol.* 225:1096–1101.

Schneirla, T. C., J. S. Rosenblatt, and E. Tobach. 1963. Maternal behavior in the cat. In *Maternal Behavior in Mammals.* H. L. Rheingold, ed. New York: John Wiley and Sons. Pp. 122–168.

Schoen, A.M.S., S. E. Curtis, E. M. Banks, and H. W. Norton. 1974. Behavior and performance of swine subjected to preweaning handling. *J. Anim. Sci.* (abstr.) 39:136–137.

Schoen, A. M. S., E. M. Banks, and S. E. Curtis. 1976. Behavior of young Shetland and Welsh ponies (*Equus caballus*). *Biol. Behav.* 1:199–216.

Schryver, H. F., S. VanWie, P. Daniluk, and H. F. Hintz. 1978. The voluntary intake of calcium by horses and ponies fed a calcium deficient diet. *J. Equine Med. Surg.* 2:337–340.

Schryver, H. F., M. T. Parker, P. D. Daniluk, K. I. Pagan, J. Williams, L. V. Soderholm, and H. F. Hintz. 1987. Salt consumption and the effect of salt on mineral metabolism in horses. *Cornell Vet.* 77:122–131.

Scott, D. W., R. W. Kirk, and J. Bentinck-Smith. 1979. Some effects of short-term methyl-prednisolone therapy in normal cats. *Cornell Vet.* 69:104–115.

Scott, J. P. 1945. Social behavior, organization and leadership in a small flock of domestic sheep. *Comp. Psychol. Monogr.* 18:1–29.

_____. Dominance reaction in a small flock of goats. *Anat. Rec.* 94:380–381.

_____. 1948. Dominance and the frustration-aggression hypothesis. *Physiol. Zool.* 21:31–39.

Scott, J. P. 1958a. *Aggression.* Chicago: University of Chicago Press.

_____. 1958b. *Animal behavior.* Chicago: University of Chicago Press.

_____. 1962. Critical periods in behavioral development. *Science* 138:949–958.

Scott, J. P., and J. L. Fuller. 1974. *Dog behavior. The genetic basis.* Chicago: University of Chicago Press.

Scott, J. P., and M-V. Marston. 1950. Critical periods affecting the development of normal and maladjustive social behavior in puppies. *J. Genet. Psychol.* 77:25–60.

Scott, M. D., and K. Causey. 1973. Ecology of feral dogs in Alabama. *J. Wildl. Manage.* 37:253–265.

Scott, P. P. 1970. Cats. In *Reproduction and breeding techniques for laboratory animals.* E. S. E. Hafez, ed. Philadelphia: Lea and Febiger. Pp. 192–208.

Seabrook, M. K. 1972. A study to determine the influence of the herdsman's personality on milk yield. *J. Agric. Labour Sci.* 1:45–59.

Seath, D. M., and G. D. Miller. 1946. Effect of warm weather on grazing performance of milking cows. *J. Dairy Sci.* 29:199–206.

Sechzer, J. A. 1970. Prolonged learning and split-brain cats. *Science* 169:889–892.

Sechzer, J. A., and J. L. Brown. 1964. Color discrimination in the cat. *Science* 144:427–429.

Sechzer, J. A., S. E. Folstein, E. H. Geiger, and R. F. Mervis. 1976. The split-brain neonate: A surgical method for corpus callosum section in newborn kittens. *Dev. Psychobiol.* 9:377–388.

Seidel, W. F., T. Roth, T. Roehrs, F. Zorick, and W. Dement. 1984. Treatment of a 12–hour shift of sleep schedule with benzodiazepines. *Science* 224:1262–1264.

Seitz, P.F.D. 1959. Infantile experiences and adult behavior in animal subjects. II. Age of separation from the mother and adult behavior in the cat. *Psychosom. Med.* 21:353–378.

Selman, I. E., A. D. McEwan, and E. W. Fisher. 1970a. Studies on natural suckling in cattle during the first eight hours post partum. I. Behavioural studies (dams). *Anim. Behav.* 18:276–283.

_____. 1970b. Studies on natural suckling in cattle during the first eight hours post partum. II. Behavioural studies (calves). *Anim. Behav.* 18:284–289.

Senn, C. L., and J. D. Lewin. 1975. Barking dogs as an environmental problem. *J. Am. Vet. Med. Assoc.* 166:1065–1068.

Seoane, J. R., and C. A. Baile. 1973a. Feeding behavior in sheep as related to the hypnotic activities of barbiturates injected into the third ventricle. *Pharm. Biochem. Behav.* 1:47–53.

_____. 1973b. Feeding elicited by injections of Ca^{++} and Mg^{++} into the third ventricle of sheep. *Experientia* 29:61–62.

Seoane, J. R., C. A. Baile, and F. H. Martin. 1972. Humoral factors modifying feeding behavior of sheep. *Physiol. Behav.* 8:993–995.

Setchell, B. P. 1978. *The mammalian testis.* Ithaca, N.Y.: Cornell University Press.

Shackleton, D. M., and C. C. Shank. 1984. A review of the social behavior of feral and wild sheep and goats. *J. Anim. Sci.* 58:500–509.

Shank, C. C. 1972. Some aspects of social behaviour in a population of feral goats (*Capra hircus* L.). *Z. Tierpsychol.* 30:488–528.

Share, I., E. Martyniuk, and M. I. Grossman. 1952. Effect of prolonged intragastric feeding on oral food intake in dogs. *Am. J. Physiol.* 169:229–235.

Shaw, E. B., and K. A. Houpt. 1985. Pre- and post-partum behaviour in mules impregnated by embryo transfer. *Equine Vet. J.* 17(suppl. 3): 73.

Shaw, E. B., K. A. Houpt, and D. F. Holmes. 1988. Body temperature and behavior of mares during the last two weeks of pregnancy. *Equine Vet.* J. 20:199–202.

Shaw, R. A. 1978. A time-controlled feeding system for cattle. *Anim. Prod.* 27: 277–284.

Sheppard, A. J., R. E. Blaser, and C. M. Kincaid. 1957. The grazing habits of beef cattle on pasture. *J. Anim. Sci.* 16:681–687.

Shillito, E. E. 1975. A comparison of the role of vision and hearing in lambs finding their own dams. *Appl. Anim. Ethol.* 1:369–377.

Shillito, E. E., and G. Alexander. 1975. Mutual recognition amongst ewes and lambs of four breeds of sheep (*Ovis aries*). *Appl. Anim. Ethol.* 1:151–165.

Shillito, E. E., and V. J. Hoyland. 1971. Observations on parturition and maternal care in Soay sheep. *J. Zool.* 165:509–512.

Shillito Walser, E. E. l986. Recognition of the sow's voice by neonatal piglets. *Behaviour* 99:177–187.

Shillito Walser, E., and P. Hague. 1980. Variations in the structure of bleats from sheep of four different breeds. *Behaviour* 75:22–35.

———. 1981. Field observations on a flock of ewes and lambs made of Clun Forest, Dalesbred and Jacob sheep. *Appl. Anim. Ethol.* 7:175–178.

Shillito Walser, E. E., E. Walters, and P. Hague. 1981. A statistical analysis of the structure of bleats from sheep of four different breeds. *Behaviour* 77:67–76.

Shillito Walser, E. E., S. Willadsen, and P. Hague. l982. Maternal vocal recognition in lambs born to Jacob and Dalesbred ewes after embryo transplantation between breeds. *Appl. Anim. Ethol.* 8:479–486.

Shreffler, C., and W. D. Hohenboken. 1974. Dominance and mating behavior in ram lambs. *J. Anim. Sci.* 39:725–731.

Shuleikina, K. V. 1976. Sensory mechanisms of learning in the behaviour of newborn kittens. *Act. Nerv. Super.* 18:48–50.

Shurrager, P. S., and E. A. Culler. 1938. Phenomena allied to conditioning in the spinal dog. *Am. J. Physiol.* 123:186–187.

Siegel, A., H. Edinger, and A. Koo. 1977. Suppression of attack behavior in the cat by the prefrontal cortex: role of the mediodorsal thalamic nucleus. *Brain Res.* 127:185–190.

Signoret, J.-P. 1967. Durée du cycle oestrien et de l'oestrus chez la truie. Action du benzoate d'oestradiol chez la femelle ovariectomisée (in French). *Ann. Biol. Anim. Biochem. Biophys.* 7:407–421.

———. 1970. Sexual behaviour patterns in female domestic pigs (*Sus scrofa* L.) reared in isolation from males. *Anim. Behav.* 18:165–168.

———. 1975. Influence of the sexual receptivity of a teaser ewe on the mating preference in the ram. *Appl. Anim. Ethol.* 1:229–232.

Signoret, J.-P., and P. Mauleon. 1962. Action de l'ablation des bulbes olfactifs sur les mecanismes de la reproduction chez la truie (in French). *Ann. Biol. Anim. Biochem. Biophys.* 2:167–174.

Signoret, J.-P., B. A. Baldwin, D. Fraser, and E.S.E. Hafez. 1975. The behaviour of swine. In *The behaviour of domestic animals*. 3d ed. E. S. E. Hafez, ed. Baltimore: Williams and Wilkins. Pp. 295–329.

Simpson, C. W., C. A. Baile, and L. F. Krabill. 1975. Neuro-chemical coding for feeding in sheep and steers. *J. Comp. Physiol. Psychol.* 88:176–182.

Sitaram, N., R. J. Wyatt, S. Dawson, and J. C. Gillin. 1976. REM sleep induction by physostigmine infusion during sleep. *Science* 191:1281–1283.

Sivak, J., and D. B. Allen. 1975. An evaluation of the "ramp" retina of the horse eye. *Vision Res.* 15:1353–1356.

Skinner, B. F. 1938. *The behavior of organisms*. New York: D. Appleton-Century.

Slotnick, B. M. 1967. Disturbances of maternal behavior in the rat following lesions of the cingulate cortex. *Behaviour* 29:204–236.

Slotnick, B. M., and B. J. Nigrosh. 1975. Maternal behavior of mice with cingulate cortical, amygdala, or septal lesions. *J. Comp. Physiol. Psychol.* 88:1181–127.

Sly, J., and F. R. Bell. 1979. Experimental analysis of the seeking behaviour observed in ruminants when they are sodium deficient. *Physiol. Behav.* 22:499–505.

Smith, C. 1985. Sleep states and learning: A review of the animal literature. *Neurosci. Biobehav. Rev.* 9:157–168.

Smith, F. V. 1965. Instinct and learning in the attachment of lamb and ewe. *Anim. Behav.* 13:84–86.

Smith, F. V., C. Van-Toller, and T. Boyes. 1966. The "critical period" in the attachment of lambs and ewes. *Anim. Behav.* 14:120–125.

Sneva, F. A. 1970. Behavior of yearling cattle on eastern Oregon range. *J. Range Manage.* 23:155–158.

Sobocinska, J. 1978. Gastric distention and thirst: Relevance to the osmotic thirst threshold and metering of water intake. *Physiol. Behav.* 20:497–501.

Soffie, M., G. Thines, and G. De Marneffe. 1976. Relation between milking order and dominance value in a group of dairy cows. *Appl. Anim. Ethol.* 2:271–276.

Solomon, R. L., and L. C. Wynne. 1953. Traumatic avoidance learning: Acquisition in normal dogs. *Psychol. Monogr.* 67(4):1–19.

Soltysik, S., and B. A. Baldwin. 1972. The performance of goats in triple choice delayed response tasks. *Acta Neurobiol. Exp.* 32:73–86.

Soltysik, S., and K. Jaworska. 1967. Prefrontal cortex and fear-motivated behaviour. *Acta Biol. Exp.* 27:429–448.

Someville, S. H., and B. G. Lowman. 1979. Observations on the nursing behaviour of beef cows suckling Charolais cross calves. *Appl. Anim. Ethol.* 5:369–373.

Sperry, R. W. 1964. The great cerebral commissure. *Sci. Am.* 210:42–52.

Sprague, R. H., and J. J. Anisko. 1973. Elimination patterns in the laboratory beagle. *Behaviour* 47:257–267.

Sprott, R. L. 1967. Barometric pressure fluctuations: Effects on the activity of laboratory mice. *Science* 157:1206–1207.

Squire, L. R. 1986. Mechanisms of memory. *Science* 232:1612–1619.

Squires, V. R. 1974. Grazing distribution and activity patterns of Merino sheep on a saltbush community in south-east Australia. *Appl. Anim. Ethol.* 1:17–30.

Squires, V. R., and G. T. Daws. 1975. Leadership and dominance relationships in Merino and border Leicester sheep. *Appl. Anim. Ethol.* 1:263–274.

Stahlbaum, C. C., and K. A. Houpt. 1989. The role of the flehmen response in the behavioral repertoire of the stallion. *Physiol. Behav.* 45:1207–1214.

Stanford, T. L. 1981. Behavior of dogs entering a veterinary clinic. *Appl. Anim. Ethol.* 7:271–279.

Stanley, W. C., and O. Elliot. 1962. Differential human handling as reinforcing events and as treatments influencing later social behavior in basenji puppies. *Psychol. Rep.* 10:775–788.

Stanley, W. C., W. E. Bacon, and C. Fehr. 1970. Discriminated instrumental learning in neonatal dogs. *J. Comp. Physiol. Psychol.* 70:335–343.

Stanley, W. C., J. E. Barrett, and W. E. Bacon. 1974. Conditioning and extinction of avoidance and escape behavior in neonatal dogs. *J. Comp. Physiol. Psychol.* 87:163–172.

Stark, P., and E. S. Boyd. 1963. Effects of cholinergic drugs on hypothalamic self-stimulation response rates of dogs. *Am. J. Physiol.* 205:745–748.

Stebbins, M. C. 1974. Social organization in free-ranging Appaloosa horses. Master's thesis, Idaho State University, Pocatello.

Stephens, D. B. 1974. Studies on the effect of social environment on the behaviour and growth rates of artificially reared British Friesian male calves. *Anim. Prod.* 18:23–24.

_____. 1975. Effects of gastric loading on the sucking response and voluntary milk intake in neonatal piglets. *J. Comp. Physiol. Psychol.* 88:796–805.

_____. 1980. The effects of 2-deoxy-D-glucose given via the jugular or hepatic portal vein on food intake and plasma glucose levels in pigs. *Physiol. Behav.* 25:691–697.

Stephens, D. B., and B. A. Baldwin. 1971. Observations on the behavior of groups of artificially reared lambs. *Res. Vet. Sci.* 12:219–224.

Stephens, D. B., and J. L. Linzell. 1974. The development of sucking behaviour in the newborn goat. *Anim. Behav.* 22:628–633.

Stephens, D. B., D. L. Ingram, and D. F. Sharman. 1983. An investigation into some cerebral mechanisms involved in schedule-induced drinking in the pig. *Q. J. Exp. Physiol.* 68:653–660.

Sterman, M. B., T. Knauss, D. Lehmann, and C. D. Clemente. 1965. Circadian sleep and waking patterns in the laboratory cat. *Electroencephalogr. Clin. Neurophysiol.* 19:509–517.

Stern, W. C., and P. J. Morgane. 1974. Theoretical view of REM sleep function: Maintenance of catecholamine systems in the central nervous system. *Behav. Biol.* 11:1–32.

Stetson, M. H., and M. Watson-Whitmyre. 1976. Nucleus suprachiasmaticus: The biological clock in the hamster? *Science* 191:197–199.

Stewart, J. C., and J. P. Scott. 1947. Lack of correlation between leadership and dominance relationships in a herd of goats. *J. Comp. Physiol. Psychol.* 40:255–264.

Stone, C. C., M. S. Brown, and G. H. Waring. 1974. An ethological means to improve swine production. *J. Anim. Sci.* (abstr.) 39:137.

Strasia, C. A., M. Thorn, R. W. Rice, and D. R. Smith. 1970. Grazing habits, diet and performance of sheep on alpine ranges. *J. Range Manage.* 23:201–208.

Stricklin, W. R., and H. Gonyou. 1981. Dominance and eating behavior of beef cattle fed from a single stall. *Appl. Anim. Behav.* 7:135–140.

Stricklin, W. R., C. C. Kautz-Scanavy, and D. L. Greger. 1985. Determination of dominance-subordinance relationships among beef heifers in a dominance tube. *Appl. Anim. Behav.* 14:111–116.

Stroup, W. W., M. K. Nielsen, and J. A. Gosey. 1987. Cyclic variation in cattle feed intake data: Characterization and implications for experimental design. *J. Anim. Sci.* 64:1638–1647.

Sturgeon, R. D., P. D. Brophy, and R. A. Levitt. 1973. Drinking elicited by intracranial microinjection of angiotensin in the cat. *Pharm. Biochem. Behav.* 1:353–355.

Sufit, E., K. A. Houpt, and M. Sweeting. 1985. Physiological stimuli of thirst and drinking patterns in ponies. *Equine Vet. J.* 17:12–16.

Sutherland, G. F. 1939. Salivary conditioned reflexes in swine. *Am. J. Physiol.* (abstr.) 126:P640–641.

Sutherland, N. S. 1961. The methods and findings of experiments on the visual discrimination of shape by animals. *Exp. Psychol. Soc. Monogr.* 1:1–68.

Sweeting, M. P., C. E. Houpt, and K. A. Houpt. 1985. Social facilitation of feeding and time budgets in stabled ponies. *J. Anim. Sci.* 160:369–374.

Sweetwood, H. L., D. F. Kripke, I. Grant, J. Yager, and M. S. Gerst. 1976. Sleep disorder and psychobiological symptomatology in male psychiatric outpatients and male nonpatients. *Psychosom. Med.* 38:373–378.

Swenson, R. M., and W. Randall. 1977. Grooming behavior in cats with pontile lesions and cats with tectal lesions. *J. Comp. Physiol. Psychol.* 91:313–326.

Syme, G. J., L. A. Syme, and T. P. Jefferson. 1974. A note on variations in the level of aggression within a herd of goats. *Anim. Prod.* 18:309–312.

Syme, L. A., G. J. Syme, T. G. Waite, and A. J. Pearson. 1975. Spatial distribution and social status in a small herd of dairy cows. *Anim. Behav.* 23:609–614.

Symoens, J., and M. van den Brande. 1969. Prevention and cure of aggressiveness in pigs using the sedative azaperone. *Vet. Rec.* 85:64–67.

Symons, L.E.A., and D. R. Hennessy. 1981. Cholecystokinin and anorexia in sheep infected by the intestinal nematode *Trichostrongylus colubriformis*. *Int. J. Parasitol.* 11:55–58.

Tanzer, H., and N. Lyons. 1977. *Your pet isn't sick. He just wants you to think so.* New York: Thomas Congdon Books.

Teitelbaum, P. 1967. *Physiological psychology: Fundamental principles.* Englewood Cliffs, N.J.: Prentice-Hall.

Tennessen, T., M. A. Price, and R. T. Berg. 1985. The social interactions of young bulls and steers after re-grouping. *Appl. Anim. Behav.* 14:37–47.

Terman, M., and J. S. Terman. 1970. Circadian rhythm of brain self-stimulation behavior. *Science* 168:1242–1244.

Ternouth, J. H., and A. W. Beattie. 1970. A note on the voluntary food consumption and the sodium-potassium ratio of sheep after shearing. *Anim. Prod.* 12:343–346.

Thiery, J. C., and J. P. Signoret. 1978. Effect of changing the teaser ewe on the sexual activity of the ram. *Appl. Anim. Ethol.* 4:87–90.

Thinus-Blanc, C., B. Poucet, and N. Chapuis. 1982. Object permanence in cats: Analysis in locomotor space. *Behav. Proc.* 7:81–86.

Thomas, G. J., W. J. Fry, F. J. Fry, B. M. Slotnick, and E. E. Krieckhaus. 1963. Behavioral effects of mammilothalamic tractotomy in cats. *J. Neurophysiol.* 26:857–876.

Thompson, D. S., and P. G. Schinckel. 1952. Incidence of oestrus in ewes. *Empire J. Exp. Agric.* 20:77–79.

Thompson, L. H., and J. S. Savage. 1978. Age at puberty and ovulation rate in gilts in confinement as influenced by exposure to a boar. *J. Anim. Sci.* 47:1141–1144.

Thompson, R. 1958. The effect of intracranial stimulation on memory in cats. *J. Comp. Physiol. Psychol.* 51:421–426.

Thompson, V. E. 1968. Neonatal orbitofrontal lobectomies and delayed response behavior in cats. *Physiol. Behav.* 3:631–635.

Thompson, W. R., and W. Heron. 1954b. The effects of early restriction on activity in dogs. *J. Comp. Physiol. Psychol.* 47:77–82.

_____. 1954b. The effects of restricting early experience on the problem-solving capacity of dogs. *Can. J. Psychol.* 8:17–31.

Thompson, W. R., R. Melzack, and T. H. Scott. 1956. "Whirling behavior" in dogs as related to early experience. *Science* 123:939.

Thorndike, E. L. 1911. *Animal Intelligence. Experimental Studies.* New York: Macmillan Co.

Thorpe, W. H. 1963. *Learning and Instinct in Animals.* Cambridge, Mass.: Harvard University Press.

Thrasher, T. N., C. J. Brown, L. C. Keil, and D. J. Ramsay. 1980. Thirst and vasopressin release in the dog: An osmoreceptor or sodium receptor mechanism? *Am. J. Physiol.* 238:R333–R339.

Tilbrook, A. J. 1987. Physical and behavioural factors affecting the sexual "attractiveness" of the ewe. *Appl. Anim. Behav.* 17:109–115.

Tischner, M. 1982. Patterns of stallion sexual behaviour in the absence of mares. *J. Reprod. Fert. Suppl.* 32:65–70.

Tischner, M., K. Kosiniak, and W. Bielanski. 1974. Analysis of the pattern of ejaculation in stallions. *J. Reprod. Fertil.* 41:329–335.

Titterington, R. W., and D. Fraser. 1975. The lying behaviour of sows and piglets during early lactation in relation to the position of the creep heater. *Appl. Anim. Ethol.* 2:47–53.

Tobach, E., L. R. Aronson, and E. Shaw. 1971. *The biopsychology of development.* New York: Academic Press.

Todd, N. B. 1963. The catnip response. Ph.D. diss., Harvard University, Cambridge, Mass.

Tomkins, T., and M. J. Bryant. 1974. Oestrous behaviour of the ewe and the influence of treatment with progestagen. *J. Reprod. Fertil.* 41:121–132.

Tomlinson, K. A., E. O. Price, and D. T. Torell. 1982. Responses of tranquilized post-partum ewes to alien lambs. *Appl. Anim. Ethol.* 8:109–117.

Topel, D. G., G. M. Weiss, D. G. Siers, and J. H. Magilton. 1973. Comparison of blood source and diurnal variation on blood hydrocortisone, growth hormone, lactate, glucose and electrolytes in swine. *J. Anim. Sci.* 36:531–534.

Tortora, D. F. 1977. *Help! This animal is driving me crazy. Solutions to your dog's behavior problems.* Chicago: Playboy Press.

_____. 1980a. Animal behavior therapy: The behavioral diagnosis and treatment of dominance-motivated aggression in canines. Pt. I. *Canine Pract.* 7:10–19.

_____. 1980b. Animal behavior therapy: The behavioral diagnosis and treatment of dominance-motivated aggression in canines. Pt. II. *Canine Pract.* 8:13–28.

Toutain, P.-L., C. Toutain, A. J. F. Webster, and J. D. McDonald. 1977. Sleep and activity, age and fatness, and the energy expenditure of confined sheep. *Br. J. Nutr.* 38:445–454.

Towbin, E. J. 1949. Gastric distention as a factor in the satiation of thirst in esophagostomized dogs. *Am. J. Physiol.* 159:533–541.

Tribe, D. E. 1949. The importance of the sense of smell to the grazing sheep. *J. Agric. Sci.* (Camb.) 39:309–312.

Trout, W. E., J. C. Pekas, and B. D. Schanbacher. 1989. Immune, growth and carcass responses of ram lambs to active immunization against desulfated cholecystokinin (CCK-8). *J. Anim. Sci.* 67:2709–2714.

Trumler, E. 1959. Das "Rossigkeitsgesicht" und ähnliches Ausdrucksverhalten bei Einhufern (in German). *Z. Tierpsychol.* 16:478–488.

Tulloh, N. M. 1961a. Behaviour of cattle in yards. I. Weighing order and behaviour before entering scales. *Anim. Behav.* 9:20–24.

_____. 1961b. Behaviour of cattle in yards. II. A study of temperament. *Anim. Behav.* 9:25–30.

Turek, F. W., and S. Losee-Olson. 1986. A benzodiazepine used in the treatment of insomnia phase-shifts the mammalian circadian clock. *Nature* 321:167–168.

Turner, D. C., and P. Bateson. 1988. *The domestic cat: The biology of its behaviour.* Cambridge, England: Cambridge University Press.

Turner, A. S., N. White II., and J. Ismay. 1984. Modified Forssell's operation for crib biting in the horse. *J. Amer. Vet. Med. Assoc.* 184:309–312.

Turner, D. C., J. Feaver, M. Mendl, and P. Bateson. 1986. Variation in domestic cat behaviour towards humans: A paternal effect. *Anim. Behav.* 34:1890–1892.

Turner, R. R. 1961. Silage self-feeding. *Vet. Rec.* 73:1432–1436.

Tyler, S. J. 1972. The behaviour and social organization of the New Forest ponies. *Anim. Behav. Monogr.* 5:85–196.

Uretsky, E., and R. A. McCleary. 1969. Effect of hippocampal isolation on retention. *J. Comp. Physiol. Psychol.* 68:1–8.

Ursin, R. 1968. The two stages of slow wave sleep in the cat and their relation to REM sleep. *Brain Res.* 11:347–356.

_____. 1970. Sleep stage relations within the sleep cycles of the cat. *Brain Res.* 20:91–97.

Ursin, R., H. Cohen, S. Henriksen, G. Mitchell, and W. Dement. 1976. Effects on sleep of restricted sleep. A cat case study. *Electroencephalogr. Clin. Neurophysiol.* 41:96–101.

Vachon, L., A. Kitsikis, and A. G. Roberge. 1984. Chlordiazepoxide, go-nogo successive discrimination and brain biogenic amines in cats. *Pharmacol. Biochem. Behav.* 20:9–22.

Vandenbergh, J. G. 1979. The influence of pheromones on puberty in rodents. Presented at the symposium on Chemical Signals in Vertebrates and Aquatic Animals, 30 May–2 June 1979. State University of New York, College of Environmental Science and Forestry, Syracuse.

Van Niekerk, A. I., J. F. D. Greenhalgh, and G. W. Reid. 1973. Importance of palatability in determining the feed intake of sheep offered chopped and pelleted hay. *Br. J. Nutr.* 30:95–105.

van Putten, G. 1969. An investigation into tail-biting among fattening pigs. *Br. Vet. J.* 125:511–517.

van Putten, G., and J. Dammers. 1976. A comparative study of the well-being of piglets reared conventionally and in cages. *Appl. Anim. Ethol.* 2:339–356.

Vecchiotti, G. G., and R. Galanti. 1986. Evidence of heredity of cribbing, weaving and stall walking in Thoroughbred horses. *Livestock Prod. Sci.* 14:91–95.

Veeckman, J., and F. O. Odberg. 1978. Preliminary studies on the behavioural detection of oestrus in Belgian "warm-blood" mares with acoustic and tactile stimuli. *Appl. Anim. Ethol.* 4:109–118.

Veissier, I., and P. Le Neindre. 1989. Weaning in calves: Its effects on social organization. *Appl. Anim. Behav.* 24:43–54.

Verberne, G., and J. De Boer. 1976. Chemocommunication among domestic cats, mediated by the olfactory and vomeronasal senses. *Z. Tierpsychol.* 42:86–109.

Villablanca, J. R., and C. E. Olmstead. 1979. Neurological development of kittens. *Dev. Psychobiol.* 12:101–127.

Vince, M. A. 1984. Teat seeking or presuckling behaviour in newly born lambs: Possible effects of maternal skin temperatures. *Anim. Behav.* 32:249–254.

Vince, M. A., and T. M. Ward. l984. The responses of newly born Clun Forest lambs to odour sources in the ewe. *Behaviour* 89:117–121.

Vince, M. A., T. M. Ward, and M. Reader. 1984. Tactile stimulation and teat seeking behaviour in newly born lambs. *Anim. Behav.* 32:1179–1184.

Vitale, A. F., M. Tenucci, M. Papini, and S. Lovari. 1986. Social behaviour of the calves of semi-wild Maremma cattle, *Bos primigenius taurus. Appl. Anim. Behav Sci.* 16:217–231.

Vogel, H. H., Jr., J. P. Scott, and M.-'V. Marston. 1950. Social facilitation and allelomimetic behavior in dogs. I. Social facilitation in a non-competitive situation. *Behaviour* 2:121–134.

Voith V. L. and P. L. Borchelt. 1985a. Elimination behavior and related problems in dogs. *Compend. Cont. Educ.* 7:537–546.

_____. 1985b. Fears and phobias in companion animals. *Comp. Contin. Ed.* 7:209–218.

Wagnon, K. A., R. G. Loy, W. C. Rollins, and F. D. Carroll. 1966. Social dominance in a herd of Angus, Hereford, and shorthorn cows. *Anim. Behav.* 14:474–477.

Walker, D. E. 1962. Suckling and grazing behaviour of beef heifers and calves. *N.Z. J. Agric. Res.* 5:331–338.

Wallace, L. R. 1949. Observations of lambing behaviour in ewes. *Proc. N.Z. Soc. Anim. Prod.* 9:85–96.

Wallace, P. 1975. Neurochemistry: Unraveling the mechanism of memory. *Science* 190:1076–1078.

Waller, G. R., G. H. Price, and E. D. Mitchell. 1969. Feline attractant, cis,trans-nepetalactone: Metabolism in the domestic cat. *Science* 164:1281–1282.

Wangsness, P. J., L. E. Chase, A. D. Peterson, T. G. Hartsock, D. J. Kellmel, and B. R. Baumgardt. 1976. System for monitoring feeding behavior of sheep. *J. Anim. Sci.* 42:1544–1549.

Wardrop, J. C. 1953. Studies in the behaviour of dairy cows at pasture. *Br. J. Anim. Behav.* 1:23–31.

Waring, G. H. 1982. Onset of behavior patterns in the newborn foal. *Equine Pract.* 4:28–34.

Waring, G. H. 1983. *Horse behavior: The behavioral traits and adaptations of domestic and wild horses, including ponies.* Park Ridge, N.J.: Noyes Publications.

Waring, G. H., S. Wierzbowski, and E. S. E. Hafez. 1975. The behaviour of horses. In *The behaviour of domestic animals,* 3d ed. E. S. E. Hafez, ed. Baltimore: Williams and Wilkins. Pp. 330–369.

Warren, J. M., and A. Baron. 1956. The formation of learning sets by cats. *J. Comp. Physiol. Psychol.* 49:227–231.

Warren, J. M., H. B. Warren, and K. Akert. 1962. Orbitofrontal cortical lesions and learning in cats. *J. Comp. Neurol.* 118:17–41.

Webb, F. M., V. N. Colenbrander, T. H. Blosser, and D. E. Waldern. 1963. Eating habits of dairy cows under drylot conditions. *J. Dairy Sci.* 46:1433–1435.

Webster, A. S. F., J. S. Smith, and J. M. Brockway. 1972. Effects of isolation confinement and competition for feed on energy exchanges of growing lambs. *Anim. Prod.* 15:189–201.

Weir, W. C., and D. T. Torell. 1959. Selective grazing by sheep as shown by a comparison of the chemical composition of range and pasture forage obtained by hand clipping and that collected by esophageal-fistulated sheep. *J. Anim. Sci.* 18:641–649.

Welch, A. R., and M. R. Baxter. 1986. Responses of newborn piglets to thermal and tactile properties of their environment. *Appl. Anim. Behav.* 15:203–215.

Welch, R. A. S., and R. Kilgour. 1970. Mis-mothering among Romneys. *N.Z. J. Agric.* 121(4):26–27.

Wells, S. M., and B. von Goldschmidt-Rothschild. 1979. Social behaviour and relationships in a herd of Camargue horses. *Z. Tierpsychol.* 49:363–380.

Wesley, F., and F. D. Klopfer. 1962. Visual discrimination learning in swine. *Z. Tierpsychol.* 19:93–104.

West, M. J. 1974. Social play in the domestic cat. *Am. Zool.* 14:427–436.

_____. 1977. Exploration and play with objects in domestic kittens. *Dev. Psychobiol.* 10:53–57.

Whalen, R. E. 1963. The initiation of mating in naive female cats. *Anim. Behav.* 11:461–463.

Whatson, T. S. 1985. Development of eliminative behaviour in piglets. *Appl. Anim. Behav.* 14:365–377.

Whipp, S. C., R. L. Wood, and N. C. Lyon. 1970. Diurnal variation in concentrations of hydrocortisone in plasma of swine. *Am. J. Vet. Res.* 31:2105–2107.

White, S. D. 1990. Naltrexone for treatment of acral lick dermatitis in dogs. *J. Am. Vet. Med. Assoc.* 196:1073–1076.

Whittemore, C. T., and D. Fraser. 1974. The nursing and suckling behaviour of pigs. II. Vocalization of the sow in relation to suckling behaviour and milk ejection. *Br. Vet. J.* 130:346–356.

Whittlestone, W. G., and L. R. Cate. 1973. An animal activated feeding device for cattle. *J. Dairy Sci.* 56:1352–1353.

Whittlestone, W. G., R. Kilgour, H. de Langen, and G. Duirs. 1970. Behavioral stress and the cell count of bovine milk. *J. Milk Food Technol.* 33:217–220.

Whittlestone, W. G., M. Mullord, R. Kilgour, and L. R. Cate. 1975. Electric shocks during machine milking. *N.Z. Vet. J.* 23:105–108.

Widdowson, E. M. 1971. Food intake and growth in the newly born. *Proc. Nutr. Soc.* 30:127–135.

Widdowson, E. M., and R. A. McCance. 1963. The effect of finite periods of undernutrition at different ages on the composition and subsequent development of the rat. *Proc. Roy. Soc. B.* 158:329–342.

Wieckert, D. A. 1971. Social behavior in farm animals. *J. Anim. Sci.* 32:1274–1277.

Wieckert, D. A., and G. R. Barr. 1966. Studies of learning ability in young pigs. *J. Anim. Sci.* (abstr.) 25:1280.

Wieckert, D. A., L. P. Johnson, K. P. Offord, and G. R. Barr. 1966. Measuring learning ability in dairy cows. *J. Dairy Sci.* (abstr.) 49:729.

Wiepkema, P. R., K. K. Van Hellemond, P. Roessingh, and H. Romberg. 1987. Behaviour and abomasal damage in individual veal calves. *Appl. Anim. Behav.* 18:257–268.

Wierzbowski, S. 1959. Odruchy plciowe ogierow (in Polish). *Rocz. Nauk. Roln.* 753–788.

———. 1978. The sexual behaviour of experimentally underfed bulls. *Appl. Anim. Ethol.* 4:55–60.

Wikmark, G., and J. M. Warren. 1972. Delayed response learning by caged-reared normal and prefrontal cats. *Psychonom. Sci.* 26:243–245.

Wilcox, S., K. Dusza, and K. Houpt. 1991. The relationship between recumbent rest and masturbation in stallions. *Eq. Vet. Sci.* 11:23–26.

Willard, J. G., J. C. Willard, S. A. Wolfram, and J. P. Baker. 1977. Effect of diet on cecal pH and feeding behavior of horses. *J. Anim. Sci.* 45:87–93.

Willham, R. L., D. F. Cox, and G. G. Karas. 1963. Genetic variation in a measure of avoidance learning in swine. *J. Comp. Physiol. Psychol.* 56:294–297.

Willham, R. L., G. G. Karas, and D. C. Henderson. 1964. Partial acquisition and extinction of an avoidance response in two breeds of swine. *J. Comp. Physiol. Psychol.* 57:117–122.

Williamson, N. B., R. S. Morris, D. C. Blood, and C. M. Cannon. 1972. A study of oestrous behaviour and oestrus detection methods in a large commercial dairy herd. I. The relative efficiency of methods of oestrus detection. *Vet. Rec.* 91:50–58.

Wilson, E. O. 1975. *Sociobiology. The New Synthesis.* Cambridge, Mass.: Harvard University Press, Belknap Press.

Wilson, M., J. M. Warren, and L. Abbott. 1965. Infantile stimulation, activity, and learning by cats. *Child Dev.* 36:843–853.

Wilsson, E. 1984–85. The social interaction between mother and offspring during weaning in German shepherd dogs: Individual differences between mothers and their effects on offspring. *Appl. Anim. Behav.* 13:101–112.

Wimersma Greidanus, Tj. B. van, B. Bohus, and D. de Wied. 1975. The role of vasopressin in memory processes. In *Progress in brain research.* Vol. 42, *Hormones, homeostasis and the brain.* W. H. Gispen, Tj. B. van Wimersma Greidanus, B. Bohus, and D. de Wied, eds. Amsterdam: Elsevier Scientific Publishing. Pp. 135–141.

Winans, S. S. 1967. Visual form discrimination after removal of the visual cortex in cats. *Science* 158:944–946.

Winchester, C. F. 1943. The energy cost of standing in horses. *Science* 97:24.

Winer, J., and J. F. Lubar. 1976. Alternation behavior of cats with medial visual cortex ablation. *Physiol. Behav.* 17:635–643.

Winfield, C. G., and R. Kilgour. 1976. A study of following behaviour in young lambs. *Appl. Anim. Ethol.* 2:235–243.

Winfield, C. G., and A. W. Makin. 1978. A note on the effect of continuous contact with ewes showing regular oestrus and of post-weaning growth rate on the sexual activity of Corriedale rams. *Anim. Prod.* 27:361–364.

Winfield, C. G., and P. D. Mullaney. 1973. A note on the social behaviour of a flock of Merino and Wiltshire horn sheep. *Anim. Prod.* 17:93–95.

Winfield, C. G., P. H. Hemsworth, M. R. Taverner, and P. D. Mullaney. 1974. Observations on the sucking behviour of piglets in litters of varying size. *Proc. Aust. Soc. Anim. Prod.* 10:307–310.

Winfield, C. G., G. J. Syme, and A. J. Pearson. 1981. Effect of familiarity with each other and breed on the spatial behaviour of sheep in an open field. *Appl. Anim. Ethol.* 7:67–75.

Winkler, W. G. 1977. Human deaths induced by dog bites, United States, 1974–1975. *Public Health Rep.* 92:425–429.

Winslow, C. N. 1944a. The social behavior of cats. I. Competitive and aggressive behavior in an experimental runway situation. *J. Comp. Psychol.* 37:297–313.

_____. 1944b. The social behavior of cats. II. Competitive, aggressive, and food-sharing behavior when both competitors have access to the goal. *J. Comp. Psychol.* 37:315–326.

Wise, R. A., and V. Dawson. 1974. Diazepam-induced eating and lever pressing for food in sated rats. *J. Comp. Physiol. Psychol.* 86:930–941.

Wolf, A. V. 1950. Osmometric analysis of thirst in man and dog. *Am. J. Physiol.* 161:75–86.

Wolski, T. R. 1982. Social behavior of the cat. *Vet. Clinic N. Am. (Small Anim. Pract.)* 12:693–706.

Wolski, T. R., K. A. Houpt, and R. Aronson. 1980. The role of the senses in mare-foal recognition. *Appl. Anim. Ethol.* 6:121–138.

Wood, M. T. 1977. Social grooming patterns in two herds of monozygotic twin dairy cows. *Anim. Behav.* 25:635–642.

Wood, P. D. P., G. F. Smith, and M. F. Lisle. 1967. A survey of intersucking dairy cows in herds in England and Wales. *Vet. Rec.* 81:396–397.

Wood, R. J., B. J. Rolls, and D. J. Ramsay. 1977. Drinking following intracarotid infusions of hypertonic solutions in dogs. *Am. J. Physiol.* 232:R88–R92.

Wood, T., S. Stanley, and T. Tobin. 1989. Operant conditioning and its applications in equine pharmacology. *J. Equine Vet. Sci.* 9:124–130.

Woodbury, C. B. 1943. The learning of stimulus patterns by dogs. *J. Comp. Psychol.* 35:29–40.

Wood-Gush, D.G.M., and D. Csermely. 1981. A note on the diurnal activity of early weaned piglets in flat-deck cages at 3 and 6 weeks of age. *Anim. Prod.* 33:107–110.

Woods, G. L., and K. A. Houpt. 1986. An abnormal facial gesture in an estrous mare. *Appl. Anim. Behav.* 16:199–202.

Wright, J. C. 1980. Early development of exploratory behavior and dominance in three litters of German shepherds. In *Early experiences and early behavior.* New York: Academic Press. Pp. 181–206.

Wright, J. C., and M. S. Nesselrote. 1987. Classification of behavior problems in dogs: Distributions of age, breed, sex and reproductive status. *Appl. Anim. Behav.* 19:169–178.

Wyers, E. J., and S. A. Deadwyler. 1971. Duration and nature of retrograde amnesia produced by stimulation of caudate nucleus. *Physiol. Behav.* 6:97–103.

Wyrwicka, W. 1978. Effects of electrical stimulation within the hypothalamus on gastric acid secretion and food intake in cats. *Exp. Neurol.* 60:286–303.

Yang, T. S., B. Howard, and W. V. Macfarlane. 1981. Effects of food on drinking behaviour of growing pigs. *Appl. Anim. Ethol.* 7:259–270.

Yang, T. S., M. A. Price, and F. X. Aherne. 1984. The effect of level of feeding on water turnover in growing pigs. *Appl. Anim. Behav.* 12:103–109.

Yarney, T. A., G. W. Rahnefeld, R. J. Parker, and W. M. Palmer. 1982. Hourly distribution of time of parturition in beef cows. *Can. J. Anim. Sci.* 62:597–605.

Yerkes, R. M. 1916. The mental life of monkeys and apes: A study of ideational behavior. *Behav. Monogr.* 3:145.

Yerkes, R. M., and C. A. Coburn. 1915. A study of the behavior of the pig *Sus scrofa* by the multiple choice method. *J. Anim. Behav.* 5:185–225.

Zablocka, T. 1975. Go–no go differentiation to visual stimuli in cats with different early visual experiences. *Acta Neurobiol. Exp.* 35:399–402.

Zablocka, T., J. Konorski, and B. Zernicki. 1975. Visual discrimination learning in cats with different early visual experiences. *Acta Neurobiol. Exp.* 35:389–398.

Zagrodzka, J., and E. Fonberg. 1977. Amygdalar area involved in predatory behavior in cats. *Acta Neurobiol. Exp.* 37:131–136.

Zahorik, D. M., and K. A. Houpt. 1977. The concept of nutritional wisdom: Applicability of laboratory learning models to large herbivores. In *Learning mechanisms in food selection.* L. M. Barker, M. R. Best, and M. Domjan, eds. Waco, Tex.: Baylor University Press. Pp. 45–67.

Zahorik, D. M., and K. A. Houpt. 1981. Species differences in feeding strategies, food hazards, and the ability to learn food aversions. In *Foraging behavior.* A. C. Kamil and T. D. Sargent, eds. New York: Garland STPM Press. Pp. 289–310.

Zahorik, D. M., K. A. Houpt, and J. Swartzman-Andert. 1990. Taste-aversion learning in three species of ruminants. *Appl. Anim. Behav.* 26:27–39.

Zemo, T., and J. O. Klemmedson. 1970. Behavior of fistulated steers on a desert grassland. *J. Range Manage.* 23:158–163.

Zenchak, J. J., and G. C. Anderson. 1973. Discrimination learning in sheep. *J. Anim. Sci.* (abstr.) 37:227.

_____. 1980. Sexual performance levels of rams (*Ovis aries*) as affected by social experiences during rearing. *J. Anim. Sci.* 50:167–174.

Zepelin, H. 1989. Mammalian Sleep. In *Principles and practice of sleep medicine.* M. H. Kryger, T. Roth, and W. C. Dement, eds. Philadelphia: W. B. Saunders. Pp. 30–49.

Zimmerman, M. B., and E. M. Stricker. 1978. Water and NaCl intake after furosemide treatment in sheep (*Ovis aries*). *J. Comp. Physiol. Psychol.* 92:501–510.

Zis, A. P., H. C. Fibiger, and A. G. Phillips. 1974. Reversal by L-dopa of impaired learning due to destruction of the dopaminergic nigroneostriatal projection. *Science* 185:960–962.

Zucker, I. 1965. Effect of lesions of the septal-limbic area on the behavior of cats. *J. Comp. Physiol. Psychol.* 60:344–352.

INDEX